Texts in Theoretical Computer Science
An EATCS Series

Springer
Berlin
Heidelberg
New York
Barcelona
Budapest
Hong Kong
London
Milan
Paris
Santa Clara
Singapore
Tokyo

Alexander Leitsch

The Resolution Calculus

With 36 Figures

 Springer

Author

Prof. Dr. Alexander Leitsch
Technische Universität Wien
Institut für Computersprachen
Resselgasse 3, A-1040 Wien, Austria

Series Editors

Prof. Dr. Wilfried Brauer
Institut für Informatik, Technische Universität München
Arcisstrasse 21, D-80333 München, Germany

Prof. Dr. Grzegorz Rozenberg
Institute of Applied Mathematics and Computer Science
University of Leiden, Niels-Bohr-Weg 1, P.O. Box 9512
2300 RA Leiden, The Netherlands

Prof. Dr. Arto Salomaa
The Academy of Finland
Department of Mathematics, University of Turku
FIN-20 500 Turku, Finland

Cataloging-in-Publication Data applied for
Die Deutsche Bibliothek – CIP-Einheitsaufnahme

Leitsch, Alexander:
The resolution calculus / Alexander Leitsch. - Berlin ;
Heidelberg ; New York ; Barcelona ; Budapest ; Hong Kong ;
London ; Milan ; Paris ; Santa Clara ; Singapore ; Tokyo :
Springer, 1997
   (Texts in theoretical computer science)
   ISBN-13:978-3-642-64473-3   e-ISBN-13:978-3-642-60605-2
   DOI:10.1007/978-3-642-60605-2

ISBN-13:978-3-642-64473-3 Springer-Verlag Berlin Heidelberg New York

Typesetting: Camera ready by author using a Springer T$_E$X macro package
Cover Design: *design & production* GmbH, Heidelberg
SPIN: 10553932   45/3142-5 4 3 2 1 0 – Printed on acid-free paper.

Preface

The History of the Book

In August 1992 the author had the opportunity to give a course on resolution theorem proving at the Summer School for Logic, Language, and Information in Essex. The challenge of this course (a total of five two-hour lectures) consisted in the selection of the topics to be presented. Clearly the first selection has already been made by calling the course "resolution theorem proving" instead of "automated deduction". In the latter discipline a remarkable body of knowledge has been created during the last 35 years, which hardly can be presented exhaustively, deeply and uniformly at the same time. In this situation one has to make a choice between a survey and a detailed presentation with a more limited scope. The author decided for the second alternative, but does not suggest that the other is less valuable. Today resolution is only one among several calculi in computational logic and automated reasoning. However, this does not imply that resolution is no longer up to date or its potential exhausted. Indeed the loss of the "monopoly" is compensated by new applications and new points of view. It was the purpose of the course mentioned above to present such new developments of resolution theory. Thus besides the traditional topics of completeness of refinements and redundancy, aspects of termination (resolution decision procedures) and of complexity are treated on an equal basis. The script of the course on resolution theorem proving appeared (in an improved version) as AILA preprint [Lei93a] and represents the skeleton of this book.

How to Use this Book

Chapters 2 (basics of resolution), 3 (refinements), and 4 (deletion) can be used for a traditional graduate course on resolution. However, some results and methods already point to the more advanced Chapters 5 (resolution decision procedures) and 6 (complexity). All the methods and results needed in Chapters 5 and 6 are provided in the first part of the book. Therefore, in principle, every student familiar with the basics of first-order logic can read the whole book. There are a few points where a basic knowledge of the theory of computability is needed to (fully) understand the material. On the other hand, Chapters 5 and 6 contain results of recent research, most of it published during the last five years. Thus the book is also addressed to researchers working in the field of automated deduction; to them it should offer a source of information about new trends and paradigms in the form of a unified and systematic presentation.

Acknowledgments

First of all my thanks go to Daniele Mundici. It was he who invited me to give a course on resolution theorem proving at the Summer School LLI in Essex; later he encouraged me to transform the course script into a AILA preprint, which improved the substance of the material. Together with Cinzia Ferrari (my thanks to her too) he also read and checked a draft of this book and suggested many improvements. Reiner Hähnle thoroughly checked the first four chapters, found many errors and proposed several substantial changes and improvements. I also thank him for using the preprint mentioned above as basis of a course on automated theorem proving; his experience and observations made on this occasion influenced the book already in an early stage. Christian Fermüller read a draft of the whole book and gave many valuable hints to improve the presentation. Robert Matzinger not only checked the draft, he somehow traversed the material to its utmost depth. He spent several months on this task and his contribution to the improvement of the book cannot be overestimated. Reinhard Pichler read most of the book and made an interesting observation in Chapter 3. Matthias Baaz gave me some very valuable comments on Chapter 6 (which is based almost entirely on our common research). Special thanks go to the referees who wrote very competent and valuable reports on this book; their comments led to essential improvements of form and content. Gernot Salzer strongly assisted in LaTeX problems and substantially contributed to the physical existence of the text. I am very grateful to Franziska Gusel for typing two thirds of the manuscript – an activity extending her usual work and inflicting additional stress. Many thanks go to Springer-Verlag, in particular to Hans Wössner and J. Andrew Ross, for their assistance and guidance in the difficult matter of transforming a text into a real book. Last, but not least, I have to thank my wife Marjan and my two sons David and Emanuel for their patience and tolerance during my work on this book. During the last four years I gave more to "resolution" than to them – sometimes forgetting that we are humans rather than scientific machines.

Vienna, October 1996   Alexander Leitsch

Table of Contents

1. Introduction 1

2. The Basis of the Resolution Calculus 5
   2.1 First-Order Logic 5
   2.2 Transformation to Clause Form 11
   2.3 Term Models and Herbrand's Theorem 23
   2.4 Decision Methods for Sets of Ground Clauses 41
       2.4.1 Gilmore's Method 41
       2.4.2 The Method of Davis and Putnam 43
   2.5 The Propositional Resolution Principle 52
   2.6 Substitution and Unification 59
   2.7 The General Resolution Principle 72
   2.8 A Comparison of Different Resolution Concepts 81

3. Refinements of Resolution 89
   3.1 A Formal Concept of Refinement 89
   3.2 Normalization of Clauses 95
   3.3 Refinements Based on Atom Orderings 99
   3.4 Lock Resolution 108
   3.5 Linear Refinements 115
   3.6 Hyperresolution 130
   3.7 Refinements: A Short Overview 145

4. Redundancy and Deletion 149
   4.1 The Problem of Proof Search 149
   4.2 The Subsumption Principle 157
   4.3 Subsumption Algorithms 181
   4.4 The Elimination of Tautologies 194
   4.5 Clause Implication 199

5. Resolution as Decision Procedure 211
   5.1 The Decision Problem 211
   5.2 A-Ordering Refinements as Decision Procedures 215

5.3   Hyperresolution as Decision Procedure 230
5.4   Hyperresolution and Automated Model Building 238

6.   On the Complexity of Resolution 253
6.1   Herbrand Complexity and Proof Length 253
6.2   Extension and the Use of Lemmas 269
      6.2.1   Structural Normalization 272
      6.2.2   Functional Extension 280

References 289

Notation Index 295

Subject Index 297

1. Introduction

Logic calculi can serve for many purposes, such as reconstructing and analyzing mathematical proofs in a formal manner or automating the search for mathematical proofs. As the paradoxes of set theory struck the mathematical community around 1900, the formal and consistent representation of theories became a central issue in foundational research. In the theory of proofs fascinating results were obtained in the 1930s, which culminated in the completeness and incompleteness (or better "incompletability") results of Gödel [Göd30],[Göd31]. Somewhat later Gentzen defined a natural notion of formal proofs [Gen34], which is closer to actual mathematical deduction than the so-called Hilbert-type systems. Like Herbrand [Her30] Gentzen investigated the form and structure of mathematical proofs, while Gödel's famous theorems were more directed to provability.

In general it was not the purpose of proof theory to develop inference systems for actual deduction or proof search. Taking into account the complexity of relevant mathematical proofs this is hardly surprising.

The conceptual roots of automated deduction can be traced back to Leibniz's brave vision of a calculus ratiocinator [Leib], which would allow slution of arbitrary problems by purely mechanical computation, once they have been translated into a special formalism. Although this dream of complete mechanization cannot become reality (as we know from results Gödel and Turing obtained in the 1930s [Göd31] [Tur36]) we need not reject the whole idea. Indeed if we choose first-order predicate logic to be our special formalism, the problem of defining a (more modest) calculus ratiocinator makes sense again. As the problem of validity of formulas in first-order logic is undecidable [Chu36], we cannot hope for a *decision procedure* that always terminates and gives the right answer. However, there are complete calculi for first-order logic, i.e., calculi in which all valid formulas of first-order logic can be derived. Thus there exists a general mechanical method to *prove* the validity of first-order sentences and thus to *verify* problems formulated in first-order logic. Still the question remains whether the first-order calculi are really adequate for problem solving in a practical sense.

As computers and programming languages developed, the problem of finding mathematical proofs automatically was one of the first attacked in the field today called Artificial Intelligence. A direct application of Herbrand's proof theoretical result was given by Gilmore [Gil60] for performing automated deduction in first-order predicate logic.

The idea behind it was the following: Herbrand's theorem defines a method to reduce the provability of a formula F in predicate logic to the provability of a corresponding propositional formula F'. Because inference in propositional logic is simpler than in predicate logic this method seems to be quite natural from the point of view of logical complexity. But from the computational point of view this method has two serious defects: 1) the finding of such a formula F', 2) the costs of inference on F' itself.

Point 2) above was considerably improved by Davis and Putnam in the same year [DP60], while the search for F' and also the (possibly enormous) size of F' remained serious obstacles. The tactic of applying proof theory (which was developed for other purposes) to automated deduction directly has clearly failed. Obviously there was a need to model some natural techniques of proof finding employed by humans within a logical calculus. The invention of such a technique is the contribution of Robinson's famous paper [Rob65], which is a real landmark in automated deduction (and in some sense its very beginning). Particularly the consequent use of the unification principle created a substantially new type of logic calculus. The resolution principle (a combination of a propositional cut rule and the substitutional unification principle) leads to a spectacular improvement in performance versus Gilmore's method. While the propositional part of the resolution rule (the atomic cut) was changed in various ways and even abandoned (in Bibel's connection method [Bib82]), the unification principle is part of every relevant computational proof calculus. Though (as indicated above) resolution is no longer the only method in automated deduction today, it is still playing a central role and some features (such as the unification principle) are common to all computational calculi.

Taking into account some more recent developments in the field of automated deduction it would not be justified to call this book "Automated Theorem Proving". There are many other methods (like the tableau method, the connection method, the systems based on rewriting, etc.) which deserve a representation on an equal basis. Moreover, the term "Automated Theorem Proving" would suggest information about implementation, experiments, performance and applications; in fact many problems in program verification and knowledge representation (which are expressible in first-order logic) were successfully handled by methods of automated theorem proving. Another important area of applications is logic programming (the execution of a logic program corresponds to a resolution deduction). In order to avoid the (practically unsolvable) problem of covering all these topics and (at the same time) to give a detailed and deep analysis of the problems, we decided for a representation of the logical aspect of resolution. We will try to show that resolution is not only a powerful method in automated deduction, but also a tool to address problems in mathematical logic; particularly we will deal with the decision problem of first-order definable classes and with the complexity of first-order inference. Thus besides the issue of completeness (which is a typical one in resolution theory) we also discuss termination and complexity of resolution. As an application of resolution decision procedures we also present a method of automated model building, thus extracting information from a resolution prover even in cases where no refutation exists. The construction of such (counter-) models is of central importance to interactive theorem proving and to model-based inference. Finally, we will discuss extension methods for the purposes of formula normalization and of inference; here we will point out that preservation and creation of logical structure essentially improves the substance (and shortens the length) of proofs.

In Chapter 2 we present the basics of resolution theory by following the usual path of textbook presentations (Herbrand's theorem, the unification principle, and resolution). A little more emphasis than usual is laid on the transformation to clause form. Chapter 3 is devoted to resolution refinements; here an attempt is made to explain the concept of refinement from a more abstract point of view. Moreover, we handle clauses as disjunctions subjected to different normalization operators; this gives us the means for a unifying treatment of different formulations of resolution deductions (clauses may be lists, sets of literals, multisets of literals, etc.). In Chapter 4 we discuss the problem of redundancy and proof search. We define the concept of search complexity and analyze its behavior under different deletion methods. Special emphasis is laid on subsumption-based refinements and on subsumption algorithms. In the last subsection of Chapter 4 we present the theorem of Lee and applications to the clause implication problem. In Chapter 5 we show how some of the resolution refinements defined in Chapter 3 can be used as decision procedures for first-order classes. The method consists in selecting adequate refinements which always terminate (i.e., they produce finitely many resolvents only) on the clausal forms of formulas belonging to such a first-order decision class. Starting from results obtained by W.H.Joyner [Joy76] we also present some more recent results and methods in resolution decision theory. We also show that the satisfiable sets of clauses obtained by the resolution decision procedures can be used as raw material for an algorithmic model building procedure. This procedure does not require backtracking and constructs an atomic representation of a Herbrand model. In Chapter 6 we present a complexity theory of resolution which is based on the concept of Herbrand complexity. We distinguish proofs with and without ground projections and show that resolution refinements may strongly differ in their resolution proof complexity. We show that resolution proofs can be very long relative to shortest proofs in full logic calculi (in fact there exists no elementary bound); here the presentation is not completely self-contained and we refer to some more advanced literature. Finally, we illustrate the importance of extension methods in computational calculi. We present a structural transformation of negation normal forms to clause form (preservation of logical structure) and a method of functional extension (creation of logical structure). We show that functional extension (F-extension) is a natural technique to formulate and use lemmas in automated deduction. The strength added by F-extension to the resolution calculus may result in a nonelementary "speed-up" of proof complexity.

With the exception of some topics in Chapter 6 the book is self-contained and directed to graduate students with a standard background in predicate logic. Although all basic definitions for first-order predicate logic are given in Chapter 2, the book is not really adequate as an introduction to mathematical logic. Some remarks are addressed to readers who know some recursion theory, but they are not crucial to the understanding of the main results.

2. The Basis of the Resolution Calculus

2.1 First-Order Logic

Throughout the whole book we will focus on theorem proving in the context of first-order logic. First-order logic plays an important role in mathematical logic and computer science. First of all it is a rich language, by which algebraic theories, computational problems, and substantial knowledge representation in Artificial Intelligence can be expressed; due to its ability to represent undecidable problems (like the halting problem for Turing machines) first-order logic (or more precisely, the validity problem for first-order logic) is undecidable. On the other hand, the valid formulas of first-order logic can be obtained by logical calculi and thus are recursively enumerable. In this sense, first-order logic is mechanizable (we can find a proof for every valid sentence, but there is no decision procedure for validity).

We assume the reader to be familiar with the basics of first-order logic and do not give a motivation for its syntax and semantics. Nevertheless we need clear and exact concepts of its language for our transformation to normal forms and its semantic justification. Henceforth the language of first-order predicate logic is denoted by PL.

We assume that there we are given countably infinite sets V (individual variables), CS (constant symbols), FS (function symbols), and PS (predicate symbols). $FS = \bigcup_{i=1}^{\infty} FS_i$, where all sets FS_i are countably infinite and mutually disjoint (FS_i is the set of i-ary function symbols).

Similarly $PS = \bigcup_{i=1}^{\infty} PS_i$, PS_i countably infinite for all $i \in \mathbb{N}_+$, where PS_i is called the set of i-ary predicate symbols.

Symbols $\wedge, \vee, \neg, \rightarrow$ stand for the logical connectives and $\forall, \exists$ for the quantifiers.

Definition 2.1.1 (term). *The set of terms T is inductively defined by:*

a) $V \subseteq T$ (variables are terms),
b) $CS \subseteq T$ (constant symbols are terms),
c) If $t_1, \ldots, t_n \in T$ and $f \in FS_n$ then $f(t_1, \ldots, t_n) \in T$,
d) No other objects are terms.

Statements like in point (d) "no other objects are ..." will henceforth be omitted and will be considered as included in the concept "definition". If $t = f(t_1, \ldots, t_n)$ for an $f \in FS_n$ and terms $t_1, \ldots, t_n$ then t is called a *functional term*; the terms t_i are called the *arguments* of t. Variables and constant symbols have no arguments.

The occurrence of terms can be defined inductively: A term s *occurs* in a term t if either $s = t$ or s occurs in an argument of t.

The set of all variables occurring in a term t is denoted by $V(t)$. A term t with $V(t) = \emptyset$ is called a *ground term* .

A natural measure for the complexity of terms is *term depth*, which we denote by τ. Formally we define

$$\tau(t) = 0 \quad \text{for} \quad t \in V \cup CS,$$

$$\tau(f(t_1, \ldots, t_n)) = 1 + \max\{\tau(t_i) | i = 1, \ldots, n\}$$

for $f \in FS_n$, $t_1, \ldots, t_n \in T$.

Definition 2.1.2 (atom formula). *If $P \in PS_n$ and $t_1, \ldots, t_n \in T$ then $P(t_1, \ldots, t_n)$ is called an* atom formula *(or simply an atom); P is called the* leading symbol *and $t_1, \ldots, t_n$ are called the* arguments *of $P(t_1, \ldots, t_n)$.*

If A is an atom formula then $A, \neg A$ are called *literals*. The set of all atom formulas is denoted by AT, the set of all literals by LIT.

Definition 2.1.3 (predicate logic formula). *We define the set of predicate logic formulas PL inductively:*

a) *$AT \subseteq PL$ (atom formulas are in PL),*
b) *If $A, B \in PL$ then $(A \vee B), (A \wedge B), (A \to B) \in PL$,*
c) *If $A \in PL$ then $\neg A \in PL$,*
d) *If $A \in PL$ and $x \in V$ such that neither $(\forall x)$ nor $(\exists x)$ occurs in A then $(\forall x)A, (\exists x)A \in PL$.*

Property (d) guarantees that one does not quantify over a variable which already occurs in bounded form. Thus by our definition $(\forall x)(\exists x)P(x, x)$ is not a formula. On the other hand dummy quantifiers are allowed. Sometimes it is useful to restrict function symbols, constant symbols, and predicate symbols to the set occurring in some fixed formulas. Thus if $F \in PL$, we denote by $FS(F)$ the set of function symbols occurring in F; similarly we define $CS(F)$ and $PS(F)$, and $FS_n(F), PS_n(F)$. If Σ is a subset of $FS \cup CS$ we denote by T_Σ the set of all terms t with the property $FS(t) \cup CS(t) \subseteq \Sigma$. If $\Sigma = FS(F) \cup CS(F)$ for some formula F we also write $T(F)$ instead of T_Σ. We call $T(F)$ the set of terms over the signature of F. Similarly we define PL_Σ (the predicate logic over Σ) for every $\Sigma \subseteq PS \cup FS \cup CS$ and $PL(F)$, the set of formulas over the signature of F.

Example 2.1.1. $x \in V$. Let $f \in FS_1$; then $f(x) \in T$. If $R \in PS_1$ then $R(x), R(f(x)) \in AT$.
Thus $(R(x) \rightarrow R(f(x))) \in PL$ and $(\forall x)(R(x) \rightarrow R(f(x))) \in PL$.

Henceforth (unless stated differently) we will use the following notational conventions:

  variables: $x, y, z, u, v, w, x_1, y_1, \ldots$
  constant symbols: $a, b, c, d, e, a_1, b_1, \ldots$
  function symbols: $f, g, h, f_1, g_1, \ldots$
  predicate symbols: $P, Q, R, S, P_1, Q_1, \ldots$

Let A be a formula such that $A = A_1 \odot A_2$, $A = \neg B$, or $A = (Qx)B$ for $\odot \in \{\wedge, \vee, \rightarrow\}, x \in V$, and $Q \in \{\forall, \exists\}$; then A_1, A_2 and B are called *immediate subformulas* of A.

A formula A *occurs* in a formula B if either $A = B$ or A occurs in an immediate subformula of B. A is called *subformula* of B if A occurs in B.

We extend the concept of occurrence to that of terms in formulas:
Let s be a term and A be an atom formula. Then s occurs in A if s occurs in an argument of A. If A is an arbitrary formula then s occurs in A if it occurs in some subformula of A.

Definition 2.1.4 (free and bounded occurrences of variables). *Let A be an atom formula and x be a variable occurring in A; then x occurs free in A. If x occurs free in A and B is of the form $A \odot C$, $C \odot A$, $\neg A$, or $(Qy)A$ (for $\odot \in \{\wedge, \vee, \rightarrow\}$, $y \neq x$, $Q \in \{\forall, \exists\}$) then x occurs free in B.*
  x occurs bounded in A if there exist a subformula of A of the form $(Qx)B$ such that x occurs in B.

Example 2.1.2. Let $A = P(x) \wedge (\forall x)(\exists y)Q(x,y)$. Then x occurs free *and* bounded in A; y only occurs bounded in A. x occurs free in the subformula $(\exists y)Q(x,y)$.

For formula transformations it is convenient to have some kind of "unique" occurrences of variables, i.e., variables cannot bind different occurrences of subformulas and cannot occur bounded and free in the same formula.

Definition 2.1.5 (standard form). *A formula A is in standard form if the following property is fulfilled: If $B_1 \odot B_2$ (for $\odot \in \{\wedge, \vee, \rightarrow\}$) is a subformula of A and a variable x occurs bounded in B_1 (B_2) then x does not occur in B_2 (B_1).*

Example 2.1.3. The formulas $P(x) \wedge Q(x)$ and $(\forall x)P(x) \wedge Q(y)$ are in standard form, but $(\forall x)P(x) \wedge Q(x)$ and $(\forall x)P(x) \wedge (\forall x)Q(x)$ are not.
  A formula without free variables is called *closed* or a *sentence*. A formula without bounded variables is called *open*.

If A is an open formula containing the variables $x_1, \ldots, x_n$ then $(\forall x_1) \ldots (\forall x_n) A$ is called the *universal closure* of A. Strictly speaking the universal closure is not unique as the order of the variables is not fixed; however, all closures are semantically equivalent.

We are now in a position to define the semantics of PL. The key concept is that of an interpretation.

Definition 2.1.6 (interpretation). *An* interpretation *of a formula F in PL is a triple $\Gamma = (D, \Phi, I)$ having the following properties:*

1) D is a nonempty set, called the domain of Γ.
2) Φ is a mapping defined on $CS(F) \cup FS(F) \cup PS(F)$ such that
    2.1) $\Phi(c) \in D$ for $c \in CS(F)$.
    2.2) For $f \in FS_n(F)$, $\Phi(f)$ is a function of type $D^n \to D$.
    2.3) For $P \in PS_n(F)$, $\Phi(P)$ is a function of type $D^n \to \{\mathbf{T}, \mathbf{F}\}$ (i.e.,
    $\Phi(P)$ is an n-ary predicate over D).
3) I is a function of type $V \to D$, called the environment or variable assignment.

The range of $\Phi(P)$ for predicate symbols P consists of $\mathbf{T}, \mathbf{F}$ which clearly symbolize the truth values true and false. An alternative way to interpret predicate symbols $P \in PS_n$ is as subsets A of D^n, where A can be identified with $\Phi(P)^{-1}(\{\mathbf{T}\})$.

Interpretations Γ are the basis for the interpretation functions u_Γ for terms, and v_Γ for formulas.

Let $F \in PL$ and Γ be an interpretation of F; we define $u_\Gamma \colon T(F) \to D$ by

$$
\begin{aligned}
u_\Gamma(x) &= I(x) \text{ for } x \in V, \\
u_\Gamma(c) &= \Phi(c) \text{ for } c \in CS(F) \text{ and} \\
u_\Gamma(f(t_1, \ldots, t_n)) &= \Phi(f)(u_\Gamma(t_1), \ldots, u_\Gamma(t_n)) \text{ for } f(t_1, \ldots, t_n) \in T(F).
\end{aligned}
$$

In order to introduce an interpretation function for quantified formulas we require the concept of variable-equivalence of interpretations.

Definition 2.1.7 (equivalence of interpretations). *Two interpretations Γ, Δ of a formula F are called* equivalent modulo $x_1, \ldots, x_k$ *if there are D, Φ, I, J such that $\Gamma = (D, \Phi, I)$ and $\Delta = (D, \Phi, J)$ and $I(v) = J(v)$ for $v \in V \setminus \{x_1, \ldots, x_k\}$ (I and J differ at most on some of the x_i). If Γ is equivalent to Δ modulo x we write $\Gamma \sim_x \Delta$.*

Equivalent interpretations have the same domain and the same interpretation for constant, function, and predicate symbols; but they may differ on a finite set of variables.

We are now ready to define the evaluation of predicate logic formulas in $PL(F)$ via an interpretation Γ.

$v_\Gamma : PL(F) \to \{\mathbf{T}, \mathbf{F}\}.$

Let F be in PL and $\Gamma = (D, \Phi, I)$ be an interpretation of F. v_Γ is defined inductively over the structure of formulas in $PL(F)$:

1. If A is an atom formula in $PL(F)$ and $A = P(t_1, \ldots, t_n)$
   then $v_\Gamma(A) = \Phi(P)(u_\Gamma(t_1), \ldots, u_\Gamma(t_n))$
2. $v_\Gamma((A \wedge B)) = and(v_\Gamma(A), v_\Gamma(B))$
   $v_\Gamma((A \vee B)) = or(v_\Gamma(A), v_\Gamma(B))$
   $v_\Gamma((A \to B)) = impl(v_\Gamma(A), v_\Gamma(B))$
   $v_\Gamma(\neg C) = not(v_\Gamma(C))$
   for $A, B, C \in PL_F$ and the usual truth functions and, or and $impl$:

$\{\mathbf{T}, \mathbf{F}\}^2 \to \{\mathbf{T}, \mathbf{F}\}$ and not: $\{\mathbf{T}, \mathbf{F}\} \to \{\mathbf{T}, \mathbf{F}\}$, defined classically by the following tables:

and	T	F
T	T	F
F	F	F

or	T	F
T	T	T
F	T	F

impl	T	F
T	T	F
F	T	T

not	
T	F
F	T

It remains to define v_Γ for quantifiers:

$v_\Gamma((\forall x)A) = \mathbf{T}$ iff for all Δ such that $\Delta \sim_x \Gamma$ we have $v_\Delta(A) = \mathbf{T}$.

Similarly, $v_\Gamma((\exists x)A) = \mathbf{T}$ iff there exists a Δ such that $\Delta \sim_x \Gamma$ and $v_\Delta(A) = \mathbf{T}$.

An interpretation Γ of A verifies A if $v_\Gamma(A) = \mathbf{T}$; if $v_\Gamma(A) = \mathbf{F}$ then we say that Γ falsifies A.

Definition 2.1.8 (model). Let A be a formula containing the free variables $x_1, \ldots, x_n$ and Γ be an interpretation of A. Then Γ is called a model of A if all Δ that are equivalent to Γ modulo $\{x_1, \ldots x_n\}$ verify A. If A is a closed then Γ is a model iff Γ verifies A.

Example 2.1.4. $F = (\forall x)(P(x, a) \to Q(x, f(a)))$.
$\Gamma = (\mathbb{N}, \Phi, I)$ such that $\mathbb{N}$ is the set of natural numbers, $I(x) = 3$ and Φ is defined as:

$\Phi(a) = 0$, $\Phi(f)(n) = n + 1$ for all $n \in \mathbb{N}$, $\Phi(P) = \leq$, $\Phi(Q) = <$.

We set $\Gamma_x^* = \{\Delta \mid \Delta \sim_x \Gamma\}$ and compute $v_\Gamma(F)$.
$v_\Gamma(F) = \mathbf{T}$ iff

For all $\Delta \in \Gamma_x^* : v_\Delta((P(x, a) \to Q(x, f(a)))) = \mathbf{T}$

iff

For all $\Delta \in \Gamma_x^* : impl(v_\Delta(P(x, a)), v_\Delta(Q(x, f(a)))) = \mathbf{T}$

iff

For all $J \sim_x I : impl(J(x) \leq u_\Delta(a)$, $J(x) < u_\Delta(f(a))) = \mathbf{T}$

iff

> For all $k \in \mathbb{N} : impl(k \le 0, k < 0 + 1) = impl(k = 0, k < 1) = \mathbf{T}$.

> Because $k = 0$ really implies $k < 1$, $v_\Gamma(F) = \mathbf{T}$ and Γ is a model of F.

Definition 2.1.9 (satisfiability and validity). *Let F, G be arbitrary formulas in PL.*

a) *F is called* satisfiable *if F has a model.*
b) *F is called* valid *if every interpretation of F is a model of F.*
c) *F and G are* logically equivalent *(notation: $F \sim G$) if F and G have the same models.*
d) *F and G are called* satisfiability-equivalent *(shorthand: sat-equivalent) if: F is satisfiable iff G is satisfiable; we write $F \sim_{sat} G$.*

Note that with respect to $\sim_{sat}$ there are only two equivalence classes, the satisfiable and the unsatisfiable formulas.

Example 2.1.5. $(\forall x)P(x, a) \sim_{sat} (\forall x)P(a, x)$, as both formulas are satisfiable. But clearly $(\forall x)P(x, a) \not\sim (\forall x)P(a, x)$ (interpret a as 0 and P as $\le$ over $\mathbb{N}$).
$F_1 : (\exists x)P(x, a) \wedge \neg P(b, a)$ is satisfiable, but $F_2 : P(b, a) \wedge \neg P(b, a)$ is not. Thus F_1 is not sat-equivalent to F_2 (the existential variable x may not be replaced by the constant symbol b). Transforming F_1 into F_2 does not preserve sat-equivalence.

As opposed to classical logical calculi, where there is not much emphasis on term structures and substitutions, the concept of substitution plays a central role in resolution theory and computational logic. For this reason we have to introduce some basic notations and terminology for substitutions.

Definition 2.1.10 (substitution). *A substitution is a mapping λ of type $V \rightarrow T$ such that $\lambda(v) \ne v$ only for finitely many $v \in V$.*

If λ is a substitution, the set $\{v | v \in V, \lambda(v) \ne v\}$ is called the *domain* of λ (notation: $dom(\lambda)$).
   $\{\lambda(v) | v \in dom(\lambda)\}$ is called the *range* of λ (notation: $rg(\lambda)$). If $rg(\lambda) \subseteq T_\Sigma$ then we say that λ is a substitution over Σ. If $dom(\lambda) = \{x_1, \ldots, x_n\}$ and $\lambda(x_i) = t_i$ for $i = 1, \ldots, n$ we represent λ by the set $\{x_1 \leftarrow t_1, \ldots, x_n \leftarrow t_n\}$.
   As usual in automated theorem proving we use a post-fix notation for substitutions and define:

$$
\begin{aligned}
x\lambda &= \lambda(x) \text{ for } x \in V, \\
a\lambda &= a \text{ for } a \in CS, \\
f(t_1,\ldots,t_n)\lambda &= f(t_1\lambda,\ldots,t_n\lambda) \text{ for } f \in FS_n, \quad t_1,\ldots,t_n \in T, \\
P(t_1,\ldots,t_n)\lambda &= P(t_1\lambda,\ldots,t_n\lambda) \text{ for atom formulas } P(t_1,\ldots,t_n), \\
(A \circ B)\lambda &= (A\lambda \circ B\lambda) \text{ for } \circ \in \{\wedge,\vee,\rightarrow\}, \\
(\neg A)\lambda &= \neg(A\lambda), \\
((Qx)A)\lambda &= (Qx)A\lambda \text{ provided that } x \notin dom(\lambda), \quad Q \in \{\forall,\exists\}.
\end{aligned}
$$

On a quantified formula containing the bounded variables $x_1,\ldots x_n$ no substitution λ with $x_i \in dom(\lambda)$ for $1 \leq i \leq n$ is admissible.

Instead of $A\{x \leftarrow t\}$ we frequently write $A[\frac{x}{t}]$. A substitution λ, for which the range consists of variable-free terms only, is called a *ground substitution*.

2.2 Transformation to Clause Form

Essential to automated theorem proving is the inference on formulas of re-stricted syntax-type. On simpler formulas more efficient inference rules can be defined and it is easier to control proof search. There is the disadvantage, however, that the structure of the formulas is destroyed, which can lead to an increase in proof complexity (this problem will be analyzed in Chapter 6) and to the loss of the intuitive meaning. We will present a method of structural transformation in Section 6.2 which preserves most of the original structure of the formulas.

Before applying the inference method itself, we transform the formulas to quantifier-free conjunctive normal form. As our inference methods are based on the idea of proof by contradiction, we transform $\neg A$ to normal form (A being the PL-formula to be proved). Afterwards the normal form of $\neg A$ will be refuted by resolution.

The transformation to normal form consists of three steps:

1. Transform $\neg A$ into a formula B such that $B \sim \neg A$, B does not contain " $\rightarrow$ " anymore and "$\neg$" appears only in front of atom formulas (B is built up by literals and $\wedge,\vee,\forall,\exists$ only).

2. Eliminate all existential quantifiers; i.e., construct an $\exists$-free formula C such that $B \sim_{sat} C$ (in this step only sat-equivalence is preserved).

3. Transform C into a quantifier-free conjunctive normal form (CNF) D such that $C \sim_{sat} D$. We call D a clause form of $\neg A$.

Remark: The third step (transformation into CNF) can performed in various ways. In this section we define two methods, the first one preserving logical equivalence and the second one satisfiability equivalence only.

Step 1) : Let F, F_1, F_2 be formula variables for PL-formulas and $A \Rightarrow B$
mean "A is transformed to B". We have to apply $\Rightarrow$ until no further
reduction is possible.

1.1) $(F_1 \rightarrow F_2) \Rightarrow (\neg F_1 \vee F_2)$
1.2) $\neg(F_1 \wedge F_2) \Rightarrow (\neg F_1 \vee \neg F_2)$
1.3) $\neg(F_1 \vee F_2) \Rightarrow (\neg F_1 \wedge \neg F_2)$
1.4) $\neg\neg F \Rightarrow F$
1.5) $\neg(\forall x)F \Rightarrow (\exists x)\neg F$
1.6) $\neg(\exists x)F \Rightarrow (\forall x)\neg F$

Formally, 1.1)–1.6) define a canonical rewrite rule system; thus every order
of applications of the rules 1.1)–1.6) leads to the same irreducible form. Now
assume that $\neg A \Rightarrow^* B$ and B is irreducible under 1.1)–1.6) ($\Rightarrow^*$ denotes the
reflexive and transitive closure of $\Rightarrow$). Then it is easy to verify that B does
not contain $\rightarrow$ and $\neg$ only occurs in front of atoms.

Moreover for every reduction $X \Rightarrow Y$ in 1.1)–1.6) we have $X \sim Y$. As $\sim$
is an equivalence relation we also obtain $\neg A \sim B$; i.e., the transformation to
B is logically correct.

Example 2.2.1. $A = (\forall x)(\exists y)((P(x,y) \wedge Q(y)) \rightarrow R(y))$.

$$\neg A \overset{1.5}{\Rightarrow} (\exists x)\neg(\exists y)((P(x,y) \wedge Q(y)) \rightarrow R(y))$$
$$\overset{1.6}{\Rightarrow} (\exists x)(\forall y)\neg((P(x,y) \wedge Q(y)) \rightarrow R(y))$$
$$\overset{1.1}{\Rightarrow} (\exists x)(\forall y)\neg(\neg(P(x,y) \wedge Q(y)) \vee R(y))$$
$$\overset{1.3}{\Rightarrow} (\exists x)(\forall y)(\neg\neg(P(x,y) \wedge Q(y)) \wedge \neg R(y))$$
$$\overset{1.4}{\Rightarrow} (\exists x)(\forall y)((P(x,y) \wedge Q(y)) \wedge \neg R(y)).$$

The last formula is irreducible and of the appropriate form.

The (unique) formula obtained from a formula F after reduction via 1.1)–1.6)
will be denoted by $\alpha(F)$.

From now on, we only deal with closed PL-formulas in standard form; this
is no real restriction, as the semantics of closed formulas is invariant under
renaming of variables.

Step 2) : Eliminate the $\exists$-quantifiers. To perform this task, we introduce
some additional formalism first: Let Q_1, Q_2 be quantifiers and $F \in PL$.
We define $(Q_1 x) \prec_q (Q_2 y)$ $((Q_1 x)$ is in the scope of $(Q_2 y))$ if there is
a subformula F_1 of F such that $F_1 = (Q_2 y)G$ and $(Q_1 x)$ occurs in G.
Because the formulas are in standard form, $\prec_q$ is irreflexive, antisymmet-
ric, and transitive. As an example take $(\forall y)(P(y) \wedge (\exists x)Q(x,y))$; here we
have $(\exists x) \prec_q (\forall y)$. From now on we will call expressions of the form (Qx)
(Q being a quantifier and x a variable) quantifier expressions.

Reading the formulas from left to right (in the standard parenthesis notation), we can speak of the "first" quantifier, or of the first $\exists$-quantifier. The

idea of the following transformation is to eliminate the first $\exists$-quantifier and to insert a new term for the ($\exists$-quantified) variable.

Definition 2.2.1 (Definition of A^*). *Let A be a closed PL-formula in standard form and $\alpha(A) = A$. Let A^- be A after omission of the first existential quantifier expression $(\exists x)$ from A; if A is $\exists$-free then we define $A^- = A$.*

case a) *A is $\exists$-free: $A^* = A^- = A$*

case b) *There is no $\forall$-quantifier expression $(\forall y)$ with $(\exists x) \prec_q (\forall y)$. Then $A^* = [A^-][_a^x]$, where $a \in CS - CS(A)$ (a is a new constant symbol)*

case c) *Suppose that $(\forall y_1), \ldots, (\forall y_n)$ is the sequence of quantifier expressions (Q) (from left to right) with $(\exists x) \prec_q (Q)$ in A. Let $f \in FS_k \setminus FS(A)$ (f is a new function symbol); we define $A^* = [A^-][_{f(y_1,\ldots,y_k)}^x]$.*

Example 2.2.2.

$$
\begin{array}{rclclcl}
A & = & ((\forall x)(\exists y)P(x,y) & \vee & (\forall u)(\exists v)\neg Q(u,v)) & \wedge & (\exists z)\neg P(z,z) \\
A^* & = & ((\forall x)P(x,f(x)) & \vee & (\forall u)(\exists v)\neg Q(u,v)) & \wedge & (\exists z)\neg P(z,z) \\
(A^*)^* & = & ((\forall x)P(x,f(x)) & \vee & (\forall u)\neg Q(u,g(u))) & \wedge & (\exists z)\neg P(z,z) \\
((A^*)^*)^* & = & ((\forall x)P(x,f(x)) & \vee & (\forall u)\neg Q(u,g(u))) & \wedge & \neg P(a,a)
\end{array}
$$

We have seen in Example 2.2.2 that $*$ can be iterated until the formula is $\exists$-free; the end formula is also closed and in standard form.

Definition 2.2.2 (Definition of transformation β). *Let $B_0 = \alpha(A)$, $B_{k+1} = B_k^*$ for $k \in \mathbb{N}$ and $m = min\{l|B_l = B_{l+1}\}$ (m is the number of $\exists$-quantifiers in A). Then we define $\beta(A) = B_m$.*

If A is a closed formula in standard form then $\beta(A)$ is $\exists$-free (closed and in standard form too). It remains to show that β is semantically admissible, i.e., $\beta(A) \sim_{sat} A$ for all α-normalized A.

Note that we do not require a prenex form for A in order to eliminate the $\exists$-quantifiers. Indeed the outcome is different when we first transform A into prenex form B and then compute $\beta(B)$. Let us take A from Example 2.2.2; the possible prefix structures of a prenex normal form of A are:

$$\exists\forall\exists\forall\exists, \ \exists\forall\forall\exists\exists, \ \forall\exists\exists\forall\exists, \ \forall\exists\forall\exists\exists, \ \forall\forall\exists\exists\exists.$$

It is easy to see that for all these prenex forms F the transformed formula $\beta(F)$ contains at least one two-place function symbol; on the other hand $\beta(A)$, for A in Example 2.2.2, contains one-place function symbols only. In [BL94] it is proved that elimination of $\exists$-quantifiers without prior construction of prenex forms is always better and can lead to dramatic speed-up of proof complexity.

The elimination of $\exists$-quantifiers via β is usually called "skolemization" after the Norwegian logician Thoralf Skolem [Sko20].

The skolemization β essentially differs from the transformation α: while $\alpha(A) \sim A$ for all PL-formulas A, $\beta(A)$ usually is not logically equivalent to A. For example, for $A = (\forall x)(\exists y)P(x,y)$ we get $\beta(A) = (\forall x)P(x,f(x))$; clearly $\beta(A) \to A$ is valid, but $A \to \beta(A)$ is not. However, $A \sim_{sat} \beta(A)$. In order to show the preservation of sat-equivalence under application of β we first prove two technical lemmas:

Lemma 2.2.1. *Let A be a predicate logic formula in standard form such that $A = \alpha(A)$. Suppose that A contains a quantifier expression (Qx) such that (Qx) is maximal with respect to $\prec_q$ $((Qx)$ is not in the scope of another quantifier): We define $A_{(Qx)}$ to be A after omission of the expression (Qx). Then $A \sim (Qx)A_{(Qx)}$ (the quantifier can be shifted in front).*

Proof. By induction on the number of connectives $\land, \lor$ in A $(conn(A))$.
$conn(A) = 0$: Because $\alpha(A) = A$ we have $A = (Q_1 x_1) \ldots (Q_n x_n)F$, where F is a literal: $(Q_1 x_1)$ is the only maximal Q-expression with respect to $\prec_q$. Thus we have $(Q_1 x_1)A_{(Q_1 x_1)} = A$ and (trivially) $A \sim (Q_1 x_1)A_{(Q_1 x_1)}$.
Induction hypothesis: Suppose the assertion holds for all A (fulfilling the conditions above) such that $conn(A) \leq k$.
Case $conn(A) = k+1$:

a) $A = (Qx)F$. Again, (Qx) is the only Q-expression which is maximal. Thus $A = (Qx)A_{(Qx)}$ and $A \sim (Qx)A_{(Qx)}$.
b) $A = (F_1 \circ F_2)$ for $\circ \in \{\land, \lor\}$.
  Let (Qx) be a quantifier which is maximal and suppose that (Qx) occurs in A (the case (Qx) in B is completely analogous).

By definition of $conn$ we have $conn(F_1) \leq k$. Clearly $\alpha(F_1) = F_1$ and (Qx) is maximal in F_1. Thus we apply the induction hypothesis and get $F_1 \sim (Qx)F_{1_{(Qx)}}$. Clearly, also $A \sim ((Qx)F_{1_{(Qx)}} \circ F_2)$ holds. Because A is in standard form, x does not occur in F_2. By elementary quantifier shifting rules for $\land, \lor$ we get:

$$((Qx)F_{1_{(Qx)}} \circ F_2) \sim (Qx)(F_{1_{(Qx)}} \circ F_2)$$

But $F_2 = F_{2_{(Qx)}}$. So we obtain

$$(Qx)(F_{1_{(Qx)}} \circ F_2) = (Qx)(F_1 \circ F_2)_{(Qx)} = (Qx)A_{(Qx)}.$$

As a consequence we get $A \sim (Qx)A_{(Qx)}$.   ◊

Remark: Note that in Lemma 2.2.1 it is not required that A is closed. Thus by iterating the quantifier shifting (as indicated in Lemma 2.2.1) we get a prenex normal form; however we are not interested in a transformation to prenex normal form.

In the next theorem we prove the semantical justification of the ∃- elimination. As already mentioned, we cannot hope to preserve logical equivalence.

Lemma 2.2.2. *Let A be a closed formula in standard form such that $\alpha(A) = A$. Then $A \sim_{sat} A^*$.*

Proof. a) A is ∃-free; by $A = A^*$ we have $A \sim_{sat} A^*$.

b) A contains existential quantifiers and the first ∃-quantifier expression $(\exists x)$ is maximal in A. From Lemma 2.2.1 we conclude $A \sim (\exists x)A^-$. By definition of A^* we have $A^* = (A^-)[\begin{smallmatrix}x\\a\end{smallmatrix}]$ for $a \in CS \setminus CS(A)$. We prove $(\exists x)A^- \sim_{sat} A^*$.

   ($\Leftarrow$) b-1) If A^* is satisfiable then $(\exists x)A^-$ is satisfiable. This is trivial by the validity of $A^-[\begin{smallmatrix}x\\a\end{smallmatrix}] \to (\exists x)A^-$.

   ($\Rightarrow$) b-2) If $(\exists x)A^-$ is satisfiable then A^* is satisfiable. Suppose that $\Gamma = (D, \Phi, I)$ is a model of $(\exists x)A^-$, i.e., $v_\Gamma((\exists x)A^-) = \mathbf{T}$. By definition of v_Γ this is equivalent to: There exists an interpretation $\Gamma_1 = (D, \Phi, I_1)$ such that $\Gamma_1 \sim_x \Gamma$ and $v_{\Gamma_1}(A^-) = \mathbf{T}$. Define $\Gamma_2 = (D, \Phi_1, I_1)$ with $\Phi_1(a) = I_1(x)$ and $\Phi_1 = \Phi$ otherwise. Then Γ_2 is an interpretation of A^* and $v_{\Gamma_2}(A^*) = v_{\Gamma_1}(A^-) = \mathbf{T}$.
   We conclude that Γ_2 is a model of A^* and that A^* is satisfiable.

c) A contains ∃-quantifiers and the first ∃ -quantifier expression $(\exists x)$ is not maximal.
   Therefore there are quantifier expressions $(\forall y_1), \ldots, (\forall y_k)$ such that

$$k \geq 1 \text{ and } (\exists x) \prec_q (\forall y_1), \ldots, (\exists x) \prec_q (\forall y_k).$$

Suppose also that $(\forall y_1), \ldots, (\forall y_k)$ occur in A in the order above (it is easy to verify that the $(\forall y_i)$ themselves can be ordered by $\prec_q$.
Because A is in standard form, $\alpha(A) = A$ and $(\forall y_1)$ is maximal in A, we apply Lemma 2.2.1 and get $A \sim (\forall y_1)A_{(\forall y_1)}$. But $\alpha(A_{(\forall y_1)}) = A_{(\forall y_1)}$ and Lemma 2.2.1 can be applied again. An easy induction argument yields
   $A \sim (\forall y_1) \ldots (\forall y_k)A_0$ for $A_0 = A_{(\forall y_1) \cdots (\forall y_k)}$.
But in A_0 $(\exists x)$ is the first existential (and a maximal) quantifier expression. By $A_0 = \alpha(A_0)$ and by Lemma 2.2.1 we get $A_0 \sim (\exists x)A_0^-$.
Defining
   $A_1 = (\forall y_1) \ldots (\forall y_k)(\exists x)A_0^-$
we obtain $A \sim A_1$.

We show now that $A_1 \sim_{sat} A^*$ holds.

c1) ($\Leftarrow$) A^* satisfiable implies A_1 satisfiable. First of all we note that $A^* \sim A_1^*$; this is easy to realize, as the quantifiers $(\forall y_1), \ldots, (\forall y_k)$ can be shifted in A like in A^*, and $(\exists x)$ is in the range of the same quantifiers in A and A_1. Moreover we also have that

$$(\forall y_1) \ldots (\forall y_k)(A_0^-)[\begin{smallmatrix}x\\f(y_1,\ldots,y_k)\end{smallmatrix}] \to (\forall y_1) \ldots (\forall y_k)(\exists x)(A_0^-)$$

is valid, i.e., $A_1^* \to A_1$ is valid. Thus also $A^* \to A_1$ is valid and the satisfiability of A^* implies that of A_1.

c2) ($\Rightarrow$) A_1 satisfiable implies A^* satisfiable. Suppose that $\Gamma = (D, \Phi, I)$ is a model of A_1. By definition of v_Γ we conclude: For all Γ_1 such that

$\Gamma_1 \sim \Gamma$ mod $y_1, \ldots, y_k$ we obtain $v_{\Gamma_1}((\exists x)(A_0^-)) = \mathbf{T}$.

By the semantics of $\exists$-quantifiers we get:

For all Γ_1, such that $\Gamma_1 \sim \Gamma$ mod $y_1, \ldots, y_k$, there exists a Γ_2 such that $\Gamma_2 \sim_x \Gamma_1$ and $v_{\Gamma_2}(A_0^-) = \mathbf{T}$.

Γ_2 is of the form (D, Φ, I_2).

That means for all Γ_1 above there exist elements $\xi(\Gamma_1)$ in D such that $I_2(x) = \xi(\Gamma_1)$, particularly we get for every tuple $(I_1(y_1), \ldots, I_1(y_k))$ elements $\xi(\Gamma_1)$. By the axiom of choice there exists a function $\varphi : D^k \to D$ such that φ selects (exactly) one $\xi(\Gamma_1)$ for every tuple in D^k.

Let $A_2 = A_1^*$ for $A_1^* = (\forall y_1), \ldots, (\forall y_k) A_0^- [^x_{f(y_1, \ldots, y_k)}]$.

We define $\Gamma_3 = (D, \Phi', I)$ such that $\Phi'(f) = \varphi$ and $\Phi' = \Phi$ otherwise. By definition of φ, Γ_3 is a model of A_2. Moreover $A^* \sim A_2$ because in A_1^* and A the order relations of the quantifiers $(\forall y_1), \ldots, (\forall y_n)$ and $(\exists x)$ are the same. Thus we can obtain A_2 from A^* by shifting $(\forall y_1), \ldots, (\forall y_n)$ in front. Therefore Γ_3 is also a model of A^* and A^* is satisfiable. $\qquad \Diamond$

Theorem 2.2.1. *Let A be a closed formula in standard form. Then $\beta(A)$ is $\exists$-free and $\beta(A) \sim_{sat} A$.*

Proof. That $\beta(A)$ is $\exists$-free is trivial by definition. Because $A \sim \alpha(A)$, it is sufficient to prove that $\alpha(A) \sim_{sat} \beta(A)$. By definition of β

$\beta(A) = B_m$ for $B_0 = \alpha(A), B_{k+1} = B_k^*$ for all k and $m = \min\{k \mid B_{k+1} = B_k\}$. Thus it is sufficient to prove

$\alpha(A) \sim_{sat} B_k$ for all $k \in N$.

We proceed by induction on k.

$k = 0 : B_0 = \alpha(A)$ and thus $\alpha(A) \sim_{sat} B_0$.

(IH) Suppose that $\alpha(A) \sim_{sat} B_k$ holds.

Because A is closed, $\alpha(A)$ and B_k are closed as well; moreover $\alpha(B_k) = B_k$ (the normal form under α is not affected by quantifier eliminations). Thus Lemma 2.2.2 is applicable and we obtain $B_k \sim_{sat} B_k^*$, or $B_k \sim_{sat} B_{k+1}$. As $\sim_{sat}$ is an equivalence relation we get $\alpha(A) \sim_{sat} B_{k+1}$. $\qquad \Diamond$

In the formula $\beta(A)$ the only remaining quantifiers are universal. As connectives we only have $\wedge, \vee, \neg$, where $\neg$ only appears in front of atom formulas. Because of this syntax type, the positions of the $\forall$-quantifiers in $\beta(A)$ are irrelevant. Particularly $\beta(A)$ is equivalent to a purely universal prefix form.

Let $(\forall y_1), \ldots, (\forall y_n)$ be the quantifiers in $\beta(A)$ and $A_1 = (\forall y_1) \ldots (\forall y_n) M$ for $M = \beta(A)_{(\forall y_1) \ldots (\forall y_n)}$.

Then M is a quantifier free matrix defined by $\wedge, \vee$ over literals. Such a form is called *negation normal form*. There are several automated theorem proving methods working on negation normal forms, (e.g., path resolution [MR85]), but for the classical resolution calculus, the matrix is required to be in conjunctive normal form. Thus if $\beta(A)$ is like above we define

$$\gamma(A) = (\forall y_1) \ldots (\forall y_n) \mathrm{CNF}(M)$$

where CNF(M) is a conjunctive normal form for M. There is a well-known standard method to transform M into a logically equivalent conjunctive normal form. Let us assume that CNF(M) is such a form; we will discuss other methods of constructing CNFs (under preservation of sat-equivalence only) later.

Theorem 2.2.2. *Let A be a closed predicate logic formula in standard form. Then $A \sim_{sat} \gamma(A)$.*

Proof. By Theorem 2.2.1 we know that $A \sim_{sat} \beta(A)$ and $\alpha(\beta(A)) = \beta(A)$, $\beta(A)$ closed, $\exists$-free, and in standard form. Let $(\forall y_1) \ldots (\forall y_k)$ be all universal quantifiers in $\beta(A)$ occurring in this order. We define

$$
\begin{aligned}
C_0 &= \beta(A), \\
C_m &= (\forall y_1) \ldots (\forall y_m)\beta(A)_{(\forall y_1) \ldots (\forall y_m)} \text{ for } 1 \leq m \leq k.
\end{aligned}
$$

Then $C_k = (\forall y_1) \ldots (\forall y_k) M$ for $M = \beta(A)_{(\forall y_1) \ldots (\forall y_k)}$.

An easy induction argument yields $C_k \sim \beta(A)$:
As $\alpha(\beta(A)_{(\forall y_1) \ldots (\forall y_m)}) = \beta(A)_{(\forall y_1) \ldots (\forall y_m)}$ Lemma 2.2.1 is applicable and we get

$$\beta(A)_{(\forall y_1) \ldots (\forall y_m)} \sim (\forall y_{m+1})\beta(A)_{(\forall y_1) \ldots (\forall y_{m+1})}$$

for $m < k$.

By prefixing the quantifiers $(\forall y_1), \ldots, (\forall y_m)$ we obtain $C_m \sim C_{m+1}$.

By CNF(M) $\sim M$ we get

$$\beta(A) \sim (\forall y_1) \ldots (\forall y_k) \mathrm{CNF}(M) = \gamma(A).$$

Thus $\gamma(A)$ is sat-equivalent to A. $\Diamond$

As the quantifiers in $\beta(A)$ and $\gamma(A)$ contain no information (stored in their position within the formula) we may omit them completely. Then we are left with a quantifier-free conjunctive normal form. That means, for $\gamma(A) = (\forall y_1) \ldots (\forall y_k) M$ we omit the quantifiers and get M. Formally M is still a PL-formula, but as it is a conjunction of disjunctions we may delete all superfluous parentheses on the formula level of M. By this transformation we obtain a form:

$$F : (L_1^1 \vee \ldots \vee L_{k_1}^1) \wedge \ldots \wedge (L_1^m \vee \ldots \vee L_{k_m}^m)$$

where the L_j^i are all literals. We call such a form a *conjunctive normal form*.

Furthermore we may describe F as a list of lists and omit all $\wedge, \vee$ connectives. But (for convenience only) we keep $\vee$ as a separation symbol between literals and delete the "$\wedge$" only. Instead of F we get a set

$$\mathcal{C} = \{L_1^1 \vee \ldots \vee L_{k_1}^1, \ldots, L_1^m \vee \ldots \vee L_{k_m}^m\}.$$

The forms $L_1^i \vee \ldots \vee L_{k_i}^i$ are called *clauses*.

$\mathcal{C}$ is called the *clause form* of A. Note that $\mathcal{C}$ is only unique with respect to a specific transformation algorithm; by focusing on such an algorithm we may indeed speak about "the" clause form of A.

In computing a clause form of A, the form $\gamma(A)$ can be skipped; rather it is sufficient to do the following:

1) compute $\alpha(A)$
2) compute $\beta(A)$
3) omit all quantifiers, construct a CNF and, eventually, a clause form.

According to our concept of model (see Definition 2.1.8) the quantifier-free CNF obtained in (3) is logically eqivalent to $\beta(A)$ and thus satisfiability equivalent to A. The clause form is, strictly speaking, not a logical formula and requires some additional semantics (given in Definition 2.3.1).

Example 2.2.3 (Transformation to clause form).

$$A = ((\forall x)(\exists y)P(x,y) \wedge (\forall u)(\forall v)(P(u,v) \to R(u))) \to (\forall z)R(z).$$

In order to prove the validity of A we transform $\neg A$ to clause form.

$$\begin{aligned}
\alpha(\neg A) &= (\forall x)(\exists y)P(x,y) \wedge (\forall u)(\forall v)(\neg P(u,v) \vee R(u)) \wedge (\exists z)\neg R(z), \\
\beta(\neg A) &= (\forall x)P(x,f(x)) \wedge (\forall u)(\forall v)(\neg P(u,v) \vee R(u)) \wedge \neg R(a), \\
\text{CNF} \quad &: \quad P(x,f(x)) \wedge (\neg P(u,v) \vee R(u)) \wedge \neg R(a),
\end{aligned}$$

clause form: $\{P(x,f(x)), \neg P(u,v) \vee R(u), \neg R(a)\}$.

Definition 2.2.3 (clause).

a) $\square$ *is a clause (the empty clause).*
b) *Literals are clauses.*
c) *If C, D are clauses then $C \vee D$ is a clause.*

For clauses we define the identities $A \vee \square \vee B = A \vee B$, $\square \vee \square = \square$.

There are different definitions of the clause concept in literature. Clauses may be defined as sets [Rob65], [CL73], [Lov78] or as sequents [Llo87]. In J.A. Robinson's book [Rob79] the so-called Quad-notation is used, which is essentially sequential and separates positive from negative atoms (no negation sign is required); it is based on a more general concept of clause (there are existential and universal clauses). Our form (defined in [BL92]) is close to the sequent notation and facilitates certain types of logical analysis.

$\neg P(x) \vee Q(x, y) \vee R(x)$ can thus be represented as

$$\{\neg P(x), Q(x, y), R(x)\} \text{ or}$$

$$P(x) \vdash Q(x, y), R(x).$$

In the sequent notation we translate a clause of the form

$$\neg A_1 \vee \ldots \vee \neg A_n \vee B_1 \vee \ldots \vee B_m$$

for atoms $A_1, \ldots, A_n, \ldots, B_1, \ldots, B_m$ into

$$A_1, \ldots, A_n \vdash B_1, \ldots, B_m$$

where $\vdash$ is a metasymbol not occurring in the syntax of predicate logic.

The set notation corresponds to a normal form under idempotency, commutativity, and associativity of $\vee$. Sometimes (but not always) it may be practical to work with such normal forms. In the sequent notation it becomes transparent that a clause is a "logic-free" form ("to be negated" can be represented by standing in the antecedent of the sequent).

Definition 2.2.4 (Horn clause, Krom clause). *A* Horn *clause is a clause containing at most one positive literal. A* Krom *clause is a clause with at most two literals.*

Horn clauses can be represented as $P \vee \neg Q_1 \vee \ldots \vee \neg Q_n$ or – as usual in the theory of logic programming [Llo87] – $P \leftarrow Q_1, \ldots, Q_n$.

From the point of view of computational complexity, the transformation of A into $\alpha(A)$ and $\beta(A)$ is "harmless", i.e., performable in polynomial time. But if M denotes the quantifier-free rest of $\beta(A)$, the transformation to CNF(M) may be exponential. But we do not need a conjunctive normal form that is logically equivalent; any sat-equivalent CNF will do the job as well.

We just mention such a transformation, which is quite standard in the theory of NP-complexity [SS76]. Suppose we have a formula of the form

$$F : (\forall y_1) \ldots (\forall y_m) M$$

where M is in negation normal form. Iterate the following procedure: Select a subformula of M having the form

$$A \vee (B \wedge C) \quad \text{or} \quad (B \wedge C) \vee A$$

(if there is no such formula then M is already in CNF).

Replace $A \vee (B \wedge C)$ by
$$D : (A \vee P(y_1, \ldots, y_m)) \wedge (B \vee \neg P(y_1, \ldots, y_m)) \wedge (C \vee \neg P(y_1, \ldots, y_m)),$$
and $(B \wedge C) \vee A$ by

$D : (B \vee \neg P(y_1, \ldots, y_m)) \wedge (C \vee \neg P(y_1, \ldots, y_m)) \wedge (A \vee P(y_1, \ldots, y_m)),$
where P is a new m-place predicate symbol not occurring in M. Clearly the cases symmetric; for the sake of simplicity, we henceforth focus on the form $A \vee (B \wedge C)$.

Let $T(M)$ be the formula obtained from M by substituting D for $A \vee (B \vee C)$ in M. Because the formula $D \rightarrow (A \vee (B \wedge C))$ is valid and $\alpha(M) = M$ we obtain the validity of $T(M) \rightarrow M$. It is straightforward to show that

$$(\forall y_1) \ldots (\forall y_m) T(M) \rightarrow (\forall y_1) \ldots (\forall y_m) M$$

is valid too. Let $T(F) = (\forall y_1) \ldots (\forall y_m) T(M)$. To obtain $F \sim_{sat} T(F)$ it remains to show that the satisfiability of F implies the satisfiability of $T(F)$. By iterating the transformation T we eventually obtain a formula in CNF that is sat-equivalent to F.

Below we give a proof sketch for $F \sim_{sat} T(F)$ (an exact proof is left as an exercise).

Suppose that we have an interpretation Γ of F such that $v_\Gamma(F) = \mathbf{T}$ and $\Gamma = (\mathcal{D}, \Phi, J)$; for every $I : V \rightarrow \mathcal{D}$ we obtain interpretations $\Gamma_I = (\mathcal{D}, \Phi, I)$. Because F is a closed formula, $v_\Gamma(F) = \mathbf{T}$ iff for all I $v_{\Gamma_I}(M) = \mathbf{T}$.

For every such Γ_I we define an interpretation Δ_I of $T(M)$ such that $v_{\Gamma_I}(M) = v_{\Delta_I}(T(M))$.
Thus let I be an arbitrary environment. Then there are elements $d_1, \ldots, d_m \in \mathcal{D}$ such that $I(y_1) = d_1, \ldots, I(y_m) = d_m$. In defining Δ_I we distinguish several cases:

case a) $v_{\Gamma_I}((A \vee (B \wedge C)) = \mathbf{F}$.

We extend Φ (of Γ_I) to Φ' by $\Phi'(P)(d_1, \ldots, d_m) = \mathbf{F}$ and define $\Delta_I = (\mathcal{D}, \Phi', I)$.
Then $v_{\Delta_I}(A \vee P(y_1, \ldots y_n)) = or(v_{\Gamma_I}(A), \Phi'(P)(d_1, \ldots, d_m)) = or(\mathbf{F}, \mathbf{F}) = \mathbf{F}$.
By definition of D we conclude $v_{\Delta_I}(D) = \mathbf{F}$.
case b)
    case b1) $v_{\Gamma_I}(A) = \mathbf{T}$, $v_{\Gamma_I}(B \wedge C)) = \mathbf{F}$.
    In this case we define $\Phi'(P)(d_1, \ldots, d_m) = \mathbf{F}$. Then

$$v_{\Delta_I}(A \vee P(y_1, \ldots, y_m)) = or(v_{\Gamma_I}(A), \Phi'(P)(d_1, \ldots d_m)) = \mathbf{T}.$$

    Moreover

$$v_{\Delta_I}(B \vee \neg P(y_1, \ldots, y_m)) = or((v_{\Gamma_I}(B), \ not(\Phi'(P)(d_1, \ldots d_m))) = \mathbf{T}$$

    and (for similar reasons) $v_{\Delta_I}(C \vee \neg P(y_1, \ldots, y_m)) = \mathbf{T}$.
    case b2) $v_{\Gamma_I}(B \wedge C) = \mathbf{T}$.
    Here we set $\Phi'(P)(d_1, \ldots, d_m) = \mathbf{T}$.
    As above it is easy to see that $v_{\Delta_I}(D) = \mathbf{T}$.

Note that the definition of $\Phi'(P)$ does not depend on I (I only defines subcases for the definition of $\Phi'(P)$). As the cases (a), (b1), and (b2) exclude each other, Φ' is consistently defined. Thus there exist interpretations $\Delta_I = (\mathcal{D}, \Phi', I)$ such that

$$v_{\Gamma_I}(A \vee B \wedge C)) = v_{\Delta_I}(D) \text{ (for all } I : \mathcal{D} \to V).$$

All other subformulas of M remain unchanged under T and do not contain the predicate symbol P.

Therefore $v_{\Gamma_I}(M) = v_{\Delta_I}(T(M))$ for all I of type $V \to \mathcal{D}$. Because $v_{\Gamma_I}(M) = \mathbf{T}$ for all I we conclude $v_{\Delta_I}(T(M)) = \mathbf{T}$ for all I. Therefore $(\mathcal{D}, \Phi', J)$ (for some arbitrary $J : V \to \mathcal{D}$) is a model of $T(F)$ and we obtain

$$F \sim_{sat} T(F).$$

While, by the law of distributivity, we get $(A \vee B) \wedge (A \vee C)$ where the occurrence of A is doubled, the increase of $T(M)$ versus M is of constant length only; this is the reason for the polynomial time complexity of the transformation via T.

There exists another intuitively much more appealing tranformation into CNF, namely the structural transformation [Ede92]. As its analysis is made easier by resolution we delay its presentation and put it into the more general context of extension methods in the final chapter of this book.

Frequently it is the case that conjunctive normal forms are there at once, i.e., $\beta(A)$ without quantifiers is already in CNF. The following example illustrates that this is a quite typical case.

Example 2.2.4. Let F be a mathematical theorem of the form

$$F_1 \wedge \ldots \wedge F_n \to \text{CON}$$

where all F_i are of the form $(Q_1 x_1) \ldots (Q_m x_m)(A_1 \wedge \ldots \wedge A_k \to B_1 \vee \ldots \vee B_l)$ where A_i, B_j are atomic formulas. Suppose that the conclusion CON is of the same form. Transforming $\neg F$ into clause form we obtain

$$\alpha(\neg F) = F_1' \wedge \ldots \wedge F_i' \wedge \ldots \wedge F_n' \wedge \alpha(\neg\text{CON}).$$

Here the F_i' are of the form

$$(Q_1 x_1) \ldots (Q_m x_m)(\neg A_1 \vee \ldots \vee \neg A_k \vee B_1 \vee \ldots \vee B_l).$$

Moreover,

$$\alpha(\neg\text{CON}) = (Q_1^d x_1) \ldots (Q_m^d x_m)(P_1 \wedge \ldots \wedge P_l \wedge \neg Q_1 \wedge \ldots \wedge \neg Q_k)$$

(for CON $= (Q_1 x_1) \ldots (Q_m x_m)(P_1 \wedge \ldots \wedge P_l \to Q_1 \vee \ldots \vee Q_k)$).
Here the Q_i^d are the dual quantifiers defined as $\forall^d = \exists, \exists^d = \forall$.

$$\beta(\neg F) = A_1'' \wedge \ldots A_i'' \wedge \ldots \wedge A_n'' \wedge P_1' \wedge \ldots \wedge P_l' \wedge \neg Q_1' \wedge \ldots \wedge \neg Q_k'$$

without quantifiers where A_i'' is of the form

$$\neg A_1' \vee \ldots \vee \neg A_k' \vee B_1' \vee \ldots \vee B_l'.$$

Thus the axioms A_i are transformed into clauses A_i'' directly, while $\neg CON$ is transformed into a set of clauses. The final clause form is thus:

$$\mathcal{C} = \{A_1'', \ldots, A_n'', P_1', \ldots, P_l', \neg Q_1', \ldots, \neg Q_k'\}.$$

We have seen that the CNF can be obtained directly from $\beta(\neg F)$.

Exercises

Background on Exercises 2.2.1, 2.2.2:

Let $T_2, \ldots, T_6$ be the formula transformations defined in the first step of the normal form transformation, i.e.,

$$
\begin{array}{lll}
T_2 & : & \neg(F_1 \wedge F_2) \Rightarrow (\neg F_1 \vee \neg F_2) \\
T_3 & : & \neg(F_1 \vee F_2) \Rightarrow (\neg F_1 \wedge \neg F_2) \\
T_4 & : & \neg\neg F \Rightarrow F \\
T_5 & : & \neg(\forall x)F \Rightarrow (\exists x)\neg F \\
T_6 & : & \neg(\exists x)F \Rightarrow (\forall x)\neg F
\end{array}
$$

Let F be a PL-formula not containing "$\rightarrow$"
We define the depth d of F recursively by:

$$
\begin{array}{lll}
d(F) & = & 0 \text{ if } F \text{ is an atom,} \\
d(\neg F) & = & 1 + d(F), \\
d((\forall x)F) & = & d((\exists x)F) = 1 + d(F) \text{ and} \\
d((F_1 \circ F_2)) & = & 1 + max\{d(F_1), d(F_2)\} \text{ for } \circ \in \{\wedge, \vee\}.
\end{array}
$$

We say that F is irreducible with respect to $\{T_2, \ldots, T_6\}$ if there exists no G such that $F \Rightarrow G$.

Exercise 2.2.1. Show that $\Rightarrow$ is terminating (on PL-formulas without $\rightarrow$), i.e., there exists no infinite sequence $(F_n) \in N$ such that $F_0 \Rightarrow \ldots \Rightarrow F_n \Rightarrow F_{n+1} \ldots$ (Hint: use induction on the depth d).

Exercise 2.2.2. Show that $\Rightarrow$ defines a unique normal form, i.e, let $\Rightarrow^*$ be the reflexive, transitive closure of $\Rightarrow$ and $F \Rightarrow^* G$, $F \Rightarrow^* H$ such that G and H are both irreducible and show that $G = H$.

Exercise 2.2.3. Show that skolemization cannot be "" "parallelized", i.e., define a formula A such that $A = A_1 \wedge A_2$ and $\beta(A)$ is not sat-equivalent to $\beta(A_1) \wedge \beta(A_2)$.

2.3 Term Models and Herbrand's Theorem

In this section we present the model theoretical framework necessary to prove the completeness of the resolution calculus. In Section 2.2 we have shown that the validity problem for a closed formula A (in standard form) can be reduced first to the satisfiability problem for $\neg A$ and, eventually, to the satisfiability problem for a set of clauses $\mathcal{C}$.

In contrast to propositional logic we cannot prove unsatisfiability of $\mathcal{C}$ by checking all interpretations, because there are infinitely many (the set of all interpretations over a fixed, infinite, countable domain is not even countable). But we will show first that we can restrict the type of interpretations. Then we will define an algorithmic method to show unsatisfiability, based on tree structures for models. Finally, Herbrand's theorem will give us a propositional criterion for unsatisfiability of clause sets in first-order logic.

In clause forms quantifiers are omitted because their position does not contain any semantical information. However, we can interpret clauses as closed formulas. Formally we define PL-formulas corresponding to sets of clauses and carry over the semantics.

Definition 2.3.1. *Let* $C = L_1 \vee \ldots \vee L_n$ *be a clause and* $V(C) = \{x_1, \ldots, x_k\}$. *We define the PL-formula* $F(C) = (\forall x_1) \ldots (\forall x_k) C'$ *where* $C' = (\ldots ((L_1 \vee L_2) \vee L_3) \ldots \vee L_n)$; *for the uniqueness of* $F(C)$ *we may assume that the variables* $x_1, \ldots, x_k$ *(first) occur in this order from left to right in* C.

If $\mathcal{C} = \{C_1, \ldots, C_m\}$ *we extend the formula operator* F *to* $\mathcal{C}$ *via*

$$F(\mathcal{C}) = \bigwedge_{i=1}^{m} F(C_i)$$

(for uniqueness we may order $\mathcal{C}$ *lexicographically and then define the conjunction). That means a (finite) set of clauses* $\mathcal{C}$ *is interpreted as a conjunction of closed disjunctions of literals.*

The semantics for sets of clauses is defined via their PL-representative $F(\ )$: Γ is an interpretation of a set of clauses $\mathcal{C}$ iff Γ is an interpretation of $F(\mathcal{C})$. For the evaluation function v_Γ on sets of clauses we define:

$$v_\Gamma(\mathcal{C}) = v_\Gamma(F(\mathcal{C})).$$

By definition of v_Γ and $F(\mathcal{C})$ we get:

1. $v_\Gamma(\mathcal{C}) = \mathbf{T}$ iff for all $C \in \mathcal{C} : v_\Gamma(\{C\}) = \mathbf{T}$
2. Let $C = L_1 \vee \ldots \vee L_k$ be a clause. $v_\Gamma(\{C\}) = \mathbf{T}$ iff for all Γ' such that $\Gamma' \sim \Gamma \bmod V(C) \ or(v'_\Gamma(L_1), \ldots, v'_\Gamma(L_k)) = \mathbf{T}$

$(or(a_1, \ldots, a_k)$ is an abbreviation for $or(\ldots (or(a_1, a_2), \ldots, a_k)))$. Intuitively 1), 2) say: a set of clauses is true in Γ iff all clauses are true in Γ. A clause is true in Γ iff for all $V(C)$-equivalent interpretations at least one literal in C is true.

To investigate satisfiability we may restrict our concept of model. Herbrand models, also called term models, have the characteristic property that their domains only consist of terms; functions are interpreted as "term builders" over the term universe.

Definition 2.3.2 (Herbrand universe). *Let C be a (finite) set of clauses. We define*

$$H_0 = \begin{cases} CS(\mathcal{C}) & \text{if } CS(\mathcal{C}) \neq \emptyset \\ \{a\} & \text{if } CS(\mathcal{C}) = \emptyset, \text{ where } a \text{ is an arbitrary constant symbol.} \end{cases}$$

For $i \geq 1$ we define recursively

$$H_i = H_{i-1} \cup \{f(t_1, \ldots, t_n) \mid f \in FS_n(\mathcal{C}); t_1, \ldots, t_n \in H_{i-1}; n \in \mathbb{N}\}.$$

Let $H(\mathcal{C}) = \bigcup_{i=0}^{\infty} H_i$. $H(\mathcal{C})$ is called the Herbrand universe *of $\mathcal{C}$.*

In Definition 2.3.2, $H(\mathcal{C})$ is the set of all ground terms definable over the signature of $\mathcal{C}$ and H_i is the subset of terms having term depth $\leq i$.

Example 2.3.1. $\mathcal{C} = \{\neg P(x) \vee P(f(x)), P(h(x,x)), \neg P(h(u,v)) \vee \neg Q(v)\}$

$$\begin{aligned} H_0 &= \{a\}, \text{ as } CS(\mathcal{C}) = \emptyset. \\ H_1 &= \{a, f(a), h(a,a)\} \\ &\vdots \\ H_{i+1} &= \{f(t) \mid t \in H_i\} \cup \{h(s,t) \mid s,t \in H_i\} \quad \text{for all } i \end{aligned}$$

Obviously, $H(\mathcal{C})$ is infinite iff there are function symbols in $\mathcal{C}$. The set of all ground atoms definable over $H(\mathcal{C})$ is called the atom set of $\mathcal{C}$. Formally:

Definition 2.3.3 (atom set). *Let C be a set of clauses. The set*

$$AS(\mathcal{C}) = \{P(t_1, \ldots, t_n) \mid P \in PS_n(\mathcal{C}), t_i \in H(\mathcal{C}), n \in \mathbb{N}\}$$

is called the atom set *of $\mathcal{C}$. The corresponding set of literals is*

$$LS(\mathcal{C}) = AS(\mathcal{C}) \cup \{\neg A \mid A \in AS(\mathcal{C})\}.$$

In the last example we get $AS(\mathcal{C}) = \{P(t) \mid t \in H(\mathcal{C})\} \cup \{Q(t) \mid t \in H(\mathcal{C})\}$. $AS(\mathcal{C})$ is finite iff $H(\mathcal{C})$ is finite.

Definition 2.3.4 (ground instance). *Let C be a clause in $\mathcal{C}$ and λ be a ground substitution with $rg(\lambda) \subseteq H(\mathcal{C})$ and $V(C) \subseteq dom(\lambda)$. Then $C\lambda$ is called a ground instance of C (in $\mathcal{C}$).*

Ground instances are obtained by ground substitutions over the signature of $\mathcal{C}$ which substitute all variables in a clause.

Example 2.3.2. $\mathcal{C} = \{P(x) \vee P(f(x)), \neg P(a), \neg P(f(a))\}$.
$P(a) \vee P(f(a))$ and $P(f(a)) \vee P(f(f(a)))$ both are ground instances of $P(x) \vee P(f(x))$ in $\mathcal{C}$. $P(b) \vee P(f(b))$ is a variable-free instance, but not a ground instance in $\mathcal{C}$.

Definition 2.3.5 (Herbrand interpretation). *Let $\mathcal{C}$ be a finite set of clauses and $\Gamma = (H(\mathcal{C}), \Phi, I)$ be an interpretation of $\mathcal{C}$. Γ is called an H-interpretation (or Herbrand interpretation) if the following conditions are fulfilled:*

1) $\Phi(a) = a$ *for all* $a \in CS(\mathcal{C})$
2) *If* $f \in FS_n(\mathcal{C})$ *then*
    $\Phi(f)(h_1, \ldots, h_n) = f(h_1, \ldots, h_n)$ *for all* $h_1, \ldots, h_n \in H(\mathcal{C})$.

Note that an H-interpretation of $\mathcal{C}$ is not just an interpretation of $\mathcal{C}$ with domain $H(\mathcal{C})$. It is of central importance that the constants and function symbols get a fixed interpretation. Constant symbols are interpreted by themselves and function symbols as term builders over the universe $H(\mathcal{C})$. The interpretation of predicate symbols is not restricted, thus we still have a uncountable set of different H-interpretations for infinite $H(\mathcal{C})$.

Example 2.3.3.

$$\mathcal{C} = \{\neg P(x) \vee P(f(x)), \quad P(a), \quad \neg P(f(z)) \vee Q(u), \quad \neg Q(g(y, y))\}.$$
$$H(\mathcal{C}) = \{a, f(a), g(a, a), f(f(a)), \ldots\}.$$

Let Γ be an H-interpretation of $\mathcal{C}$. Then $\Phi(a)$, $\Phi(f)$, and $\Phi(g)$ have the following interpretations:

$$\Phi(a) = a,$$

$$\Phi(f) = \{(t, f(t)) | t \in H(\mathcal{C})\},$$

$$\Phi(g) = \{(s, t, g(s, t)) | s, t \in H(\mathcal{C})\}.$$

We are only free to define $\Phi(P)$ and $\Phi(Q)$; to this purpose we let

$$\Phi(P)(h) = \mathbf{T} \text{ for all } h \in H(\mathcal{C})$$
$$\Phi(Q)(h_1, h_2)) = \mathbf{F} \text{ for all } h_1, h_2 \in H(\mathcal{C})$$

It is easy to realize that $v_\Gamma(\neg P(x) \vee P(f(x))) = \mathbf{T}$ and $v_\Gamma(P(a)) = \mathbf{T}$, but $v_\Gamma(\neg P(f(z)) \vee Q(u)) = \mathbf{F}$. Thus Γ falsifies $\mathcal{C}$ (as does any interpretation, because $\mathcal{C}$ is unsatisfiable).

H-interpretations Γ are characterized by the interpretation of the predicate symbols; thus we can represent Γ by the set

$$\mathcal{M} = \{P(t_1, \ldots, t_{ar(P)}) \mid \Phi(P)(t_1, \ldots, t_{ar(P)}) = \mathbf{T}, t_i \in H(\mathcal{C})\},$$

where $ar(P)$ denotes the arity of P.

In the example above we obtain $\mathcal{M} = \{P(a), P(f(a)), P(g(a,a)), \ldots\}$. Because

$$v_\Gamma(P(t_1, \ldots, t_n)) = \Phi(P)(t_1, \ldots, t_n)$$

for H-interpretations, $\mathcal{M}$ is the subset of $AS(\mathcal{C})$ which is true in Γ. Because clauses are always interpreted as closed formulas, the environment I in $(H(\mathcal{C}), \Phi, I)$ is without importance.

The importance of H-interpretations is based on the fact that satisfiability in H-interpretations coincides with the concept of unrestricted satisfiability; otherwise expressed, if we can prove that a set of clauses does not have H-models then we know it is unsatisfiable.

We show this result by a construction principle which assigns an H-model Γ_H to every model Γ. The first step consists in defining a mapping $\omega : H(\mathcal{C}) \to D$ which defines a correspondence between the elements in the domain D and $H(\mathcal{C})$.

Definition 2.3.6 (mapping ω). *Let $\Gamma = (D, \Phi, I)$ be an arbitrary interpretation of $\mathcal{C}$.*

1. *For all $a \in CS(\mathcal{C})$ we define $\omega(a) = \Phi(a)$.*
   If $CS(\mathcal{C}) = \emptyset$ and $H_0(\mathcal{C}) = \{a\}$ then we define $\omega(a) = \alpha$ for some element $\alpha \in D$.
2. *For $f(t_1, \ldots, t_n) \in H(\mathcal{C})$ we define*

$$\omega(f(t_1, \ldots, t_n)) = \Phi(f)(\omega(t_1), \ldots, \omega(t_n)).$$

Note that ω coincides with u_Γ on ground terms.

Example 2.3.4. $\mathcal{C} = \{P(x,y) \vee \neg Q(y,x), \; \neg P(u,v) \vee P(f(u), f(v))\}$.

Let $\Gamma = (D, \Phi, I)$ be the following model:

$$
\begin{aligned}
D &= \mathbb{N}, \\
\Phi(f) &= \{(n, n^2) \mid n \in \mathbb{N}\}, \quad \Phi(P) = '<', \quad \Phi(Q) = '>' . \\
H(\mathcal{C}) &= \{a, f(a), f(f(a)), \ldots\}.
\end{aligned}
$$

Definition of ω:

As $a \notin CS(\mathcal{C})$ we set $\omega(a) = 2$ (arbitrarily).

$$\omega(f(a)) = \Phi(f)(\omega(a)) = \omega(a)^2 = 2^2$$
$$\vdots$$
$$\omega(f^{(n)}(a)) = \Phi(f)(\omega(f^{(n-1)}(a))) = \omega(f^{(n-1)}(a))^2.$$

By solving the recursion we obtain $\omega(f^{(n)}(a)) = 2^{2^n}$. By ω we can "translate" the predicates $\Phi(P), \Phi(Q)$ over D to predicates $\Phi_H(P), \Phi_H(Q)$ over $H(\mathcal{C})$:

$$\begin{aligned}
\Phi_H(P)(a,a) &= \Phi(P)(\omega(a),\omega(a)) &&= \Phi(P)(2,2) = (2 < 2) = \mathbf{F} \\
\Phi_H(Q)(a,a) &= \Phi(Q)(\omega(a),\omega(a)) &&= (2 > 2) = \mathbf{F} \\
\Phi_H(P)(a,f(a)) &= \Phi(P)(\omega(a),\omega(f(a))) &&= (2 < 2^2) = \mathbf{T}
\end{aligned}$$
$$\text{etc.}$$

By this definition of Φ_H on $PS(\mathcal{C})$ we obtain an H-interpretation Γ_H on $\mathcal{C}$ such that:
$$v_{\Gamma_H}(P(s,t)) = \Phi(P)(\omega(s),\omega(t)),$$
$$v_{\Gamma_H}(Q(s,t)) = \Phi(Q)(\omega(s),\omega(t)).$$

The definition of Γ_H in the example above motivates the following definition:

Definition 2.3.7. *Let $\Gamma = (D, \Phi, I)$ be an interpretation of a set of clauses $\mathcal{C}$; then $\Gamma_H = (H(\mathcal{C}), \Phi_H, J)$ (for arbitrary $J : V \to H(\mathcal{C})$) is called a corresponding H-interpretation to Γ if:*

a) Γ_H is an H-interpretation
b) $\Phi_H(P)(t_1,\ldots,t_{n(P)}) = \Phi(P)(\omega(t_1),\ldots,\omega(t_{n(P)}))$ for all $P \in PS(\mathcal{C})$ and all $t_1,\ldots t_{n(P)} \in H(\mathcal{C})$.

We are now ready to state the central results about H-interpretations.

Theorem 2.3.1. *A set of clauses is satisfiable iff it has an H-model.*

Proof. Let $\mathcal{C}$ be a set of clauses.

$\Leftarrow$: trivial, as every H-model is a model.

$\Rightarrow$: We prove the contraposition: If $\mathcal{C}$ does not have an H-model then $\mathcal{C}$ is unsatisfiable.

Thus we suppose that all H-interpretations falsify $\mathcal{C}$. Then for each Γ_H, such that Γ_H is the corresponding H-interpretation to an interpretation Γ, $v_{\Gamma_H}(\mathcal{C}) = \mathbf{F}$. Therefore the proof of "$\Rightarrow$" can be reduced to a proof of the following statement:

(*) If Γ_H is the corresponding H-interpretation to Γ and $v_{\Gamma_H}(\mathcal{C}) = \mathbf{F}$ then also $v_{\Gamma}(\mathcal{C}) = \mathbf{F}$.

As there is an interpretation Γ_H for every interpretation Γ of $\mathcal{C}$, statement (*) implies that $\mathcal{C}$ is unsatisfiable. It remains to prove (*):

Suppose that $v_{\Gamma_H}(\mathcal{C}) = \mathbf{F}$ for some interpretation Γ of $\mathcal{C}$. By the semantics of sets of clauses there exists a $C \in \mathcal{C}$ such that $v_{\Gamma_H}(\{C\}) = \mathbf{F}$.

But $\{C\}$ is interpreted as closed universal disjunction; so for $C = L_1 \vee \ldots \vee L_k$ we get:

There is a Γ'_H such that $\Gamma'_H \sim \Gamma_H \bmod V(C)$ such that for all $i = 1, \ldots, k$:
$$v_{\Gamma'_H}(L_i) = \mathbf{F}.$$

Let $L \in \{L_1, \ldots, L_k\}$. Then there is a $P \in PS(C)$ and terms $t_1, \ldots, t_n$ such that $L = P(t_1, \ldots, t_n)$ or $L = \neg P(t_1, \ldots, t_n)$. In order to prove $v_{\Gamma'}(C) = \mathbf{F}$ for a $\Gamma' \sim \Gamma \bmod V(C)$ – thus $v_\Gamma(\{C\}) = \mathbf{F}$ – it is enough to show: There is a Γ' such that $\Gamma' \sim \Gamma \bmod V(C)$ and

$$v_{\Gamma'_H}(P(t_1, \ldots, t_n)) = v_{\Gamma'}(P(t_1, \ldots, t_n)).$$

By direct inspection of $v_{\Gamma'_H}(P(t_1, \ldots, t_n))$ we find:

$$v_{\Gamma'_H}(P(t_1, \ldots, t_n)) = \Phi_H(P)(u_{\Gamma'_H}(t_1), \ldots, u_{\Gamma'_H}(t_n))$$

where $u_{(\,)}$ is the semantic function for terms, which – in this case – is of the type:

$$u_{\Gamma'_H} : T(C) \to H(C).$$

Every term is interpreted as a ground instance of itself; consequently there exists a ground substitution σ ($\sigma = \sigma(I'_H)$) such that $u_{\Gamma'_H}(t_i) = t_i\sigma$ for $i = 1, \ldots, n$. We obtain $v_{\Gamma'_H}(P(t_1, \ldots, t_n)) = \Phi_H(P)(t_1\sigma, \ldots, t_n\sigma)$.

From Definition 2.3.7 we get:

$$\Phi_H(P)(t_1\sigma, \ldots, t_n\sigma) = \Phi(P)(\omega(t_1\sigma), \ldots, \omega(t_n\sigma)).$$

Note that Γ'_H corresponds to Γ', because it corresponds to Γ, for every Γ' such that $\Gamma' \sim \Gamma \bmod V(C)$: From Lemma 2.3.1 (to be proved below) we conclude that there exists a Γ' such that $\Gamma' \sim \Gamma \bmod V(C)$ and

$$\omega(t_i\sigma) = u_{\Gamma'}(t_i) \text{ for } i = 1, \ldots, n.$$

Therefore we obtain

$$\Phi_H(P)(t_1\sigma, \ldots, t_n\sigma) = \Phi(P)(u'_\Gamma(t_1), \ldots, u'_\Gamma(t_n)).$$

and

$$v'_{\Gamma_H}(P(t_1, \ldots, t_n)) = v'_\Gamma(P(t_1, \ldots t_n)).$$

We conclude $v'_{\Gamma_H}(C) = v'_\Gamma(C) = \mathbf{F}$ and thus $v_\Gamma(C) = \mathbf{F}$.   ◊

Lemma 2.3.1. *Let* $\Gamma = (D, \Phi, I)$ *be an interpretation of a set of clauses* C *and* $\Gamma_H = (H(C), \Phi_H, J)$ *a corresponding H-interpretation. Furthermore let* σ *be a ground substitution over the signature of* C. *Then there exists a* Γ' *such that* $\Gamma' \sim \Gamma \bmod V(C)$ *and* $u_{\Gamma'}(t) = \omega(t\sigma)$ *for all* $t \in T(C)$ *with* $t\sigma \in H(C)$.

Proof. Let $\sigma = \{y_1 \leftarrow h_1, \ldots, y_m \leftarrow h_m\}$.
We define an environment I' by: $I'(y_1) = \omega(h_1), \ldots, I'(y_m) = \omega(h_m)$. Now let t be a term over C such that $V(t) \subseteq \{y_1, \ldots, y_m\}$. We show $u_{\Gamma'}(t) = \omega(t\sigma)$ by induction on $\tau(t)$ (term depth). $\tau(t) = 0$:

a) $t = a$ for $a \in CS(\mathcal{C})$:

  $\omega(a\sigma) = \omega(a) = \Phi(a) = u_{\Gamma'}(a)$ by definition of ω and $u_{\Gamma'}$.

b) $t \in \{y_1, \ldots, y_m\}$:

  $\omega(y_i\sigma) = \omega(h_i) = I'(y_i) = u_{\Gamma'}(y_i)$ by definition of I'.

(IH): Suppose that $u'_\Gamma(t) = \omega(t\sigma)$ for t with $\tau(t) \leq n$.
Let t be a term over $\mathcal{C}$ such that $V(t) \subseteq \{y_1, \ldots, y_m\}$ and $\tau(t) = n + 1$.
Then there exists a function symbol f and terms $t_1, \ldots, t_k$ such that $t = f(t_1, \ldots, t_k)$ and $\tau(t_i) \leq n$ for all $i = 1, \ldots, k$. By definition of ω and by (IH) we get:

$\omega(f(t_1, \ldots, t_k)\sigma) = \omega(f(t_1\sigma, \ldots, t_k\sigma)) = \Phi'(f)(\omega(t_1\sigma), \ldots, \omega(t_k\sigma)) = \Phi(f)(u_{\Gamma'}(t_1), \ldots, u_{\Gamma'}(t_n)) = u_{\Gamma'}(f(t_1, \ldots, t_n))$.   ◊

Theorem 2.3.1 simplifies proofs of unsatisfiabllity of sets of clauses. Since, in Herbrand interpretations, function symbols and constant symbols have a fixed semantics, one may limit attention to the interpretation of predicate symbols.

Example 2.3.5. Let $\mathcal{C} = \{C_1, C_2, C_3\}$ for

$$C_1 = P(x, f(a)), \quad C_2 = \neg P(u, v) \vee Q(f(v)) \text{ and } C_3 = \neg Q(z).$$

Let Γ be an H-interpretation of $\mathcal{C}$. Γ is characterized by the interpretation of predicate symbols and $v_\Gamma(P(t_1, \ldots, t_n)) = \Phi(P)(t_1, \ldots, t_n)$ for ground atoms $P(t_1, \ldots, t_n)$. After Example 2.3.3 we mentioned that Γ can be represented by $\mathcal{M}_\Gamma = \{A \mid A \in AS(\mathcal{C}), v_\Gamma(A) = \mathbf{T}\}$. We assume $v_\Gamma(\mathcal{C}) = \mathbf{T}$ and derive a contradiction. By $v_\Gamma(\mathcal{C}) = \mathbf{T}$ and by definition of v_Γ we obtain:

$Q(t) \notin \mathcal{M}_\Gamma$ for all $t \in H(\mathcal{C})$ (otherwise $v_\Gamma(\{C_3\}) = \mathbf{F}$).

But $v_\Gamma(\mathcal{C}) = \mathbf{T}$ and $Q(t) \notin \mathcal{M}_\Gamma$ for all $t \in H(\mathcal{C})$ implies $P(s, t) \notin \mathcal{M}_\Gamma$ for all $s, t \in H(\mathcal{C})$. For suppose $P(s, t) \in \mathcal{M}_\Gamma$ for some terms $s, t \in H(\mathcal{C})$; then $v_\Gamma(\{\neg P(u, v) \vee Q(f(v))\}) = \mathbf{F}$, as $v_\Gamma(\neg P(s, t) \vee Q(f(t))) = \mathbf{F}$. But $v_\Gamma(\{C_2\}) = \mathbf{F}$ implies $v_\Gamma(\mathcal{C}) = \mathbf{F}$. But as $P(s, t) \notin \mathcal{M}_\Gamma$ for all $s, t \in H(\mathcal{C})$, we get

$v_\Gamma(\{P(x, f(a))\}) = \mathbf{F}$ ($v_\Gamma(P(a, f(a))) = \mathbf{F}$) and $v_\Gamma(\{C_1\}) = \mathbf{F}$.

But $v_\Gamma(\{C_1\}) = \mathbf{F}$ implies $v_\Gamma(\mathcal{C}) = \mathbf{F}$. We obtain a contradiction.

The above argument yields a semantical counterpart of a resolution-based refutation. There the same problem reduction takes place in the object language (by cutting out literals). The technique of excluding interpretations from being models can be systematized and represented in the form of so-called semantic trees. Before developing the semantic tree concept we make some basic observations on H-interpretations:

Let $\Gamma = (H(\mathcal{C}), \Phi, I)$ be an H-interpretation of $\mathcal{C}$. We know that $v_\Gamma(A)$ for atoms $A \in AS(\mathcal{C})$ determines the value of $v_\Gamma(\mathcal{C})$. Because $\mathcal{C}$ represents a conjunctive normal form Γ falsifies $\mathcal{C}$ iff there exists a C in $\mathcal{C}$ such that Γ

falsifies C, i.e., $v_\Gamma(\{C\}) = \mathbf{F}$. Thus let $C = L_1 \vee \ldots \vee L_k$; in order that $v_\Gamma(\{C\}) = \mathbf{F}$, there must exist a ground instance $C' = L_1' \vee \ldots \vee L_k'$ such that $v_\Gamma(L_1' \vee \ldots \vee L_k') = \mathbf{F}$. Because C' is a disjunction we have:

$$L_i' \text{ is negative implies } v_\Gamma(L_i'^d) = \mathbf{T}, \ L_i' \text{ is positive implies } v_\Gamma(L_i') = \mathbf{F}.$$

Thus we obtain $v_\Gamma(L_i) = \mathbf{F}$ for $L_i \in AS(C), v_\Gamma(L_i^d) = \mathbf{T}$ for $L_i^d \in AS(C)$.

We conclude that a set of clauses C is unsatisfiable iff for every H-interpretation Γ there exists a ground instance C' of a clause C in C having the following property: If L is a positive literal in C' then $v_\Gamma(L) = \mathbf{F}$. If L is a negative literal in C' then $v_\Gamma(L^d) = \mathbf{T}$.

Example 2.3.6.

$$\begin{aligned}
C &= \{C_1, C_2, C_3, C_4\}, \\
C_1 &= P(x) \vee Q(f(a)), \quad C_2 = \neg P(y) \vee Q(y), \\
C_3 &= P(f(v)) \vee \neg Q(w), \quad C_4 = \neg P(z) \vee \neg Q(f(a)). \\
H(C) &= \{a, f(a), f(f(a)), \ldots\}.
\end{aligned}$$

We show that C is unsatisfiable in the following way: For every H-interpretation Γ there exists a ground instance of a clause which is falsified by Γ. Thus let Γ be an arbitrary H-interpretation. We must have $v_\Gamma(P(f(a))) = \mathbf{T}$ or $v_\Gamma(P(f(a))) = \mathbf{F}$.

case a) $v_\Gamma(P(f(a))) = \mathbf{T}$
    a1) $v_\Gamma(Q(f(a))) = \mathbf{T}$. In this case Γ falsifies $C_4\lambda$ for $\lambda = \{z \leftarrow f(a)\}$.
    a2) $v_\Gamma(Q(f(a))) = \mathbf{F}$. Γ falsifies $C_2\lambda = \neg P(f(a)) \vee Q(f(a))$
       for $\lambda = \{y \leftarrow f(a)\}$.
From case a) we conclude that every H-interpretation Γ with $v_\Gamma(P(f(a))) = \mathbf{T}$ falsifies C.
case b) $v_\Gamma(P(f(a))) = \mathbf{F}$
    b1) $v_\Gamma(Q(f(a))) = \mathbf{T}$. Let $C_3\lambda = P(f(a)) \vee \neg Q(f(a))$
       for $\lambda = \{v \leftarrow a, w \leftarrow f(a)\}$. Then Γ falsifies $C_3\lambda$.
    b2) $v_\Gamma(Q(f(a))) = \mathbf{F}$. Γ falsifies $C_1\lambda = P(f(a)) \vee Q(f(a))$
       for $\lambda = \{x \leftarrow f(a)\}$.

It follows from case b) that every interpretation Γ with $v_\Gamma(P(f(a))) = \mathbf{F}$ falsifies C. As either case a) or case b) must hold we conclude that every H-interpretation falsifies C, and – by Theorem 2.3.2 – that C is unsatisfiable.

The various alternatives in the above argument can be efficiently accomodated in a tree (see Figure 2.1).

In Figure 2.1 we use the notation "$\neg A$" to indicate $v_\Gamma(A) = \mathbf{F}$ for $A \in AS(C)$. A node crossed by "$\times$" indicates the falsification of a clause in C; the falsified clause is written below the node. Nodes of "$\times$"-type will be called failure nodes.

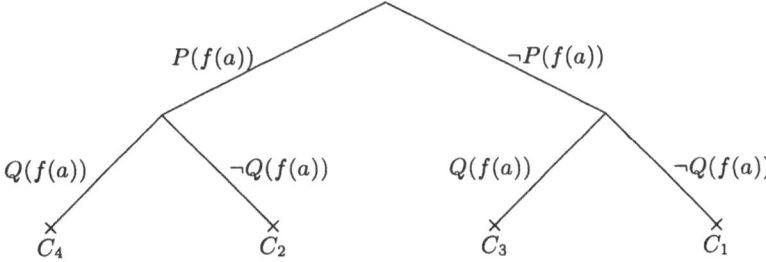

Fig. 2.1. A semantic tree

We first consider semantic trees representing (Herbrand) interpretations and focus on the relations to the falsification of sets of clauses later.

For a formal definition of the trees, we introduce the following notation: Let $\mathcal{T}$ be a tree; then $\mathrm{NOD}(\mathcal{T})$ denotes the set of nodes in $\mathcal{T}$, $E(\mathcal{T})$ the set of edges in $\mathcal{T}$ and $\xi(\mathcal{T}) : E(\mathcal{T}) \to X$ a labeling function for edges. $\mathrm{ROOT}(\mathcal{T})$ denotes the root of $\mathcal{T}$.

The triple $(\mathrm{NOD}(\mathcal{T}), E(\mathcal{T}), \xi(\mathcal{T}))$ is called a *labeled tree*.

Definition 2.3.8 (semantic tree). *A labeled tree* $\mathcal{T}$ *is called a* semantic tree *for a set of clauses* $\mathcal{C}$ *if the following conditions are fulfilled:*

1. $\xi(\mathcal{T}) : E(\mathcal{T}) \to LS(\mathcal{C})$.
2. $\mathcal{T}$ *is a binary tree (all nodes have degree two or zero).*
3. *If* e_1, e_2 *are the edges starting from a common node then* $\xi(\mathcal{T})(e_1) = \xi(\mathcal{T})(e_2^d)$.
4. *Let* N *be a node in* $\mathcal{T}$ *and* π *be the (unique) path connecting* N *with the root of* $\mathcal{T}$ *and let* $\gamma_N = \{L | (\exists e \in E(\mathcal{T}))$ *(e is an edge on* π *and* $\xi(\mathcal{T})(e) = L)\}$; *then* γ_N *does not contain complementary literals (i.e.,* γ_N – *the set of all literals appearing on* π – *is satisfiable).*

Example 2.3.7. $\mathcal{C} = \{P(x, y), \neg P((z, f(z)) \vee Q(a), \neg Q(w)\}.$

The tree in Figure 2.2 is a semantic tree, the trees in Figures 2.3 and 2.4 are not (in Figure 2.3 condition 3 is violated, in Figure 2.4 condition 4).

Every path on a semantic tree represents a (partial) truth assignment for $AS(\mathcal{C})$. Particularly $\gamma_N = \{L_1, \ldots, L_n\}$ represents all interpretations Γ with $v_\Gamma(L_i) = \mathbf{T}$ for $i = 1, \ldots, n$. The trees in Figures 2.2–2.4 are finite and thus cannot represent full (single) interpretations of $AS(\mathcal{C})$ (in these examples $AS(\mathcal{C})$ is infinite because $H(\mathcal{C})$ is infinite). To represent specific interpretations we also need infinite paths.

Definition 2.3.9. *A path in a tree* $\mathcal{T}$ *is called* maximal *if it starts in the root* $\mathrm{ROOT}(T)$ *and ends in a leaf node (note that maximal paths are always finite).*

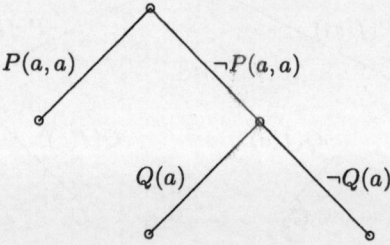

Fig. 2.2. A semantic tree

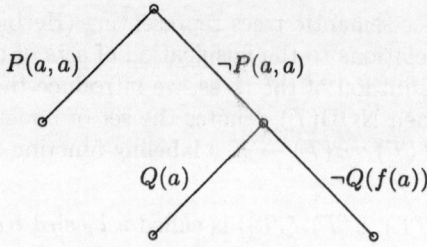

Fig. 2.3. Not a semantic tree – condition 3 violated

Definition 2.3.10 (branch). *A path π in a tree $\mathcal{T}$ is called a* branch *if the following properties are fulfilled:*

1. π starts in ROOT(T) *and*
2. π is infinite or maximal.

We extend the interpretations defined by nodes to interpretations defined by branches:
If a branch π is finite then we define $\gamma(\pi) = \gamma_N$ for the leaf node N of π; if $\pi = (N_i)_{i \in \mathbb{N}}$ then

$$\gamma(\pi) = \bigcup_{i \in \mathbb{N}} \gamma_{N_i}.$$

Definition 2.3.11. *Let $\mathcal{T}(C)$ be a semantic tree for C. $\mathcal{T}(C)$ is called* complete *if for every branch π in $\mathcal{T}(C)$ and*

for every $A \in AS(C)$ either $A \in \gamma(\pi)$ or $\neg A \in \gamma(\pi)$.

Complete trees have the following property: If $H(C)$ is infinite then every branch is infinite; if $H(C)$ is finite then also $AS(C)$ is finite and therefore every semantic tree is finite. For $H(C)$ finite and $\mathcal{T}(C)$ complete we thus get:

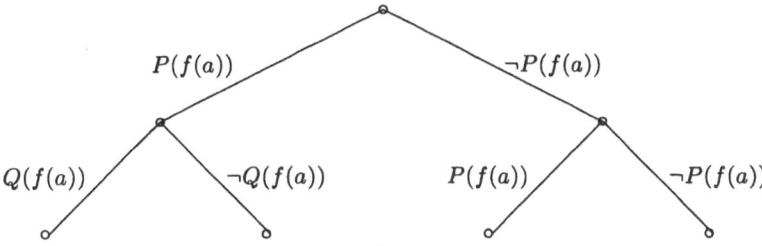

Fig. 2.4. Not a semantic tree – condition 4 violated

If N is a leaf node and A is an arbitrary atom in $AS(\mathcal{C})$ then either $A \in \gamma_N$ or $\neg A \in \gamma_N$. Thus (in both cases) every branch represents a full H-interpretation of $\mathcal{C}$.

A complete semantic tree can always be defined by the following construction: Let $\psi : \mathbb{N} \to AS(\mathcal{C})$ be an enumeration of $AS(\mathcal{C})$ (for $\| AS(\mathcal{C}) \| = k$ we define $\psi : M \to AS(\mathcal{C})$ for $M = \{0, \ldots, k - 1\}$).
We start by defining a tree $\mathcal{T}_0$ by:

$$\mathcal{T}_0 = (\mathrm{NOD}_0, E_0, \xi_0), \quad \mathrm{NOD}_0 = \{\mathrm{ROOT}(\mathcal{T}_0)\}, \quad E_0 = \emptyset, \quad \xi_0 = \emptyset.$$

Inductively suppose that $\mathcal{T}_n = (\mathrm{NOD}_n, E_n, \xi_n)$; if $n = \| AS(\mathcal{C}) \|$ then the construction is completed.

If $n < \| AS(\mathcal{C}) \|$ we continue as follows:
Let $\mathrm{FIN}(\mathcal{T}_n)$ be the set of all leaf nodes in NOD_n. For every $N \in FIN(\mathcal{T}_n)$ we define two new nodes $\alpha_1(N), \alpha_2(N)$ (which are different from each other) and

$$\mathrm{NOD}_{n+1} = \mathrm{NOD}_n \cup \bigcup_{N \in \mathrm{FIN}(\mathcal{T}_n)} \{\alpha_1(N), \alpha_2(N)\},$$
$$E_{n+1} = E_n \cup \bigcup_{N \in \mathrm{FIN}(\mathcal{T}_n)} \{(N, \alpha_1(N)), (N, \alpha_2(N))\}$$
$$\xi_{n+1} = \xi_n \cup \{(N, \alpha_1(N), \psi(n)), (N, \alpha_2(N), \neg\psi(n)) \mid N \in \mathrm{FIN}(\mathcal{T}_n)\}.$$

The new tree $\mathcal{T}_{n+1}$ is then defined as

$$\mathcal{T}_{n+1} = (\mathrm{NOD}_{n+1}, E_{n+1}, \xi_{n+1}).$$

Eventually we define the "limit" tree:

$$\hat{\mathcal{T}}(\mathcal{C}) = (\hat{\mathrm{NOD}}, \hat{E}, \hat{\xi}) \text{ with}$$
$$\hat{\mathrm{NOD}} = \bigcup_{i=0}^{\alpha} \mathrm{NOD}_i,$$
$$\hat{E} = \bigcup_{i=0}^{\alpha} E_i,$$
$$\hat{\xi} = \bigcup_{i=0}^{\alpha} \xi_i,$$

for $\alpha =$ cardinality of $AS(\mathcal{C})$.

If ψ is a computable function then $\hat{\mathcal{T}}(\mathcal{C})$ can be constructed effectively.
If $\| AS(\mathcal{C}) \| = k$ then clearly $\hat{\mathcal{T}}(\mathcal{C}) = \mathcal{T}_k$. Suppose now that $\mathcal{C}$ is unsatisfiable;

the question is, how do we recognize unsatisfiability in $\hat{\mathcal{T}}(\mathcal{C})$? The basic idea in solving this problem is the following: In constructing the trees $\mathcal{T}_n$ stop generation of further nodes and edges on nodes where a clause in $\mathcal{C}$ is falsified. The following example illustrates this procedure.

Example 2.3.8. $\mathcal{C} = \{C_1, C_2, C_3\}$
  $C_1 = P(x, f(x)),\quad C_2 = \neg P(a, f(y)) \vee R(y),\quad C_3 = \neg R(z),$
  $\psi(0) = P(a,a),\ \ \psi(1) = R(a),\ \ \psi(2) = P(a, f(a)),$ etc.

In Figures 2.5 – 2.7, 3 subtrees of $\hat{\mathcal{T}}(\mathcal{C})$ are constructed in order to get a final tree in which every leaf node falsifies a clause in $\mathcal{C}$.

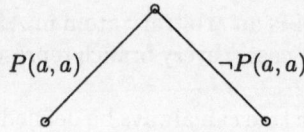

Fig. 2.5. $\mathcal{T}_1$

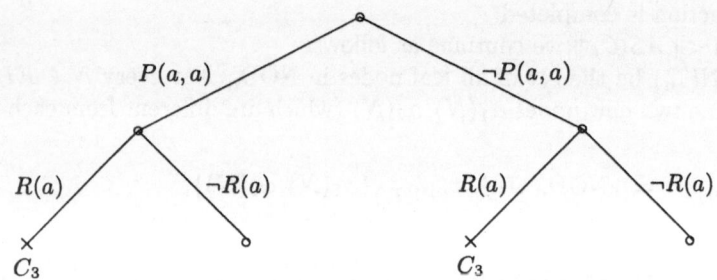

Fig. 2.6. $\mathcal{T}_2$

As every leaf node in $\mathcal{T}_3'$ falsifies a clause in $\mathcal{C}$, we stop the construction of subtrees of $\hat{\mathcal{T}}(\mathcal{C})$ completely. From $\mathcal{T}_3'$ we can extract the information that $\mathcal{C}$ is unsatisfiable; the reason is that every H-interpretation of $\mathcal{C}$ must be an extension of one of the partial interpretations appearing in the branches of $\mathcal{T}_3'$.

The example above motivates the following definitions:

Definition 2.3.12. *Let $\mathcal{T}(\mathcal{C})$ be a semantic tree for $\mathcal{C}$ and $N \in \mathrm{NOD}(\mathcal{T}(\mathcal{C}))$. N falsifies a clause C in $\mathcal{C}$ if there exists a ground instance C' of C such that for all L in C', L^d is contained in γ_N.*

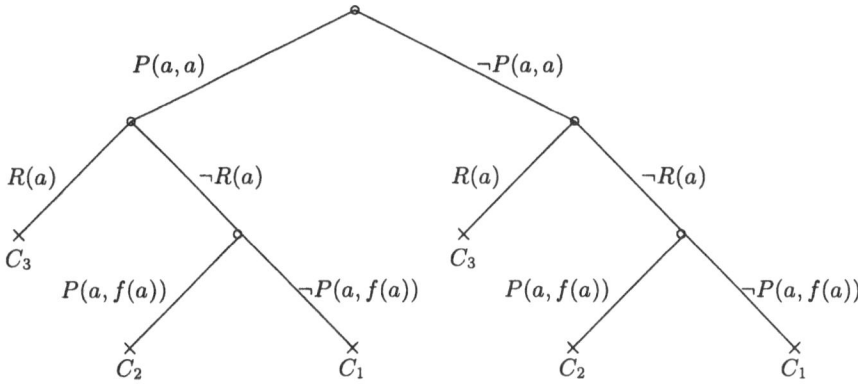

Fig. 2.7. $\mathcal{T}_3'$ (reduced $\mathcal{T}_3$)

Definition 2.3.13 (failure node). *Let $\mathcal{T}(\mathcal{C})$ be a semantic tree for $\mathcal{C}$ and $N \in \mathrm{NOD}(\mathcal{T}(\mathcal{C}))$. N is called a* failure node *if N falsifies a clause in $\mathcal{C}$, but no ancestor node of N falsifies any clause in $\mathcal{C}$.*

In the construction indicated in the last example, every node that falsifies a clause is also a failure node (as branches with failure nodes are not extended).

Definition 2.3.14. *A semantic tree is called* closed *if on every branch there is a failure node.*

In the last example, $\hat{\mathcal{T}}(\mathcal{C}), \mathcal{T}_3(\mathcal{C})$, and $\mathcal{T}_3'$ are all closed. $\mathcal{T}_3'$ is a minimal closed tree in the sense that falsifying nodes are all leaf nodes. Closed semantic trees represent unsatisfiable sets of clauses.

Theorem 2.3.2. *A set of clauses $\mathcal{C}$ is unsatisfiable iff $\hat{\mathcal{T}}(\mathcal{C})$ is closed.*

Proof. 1. $\mathcal{C}$ is unsatisfiable $\Rightarrow$:
  We have to show that on every branch there exists a failure node. Let $B = (N_i)_{i \in M}$ be a branch in $\hat{\mathcal{T}}(\mathcal{C})$ (M being $\mathbb{N}$ or an initial segment of $\mathbb{N}$). As $\hat{\mathcal{T}}(\mathcal{C})$ is complete, we find for every $A \in AS(\mathcal{C})$ an edge e on B such that $\xi(e) = A$ or $\xi(e) = \neg A$. We extend B to an H-interpretation $\Gamma_B = (H(\mathcal{C}), \Phi, I)$, where I can be chosen arbitrarily. Γ_B is determined by the values of v_{Γ_B} on $AS(\mathcal{C})$. We define

$$v_{\Gamma_B}(A) \quad = \quad \mathbf{T} \quad \text{if there is an } e \text{ on } B \text{ such that } \xi(e) = A,$$
$$= \quad \mathbf{F} \quad \text{otherwise}$$

Because $\mathcal{C}$ is unsatisfiable $v_{\Gamma_B}(\mathcal{C}) = \mathbf{F}$. Thus there exists a ground instance C' of a clause C in $\mathcal{C}$ such that for $C' = L_1' \vee \ldots \vee L_m'$ we obtain

$$v_{\Gamma_B}(L_i') \quad = \quad \mathbf{F} \quad \text{for positive } L_i'$$
$$v_{\Gamma_B}(L_i'^d) \quad = \quad \mathbf{T} \quad \text{for negative } L_i'.$$

By definition of v_{Γ_B} there must exist edges $e_1, \ldots, e_m$ on B such that $\xi(e_i) = L_i'^d$ for $i = 1, \ldots, m$.

Now let $e_k \in \{e_1, \ldots, e_m\}$ be the node among the $e_1, \ldots, e_m$ having the maximal depth (e_k is determined uniquely). Let $e_k = (N_1, N_2)$, then obviously

$$\{L_1'^d, \ldots, L_m'^d\} \subseteq \gamma_{N_2}$$

and N_2 falsifies C. If N_2 is not a failure node, there is an ancestor node N of N_2 which is a failure node and N is on B. We conclude that there exists a failure node on B.

2. $\hat{\mathcal{T}}(C)$ is closed $\Rightarrow$:

We show that C is unsatisfiable. For this purpose it is enough to show that C does not possess an H-model. Let $\Gamma = (H(C), \Phi, I)$ be an H-interpretation on C and let $\Psi : \mathbb{N} \to AS(C)$ be the enumeration used for the definition of $\hat{\mathcal{T}}(C)$. We define a mapping $\lambda : M \to LS(C)$ (M being an initial segment of $\mathbb{N}$ or $\mathbb{N}$ itself) by

$$\begin{aligned} \lambda(i) &= \Psi(i) \quad \text{for} \quad v_\Gamma(\Psi(i)) = \mathbf{T} \\ &= \neg\Psi(i) \quad \text{for} \quad v_\Gamma(\Psi(i)) = \mathbf{F} \end{aligned}$$

Because $\hat{\mathcal{T}}(C)$ is complete there exists a branch $B = (e_i)_{i \in M}$ such that $\xi(e_i) = \lambda(i)$ for all $i \in M$. For $M = \mathbb{N}$ the existence of the required branch B can be shown by induction. Because $\hat{\mathcal{T}}(C)$ is also closed there exists a ground instance $C' = L_1' \vee \ldots \vee L_m'$ of a clause $C \in \mathcal{C}$ such that $\{L_1'^d, \ldots, L_m'^d\} \subseteq \gamma_N$ for a failure node N appearing on B. By definition of λ there exist $i_1, \ldots, i_m \in \mathbb{N}$ such that $\lambda(i_k) = L_k'^d$ for $k = 1, \ldots, m$. As a consequence we get

$$\begin{aligned} v_\Gamma(\Psi(i_k)) = \mathbf{T} &\Rightarrow \Psi(i_k) = \lambda(i_k) = L_k'^d \Rightarrow \neg\Psi(i_k) = L_k', \\ v_\Gamma(\Psi(i_k)) = \mathbf{F} &\Rightarrow \Psi(i_k) = \lambda(i_k)^d = L_k' \Rightarrow \Psi(i_k) = L_k'. \end{aligned}$$

Therefore Γ falsifies C and thus also $\mathcal{C}$. $\diamond$

Let $\mathcal{T}$ be a semantic tree for $\mathcal{C}$. Starting from $\mathcal{T}$ we define a semantic tree $\mathrm{Clos}(\mathcal{T})$, which is constructed from $\mathcal{T}$ by omitting all paths which start from failure nodes. If $\mathcal{T}$ is closed then all leaf nodes of $\mathrm{Clos}(\mathcal{T})$ are failure nodes.

Lemma 2.3.2. *Let $\mathcal{T}$ be a closed semantic tree for a set of clauses C, then $\mathrm{Clos}(\mathcal{T})$ is finite.*

Proof. $\mathrm{Clos}(\mathcal{T})$ is (like $\mathcal{T}$ itself) a binary tree. Suppose now that $\mathrm{Clos}(\mathcal{T})$ is infinite; then by König's lemma [KD68] $\mathrm{Clos}(\mathcal{T})$ must possess an infinite path, as the degree of all nodes in $\mathrm{Clos}(\mathcal{T})$ is ≤ 2 (and thus finite). But if $\mathrm{Clos}(\mathcal{T})$ possesses an infinite path, there is a branch in $\mathcal{T}$ without failure node; but then $\mathcal{T}$ is not closed, contrary to the assumption. $\diamond$

The existence of a a closed finite tree is a property of central importance in mathematical logic as it guarantees *compactness*. We only mention this connection here, but refer to the excellent and deep presentation of this problem area in [Rob79](chapters 3 and 4).

We are now in a position to define a complete refutation calculus for clause logic. Given a set of clauses $\mathcal{C}$, we start the construction of $\hat{\mathcal{T}}(\mathcal{C})$, where $\hat{\mathcal{T}}(\mathcal{C})$ is defined via a recursive enumeration of the atom set $AS(\mathcal{C})$; each time we have found a failure node, we stop further node and edge construction below this node. It is obvious that this construction must result in $\mathrm{Clos}(\hat{\mathcal{T}}(\mathcal{C}))$; but $\mathrm{Clos}(\hat{\mathcal{T}}(\mathcal{C}))$ is finite if $\hat{\mathcal{T}}(\mathcal{C})$ is closed. By Theorem 2.3.2 we have a guarantee that $\hat{\mathcal{T}}(\mathcal{C})$ is closed if $\mathcal{C}$ is unsatisfiable. Thus for every unsatisfiable set of clauses $\mathcal{C}$ we can effectively generate a finite, closed semantic tree $\mathrm{Clos}(\hat{\mathcal{T}}(\mathcal{C}))$. The tree $\mathrm{Clos}(\hat{\mathcal{T}}(\mathcal{C}))$ can be considered as refutation of $\mathcal{C}$. More formally we get the refutation procedure shown in Figure 2.8.

procedure REFUTE;
  {input is a set of clauses $\mathcal{C}$}
begin
  if $\square \in \mathcal{C}$ then
    contr $\leftarrow$ TRUE
  elsebegin
    $n \leftarrow 0$;
    contr $\leftarrow$ FALSE;
    $\mathcal{T}_0 \leftarrow (\{\mathrm{ROOT}(\mathcal{T}_0)\}, \emptyset, \emptyset)$;
    while $\neg$ contr and $n \leq |AS(\mathcal{C})|$ do
      $\mathcal{T}_{n+1} \leftarrow \mathcal{T}_n$ extended   by   the   nodes
          $(N, \Psi^{(n)}, N_1)$ and $(N, \neg\Psi(n), N_2)$
          for all $N \in \mathrm{FIN}(\mathcal{T}_n)$ which are not
          failure nodes;
      contr $\leftarrow$ all nodes in $\mathrm{FIN}(\mathcal{T}_{n+1})$ are failure
          nodes;
      $n \leftarrow n + 1$
    end while
  end if;
  if contr then
    "satisfiable"
  else
    "unsatisfiable"
  end if
end refute.

Fig. 2.8. Refutation procedure REFUTE

The procedure REFUTE is only of theoretical interest (the algorithm is highly inefficient). Note that the predicate "N is failure node" is decidable. If $AS(\mathcal{C})$ is finite then REFUTE halts and decides the satisfiability of $\mathcal{C}$. If, on the other hand, $AS(C)$ is infinite and C is satisfiable then REFUTE does not halt. If C is unsatisfiable and $AS(C)$ is infinite, REFUTE must halt because $\text{Clos}(\hat{T}(C))$ is finite.

REFUTE is a quite crude method, as the construction of $\text{Clos}(\hat{T}(C))$ is based on the Herbrand universe and on the atom set defined by C; it does not take into account the actual structure of the atoms appearing in C. The following example shows that with respect to computational efficiency, the enumeration Ψ of $AS(C)$ is crucial.

Example 2.3.9. $C = \{C_1, C_2\}$ for $C_1 = P(h(y, h(a, b)))$, $C_2 = \neg P(h(z, z))$. C is unsatisfiable. If we choose the enumeration Ψ such that $\Psi(0) = P(h(h(a, b), h(a, b)))$ then the tree shown in Figure 2.9 is $\text{Clos}(\hat{T}(C))$, which refutes C.

Fig. 2.9. $\Psi(0) = P(h(h(a, b), h(a, b)))$

But choosing a straightforward enumeration Ψ (by ordering the atoms according to their size) as:

$$\Psi(0) = P(a), \ \Psi(1) = P(b), \ \Psi(2) = P(h(a, a)), \ldots,$$

$$\Psi(5) = P(h(b, b)), \Psi(6) = P(h(a, h(a, a))), \ldots, \ \Psi(21) = P(h(h(b, b), b)), \ldots$$

the construction of $\text{Clos}(\hat{T}(C))$ becomes very expensive. Note that finding $\text{Clos}(\hat{T}(C))$ for this Ψ requires (at least) the construction of T_{22}, a tree with 2^{22} leaf nodes.

The following theorem shows that unsatisfiability of sets of clauses C can be characterized by finite sets of ground clauses C' obtained from C by ground instantiation. As ground instances of clauses have a structure similar to the original this last characterization of unsatisfiablility is closer to common inference concepts.

Theorem 2.3.3 (Herbrand's theorem). *A set of clauses C is unsatisfiable iff there exists a finite unsatisfiable set of clauses C' such that C' consists of ground instances of clauses in C.*

Proof. 1. Suppose that there exists a finite, unsatisfiable set of ground clauses C' defined by C.

If $C' = \{C'_1, \ldots, C'_k\}$ then $F(C') = \bigwedge_{i=1}^k C'_i$ (note that for ground clauses C' $F(\{C'\}) = C'$). Because $F(\mathcal{C}) \to F(\{C\})$ is valid for every $C \in \mathcal{C}$ and $F(\{C\}) \to C'$ is valid for every ground instance C' of C, $F(\mathcal{C}) \to C'_i$ is valid for every C'_i above; but then, clearly, $F(\mathcal{C}) \to F(C')$ is valid too. We conclude that $\mathcal{C}$ must be unsatisfiable (as every model of $\mathcal{C}$ is a model of C').

2. Suppose that $\mathcal{C}$ is unsatisfiable. By Theorem 2.3.2 and Lemma 2.3.2 we conclude that $\mathrm{Clos}(\hat{\mathcal{T}}(\mathcal{C}))$ is finite. We define

$C' =$ the set of all ground instances of clauses in $\mathcal{C}$ which are falsified by a leaf node in $\mathrm{Clos}(\hat{\mathcal{T}}(\mathcal{C}))$.

As $\mathrm{Clos}(\hat{\mathcal{T}}(\mathcal{C}))$ is finite and every failure node falsifies at most finitely many ground instances, C' must be finite. We have to show that C' is unsatisfiable.

Let Γ be an H-interpretation of C'. Because $AS(C') \subseteq AS(\mathcal{C})$ there is an H-interpretation Δ of $\mathcal{C}$ such that $v_\Delta(A) = v_\Gamma(A)$ for all $A \in AS(C')$. Because $\mathcal{C}$ is unsatisfiable, Δ falsifies $\mathcal{C}$. Let I_Δ be a branch in $\hat{\mathcal{T}}(\mathcal{C})$ corresponding to Δ; then there is a failure node N on I_Δ which is leaf node in $\mathrm{Clos}(\hat{\mathcal{T}}(\mathcal{C}))$. By definition of C', there is a C' which is falsified on N and thus $v_\Delta(C') = \mathbf{F}$. By definition of Δ we have $v_\Gamma(C') = v_\Delta(C') = \mathbf{F}$ and Γ falsifies C'. Therefore Γ also falsifies C'. We conclude that C' is unsatisfiable. ◊

In its original form, Herbrand's theorem [Her31] was not formulated for sets of clauses but for existential prenex forms in predicate logic; it expressed the fact that such an existential form is provable in PL iff there is a (finite) disjunction, defined out of ground instances of the matrix, which is provable in PL. Thus the essence of Herbrand's theorem was the characterization of predicate logic provability by provability in propositional logic. This is also the essential aspect of Theorem 2.3.3, because the set of ground instances C' is unsatisfiable if it is unsatisfiable propositionally. The essential point in the reduction to C' is the decidability of the satisfiability problem in propositional logic (even the primitive truth-table method will do the job).

Example 2.3.10. We refute a clause set $\mathcal{C}$ by giving a finite set of ground instances C' and by proving that C' is unsatisfiable.

$$\mathcal{C} = \{P(x), \neg P(y) \vee Q(y), \neg Q(f(z)) \vee \neg Q(g(z))\}$$

$$C' = \{P(f(a)), P(g(a)), \neg P(f(a)) \vee Q(f(a)), \neg P(g(a)) \vee Q(g(a)),$$
$$\neg Q(f(a)) \vee \neg Q(g(a))\}.$$

The problem of proving unsatisfiability of C' can be reduced to showing unsatisfiability of the following conjunctive normal form F in propositional logic:

$$F = X \wedge Y \wedge (\neg X \vee Z) \wedge (\neg Y \vee U) \wedge (\neg Z \vee \neg U).$$

F corresponds to a set of propositional clauses

$$C^* = \{X,\ Y,\ \neg X \vee Z,\ \neg Y \vee U,\ \neg Z \vee \neg U\}.$$

We may refute F by truth-tables or C^* by a (finite) semantic tree. Later we will see that the Davis–Putnam method and resolution are superior to the rather rough expansion method REFUTE presented in this section.

Exercises

Background on Exercise 2.3.1:

Definition 2.3.15 (prenex form). *Let M be an open formula containing the variables $x_1, \ldots x_n$ and $Q_i \in \{\forall, \exists\}$ for $i = 1, \ldots n$. Then the formula $F :\ (Q_1 x_1) \ldots (Q_n x_n) M$ is in* prenex form; *M is called the* matrix *of F and $(Q_1 x_1) \ldots (Q_n x_n)$ the* prefix *of F. If the quantifiers in the prefix are all universal we call F a* universal prenex form; *if the quantifiers are all existential we speak about an* existential prenex form.

Every closed formula can be transformed into a logically equivalent prenex normal form; this can easily be achieved by variable renamings and by quantifier shifting rules. In particular every clause form is logically equivalent to a universal prenex form. If F is such a form we can define the Herbrand universe $H(F)$ in the same way as for sets of clauses; also the concept of H-interpretation can be carried over to universal prenex forms – even if the matrix is not in CNF (it is in fact like in Definition 2.3.5).

Exercise 2.3.1. Let F be a universal prenex formula. Show that F is satisfiable iff there exists an H-model of F.

Exercise 2.3.2. Let $\mathcal{T}$ be a semantic tree for a (finite) set of clauses $\mathcal{C}$ and N be failure node in $\mathcal{T}$. Prove that N can falsify only finitely many ground instances of clauses in $\mathcal{C}$.

Background on Exercise 2.3.3:

Let A, B be two closed formulas of predicate logic. A is called *taut-equivalent* to B (notation $A \sim_{taut} B$) if the validity of A is equivalent to the validity of B. Note that $A \sim_{taut} B$ does not imply $A \sim B$. For $\sim_{taut}$ there are results completely analogous (and dual) to those for $\sim_{sat}$.

Exercise 2.3.3. Let A be a prenex formula of the form $(Q_1 x_1) \ldots (Q_n x_n) M$. Prove that there exists a substitution instance M' of M and a set $\{y_1, \ldots, y_m\} \subseteq \{x_1, \ldots, x_n\}$ such that

$$A \sim_{taut} (\exists y_1) \ldots (\exists y_m) M'$$

(the right hand side is called the existential form of A).

Background on Exercise 2.3.4:

Herbrand's theorem is not restricted to clausal form. One can characterize the unsatisfiability of every universal prenex form A by a finite conjunction of ground instances of the matrix of A.

Exercise 2.3.4 (Herbrand's theorem for universal prenex forms).
Let $F: (\forall x_1)\ldots(\forall x_n)M$ be a universal prenex form. Show that F is unsatisfiable iff there exists a finite set of ground instances $\{M'_1,\ldots,M'_m\}$ of M such that $M'_1 \wedge \ldots \wedge M'_m$ is unsatisfiable.

Background on Exercise 2.3.5:

Herbrand's theorem can be formulated for existential prenex forms as well (this is also the original formulation). We only have to replace satisfiability by validity and conjunction by disjunction. Together with prenexing and the transformation in Exercise 2.3.3 this gives a method to characterize the validity of closed formulas by disjunctions of ground formulas.

Exercise 2.3.5 (Herbrand's theorem for existential prenex forms).

Let $F: (\exists x_1)\ldots(\exists x_n)M$ be an existential prenex form. Show that F is valid iff there exists a finite set of ground instances $\{M'_1,\ldots,M'_m\}$ of M such that $M'_1 \vee \ldots \vee M'_m$ is valid.

2.4 Decision Methods for Sets of Ground Clauses

2.4.1 Gilmore's Method

The first automated theorem prover for first-order logic was written by Gilmore in 1960 [Gil60]; it was essentially based on Herbrand's theorem and reduced unsatisfiability to propositional unsatisfiability. In order to get a simple method to determine the unsatisfiability of the set of ground clauses $\mathcal{C}'$ corresponding to $\mathcal{C}$, he transformed $\mathcal{C}'$ to disjunctive normal form. In getting the "candidates" $\mathcal{C}'$, Gilmore used saturation by ground instances from $H_n(\mathcal{C})$ for $n = 0, 1, \cdots$. More formally, let $\mathcal{C}$ be a set of clauses and

$\mathcal{C}'_n$ = set of all ground instances $C\lambda$ for $C \in \mathcal{C}$ and $rg(\lambda) \subseteq H_n(\mathcal{C})$.

Clearly, $\mathcal{C}'_n$ is finite for all n. By Herbrand's theorem we know that, in case $\mathcal{C}$ is unsatisfiable, there must be an n such that $\mathcal{C}' \subseteq \mathcal{C}'_n$. Thus one gets a refutational method by successively generating the sets $\mathcal{C}'_n$, testing them for satisfiability, and stopping if an unsatisfiable $\mathcal{C}'_n$ is found. If such a method of level saturation is chosen, it is crucial to have an adequate algorithm for deciding satisfiability of propositional sets of clauses.

Gilmore's technique was the following:

Transform C'_n into disjunctive normal form DNF(C'_n) and test DNF(C'_n) for satisfiability.

The reason for such a transformation can be found in the fact that satisfiability of disjunctive normal forms can be decided in deterministic time $O(n \log n)$, n being the length of the formula:

we only have to order the constituents by an atom ordering and then look whether "$A \wedge \neg A$" appears in constituents. If every constituent contains such a complementary pair then DNF(C'_n) (and thus C'_n) is unsatisfiable.

We give Gilmore's method in pseudocode (Figure 2.10), considering the case of an infinite Herbrand universe.

```
begin
    contr ← FALSE;
    n ← 0;
    while not contr do
        D' ← DNF(C'_n);
        contr ← all constituents in D' contain complementary literals;
        n ← n + 1
    end while
end
```

Fig. 2.10. Gilmore's method

By the comments given above we realize that Gilmore's method is indeed a refutational method for clause logic. but there are two weak points:

1) The generation of the C'_n and
2) the disjunctive normal form.

Point 1) is germane to all "direct" applications of Herbrand's theorem, while point 2) concerns propositional logic only.

Indeed, the transformation of a conjunctive into a disjunctive normal form is (almost always) exponential.

If we start with a set of clauses $\{L^i_1 \vee L^i_2 / i = 1, \cdots, n\}$ (containing $2n$ literal occurrences only) the disjunctive normal form will contain $n2^n$ literal occurrences.

Moreover, we cannot hope to get a fast (i.e., polynomial) transformation of the C'_n into sat-equivalent disjunctive normal forms D_n'; such a transformation would give no less than the result P = NP (for the NP-problem see e.g. [GJ79], because (by such a transformation) the NP-complete satisfiability problem for conjunctive normal forms would be solvable in deterministic polynomial time.

Gilmore's pioneering implementation did not yield actual proofs of even quite simple predicate logic formulas. One possible improvement of Gilmore's method consists in avoiding the transformation to DNF by developing decision methods for satisfiability directly on conjunctive normal forms. Such an improvement was achieved by Davis and Putnam [DP60] shortly after Gilmore's implementation.

2.4.2 The Method of Davis and Putnam

Like Gilmore's method, the method of Davis and Putnam is based on the successive production of ground clause sets C'_n and testing C'_n for unsatisfiability. But the propositional decision procedure of Davis and Putnam is much more efficient than Gilmore's. Although it is no longer up to date for predicate clause logic, it is still a very efficient method for testing satisfiability of propositional conjunctive normal forms; moreover we will use the method of Davis and Putnam as a proof technique (at the metalevel) in later chapters. The method we describe in this section is in fact a variant defined by M. Davis, G. Logemann, and D. Loveland in [DLL62]. We start with a motivating example.

Example 2.4.1. Let C be the set of clauses

$$C = \{P(f(x)) \vee R(y),\ \neg P(u) \vee R(u),\ \neg R(f(z)) \vee \neg R(w)\}.$$

We consider the following set of ground instances

$$C' = \{P(f(a)) \vee R(f(a)),\ \neg P f(a)) \vee R(f(a)),\ \neg R(f(a)) \vee \neg R(f(a))\}.$$

C' is a subset of C'_1.

Our first reduction consists in omitting multiple literals within clauses. By this step we reduce C' to

$$C'' = \{P(f(a)) \vee R(f(a)),\ \neg P f(a)) \vee R(f(a)),\ \neg R(f(a))\}.$$

Suppose now that C'' is satisfiable, i.e., $v_\Gamma(C'') = \mathbf{T}$ for some H-interpretation Γ. But $v_\Gamma(C'') = \mathbf{T}$ only if all clauses in C'' evaluate to $\mathbf{T}$, thus $v_\Gamma(R(f(a)))$ must be $\mathbf{F}$.

If $v_\Gamma(R(f(a))) = \mathbf{F}$, then $v_\Gamma(C'') = v_\Gamma(C^{(3)})$ for $C^{(3)} = \{P(f(a)), \neg P(f(a))\}$.

But $C^{(3)}$ is clearly contradictory and we obtain $v_\Gamma(C^{(3)}) = \mathbf{F}$. Thus we get a contradiction and have to conclude that C'' – and also C' – is unsatisfiable.

For computational purposes it is practical to remove multiple occurrences of literals in clauses.

Definition 2.4.1. *A clause C is called* reduced *if every literal in C occurs at most once. If C is an arbitrary clause we write C_r for the reduced clause derived from C by keeping the leftmost occurrence of all multiple literals.*

Definition 2.4.2 (The rules of Davis and Putnam). *Let C' be a set of reduced ground clauses. We define the following rules on C':*

1. *Tautology rule:*
   Delete all clauses in C' containing complementary literals.

2. *One-Literal-Rule (C' does not contain tautologies):*
   Let $C \in C'$ and $C = L$ for a literal L.
   a) Remove all clauses D from C' which contain L.
   b) Delete L^d in the remaining clauses.

3. *Pure literal rule (C' does not contain tautologies):*
   Let $\mathcal{D}'$ be a subset of C' with the following property: There exists a literal L appearing in all clauses of $\mathcal{D}'$, but L^d does not appear in C'.
   *   Rule: Replace C' by $C' - \mathcal{D}'$.*

4. *Splitting rule (C' does not contain tautologies).*
   Let $C' = \{A_1, \ldots, A_n, B_1, \ldots, B_m\} \cup R$ such that
   a) R neither contains L nor L^d.
   b) All A_i contain L, but not L^d.
   c) All B_j contain L^d, but not L.

Let

$$A_i' = A_i \text{ after deletion of } L, \ B_j' = B_j \text{ after deletion of } L^d.$$

Then the rule consists in splitting C' into C_1', C_2' for

$$C_1' = \{A_1', \ldots, A_n'\} \cup R,$$

$$C_2' = \{B_1', \ldots, B_m'\} \cup R,$$

We now introduce some useful notation: If $C = A \lor L \lor B$ is a reduced clause then $C \setminus L = A \lor B$. If $\mathcal{D}$ is a set of ground clauses we write $\mathcal{D}^\sim$ for $\mathcal{D}$ after application of the tautology (one-literal, pure literal) rule and $\mathcal{D}_1, \mathcal{D}_2$ for the clauses obtained from $\mathcal{D}$ by the splitting rule. If L is removed from L we obtain $\square$ (note that $L = L \lor \square$).

Theorem 2.4.1. *The rules of Davis and Putnam are correct. That means for every reduced set of ground clauses $\mathcal{D}$ we have:*

1. *$\mathcal{D} \sim_{sat} \mathcal{D}^\sim$ if one of the rules 1–3 is applicable,*
2. *$\mathcal{D}$ is satisfiable iff $\mathcal{D}_1$ is satisfiable or $\mathcal{D}_2$ is satisfiable, if rule 4 is applicable.*

Proof.

1. The tautology rule:
   Let C be a tautological clause in a set of reduced ground clauses $\mathcal{D}$. By $F(\mathcal{D}) \sim F(\mathcal{D} - \{C\})$ we get $\mathcal{D} \sim \mathcal{D} - \{C\}$ and thus $\mathcal{D} \sim_{sat} \mathcal{D} - \{C\}$. By iterating this argument we eventually obtain $\mathcal{D} \sim \mathcal{D}^{\sim}$.

2. The one-literal rule:
   Let $\mathcal{D}$ be a reduced set of clauses of the form
   $$\mathcal{D} = \{L, A_1, \ldots, A_n, B_1, \ldots, B_m\} \cup R$$
   such that L occurs in the A_i, L^d occurs in the B_j and R neither contains L nor L^d. Using merely commutativity and associativity of $\vee$ we get that $\mathcal{D}$ is logically equivalent to

   $$\mathcal{D}' = \{L, L \vee A_1', \ldots, L \vee A_n', L^d \vee B_1', \ldots, L^d \vee B_m'\} \cup R.$$

   for $A_i' = A_i \setminus L, B_j' = B_j \setminus L$.

   2a) Suppose that $\mathcal{D}'$ is satisfiable. Then there exists an H-interpretation Γ such that $v_\Gamma(\mathcal{D}') = \mathbf{T}$. By definition of v_Γ, $v_\Gamma(\{L\})$ must be true. By definition of "or" we obtain

   $$v_\Gamma(\mathcal{D}') = v_\Gamma(\{B_1', \ldots, B_m'\} \cup R) = v_\Gamma(\mathcal{D}'^{\sim}).$$

   We conclude that $\mathcal{D}'^{\sim}$ is satisfiable.

   2b) Suppose that $\mathcal{D}'^{\sim}$ is satisfiable and $v_\Gamma(\mathcal{D}'^{\sim}) = \mathbf{T}$. Let Γ' be like Γ with the exception $v_{\Gamma'}'(L) = \mathbf{T}$. Because neither L nor L^d occurs in $\mathcal{D}'^{\sim}$ we get $v_{\Gamma'}'(\mathcal{D}'^{\sim}) = v_\Gamma(\mathcal{D}')$, but also $v_{\Gamma'}'(\{L, L \vee A_1', \ldots, L \vee A_n'\}) = \mathbf{T}$. We conclude $v_{\Gamma'}'(\mathcal{D}') = \mathbf{T}$ and thus that $\mathcal{D}$ is satisfiable.

3. Pure literal rule: Let $\mathcal{D}$ be a reduced set of ground clauses such that $\mathcal{D} = \{A_1, \ldots, A_n\} \cup R$ and L occurs in every A_i, but neither L nor L^d occurs in R. By definition of the pure literal rule we have $\mathcal{D}^{\sim} = R$.

   3a) Suppose that $\mathcal{D}$ is satisfiable. By the validity of $F(\mathcal{D}) \to F(R)$ we conclude that also R (and thus $\mathcal{D}^{\sim}$) is satisfiable.

   3b) Suppose that $\mathcal{D}^{\sim}$ is satisfiable. Then there is an H-interpretation Γ such that $v_\Gamma(\mathcal{D}^{\sim}) = \mathbf{T}$. As in 2b) we define a Γ' such that $v_{\Gamma'}'(L) = \mathbf{T}$ and $v_\Gamma(L') = v_\Gamma(L')$ for all other ground literals L'. Then clearly $v_{\Gamma'}'(\mathcal{D}) = v_{\Gamma'}'(\{A_1, \ldots, A_n\} \cup R) = \mathbf{T}$.

4. The splitting rule:

Let $\mathcal{D}$ be a set of reduced ground clauses such that $\mathcal{D} = \{A_1, \ldots, A_n, B_1, \ldots, B_m\} \cup R$, where the A_i contain a literal L, the B_j contain L^d and R neither contains L nor L^d.

4a) Suppose that $\mathcal{D}$ is satisfiable. Then there exists an H-interpretation Γ such that $v_\Gamma(\mathcal{D}) = \mathbf{T}$. For Γ there are two possibilities, $v_\Gamma(L) = \mathbf{T}$ or $v_\Gamma(L) = \mathbf{F}$. If $v_\Gamma(L) = \mathbf{T}$ then $v_\Gamma(A_i) = \mathbf{T}$ for all $i = 1, \ldots, n$ and $v_\Gamma(B_j) = v_\Gamma(B_j \setminus L^d)$ for all $j = 1, \ldots, m$. Therefore

$$v_\Gamma(\mathcal{D}) = v_\Gamma(\{B_1 \setminus L^d, \ldots B_m \setminus L^d\} \cup R) = v_\Gamma(\mathcal{D}_2) = \mathbf{T}.$$

Similarly if $v_\Gamma(L) = \mathbf{F}$ we obtain

$$v_\Gamma(\mathcal{D}) = v_\Gamma(\{A_1 \setminus L, \ldots A_n \setminus L\} \cup R) = v_\Gamma(\mathcal{D}_1) = \mathbf{T}.$$

Therefore we obtain the result that either $\mathcal{D}_1$ is satisfiable or $\mathcal{D}_2$ is satisfiable.

4b) Suppose that either $\mathcal{D}_1$ is satisfiable or $\mathcal{D}_2$ is satisfiable. If $\mathcal{D}_1$ is satisfiable via Γ then define $\Gamma' = \Gamma$, except $v'_\Gamma(L) = \mathbf{F}$. Then, clearly $v'_\Gamma(\{B_1, \ldots, B_m\}) = \mathbf{T}$ and thus $v'_\Gamma(\mathcal{D}) = \mathbf{T}$. It follows that $\mathcal{D}$ is satisfiable. If $\mathcal{D}_2$ is satisfiable via Γ then define $\Gamma' = \Gamma$, except $v'_\Gamma(L) = \mathbf{T}$. Then we obtain $v'_\Gamma(\{A_1, \ldots, A_n\}) = \mathbf{T}$ and $v'_\Gamma(\mathcal{D}) = \mathbf{T}$; again, $\mathcal{D}$ is satisfiable. By definition 2.4.2 it follows that the splitting rule is correct.   ◊

By following the arguments in the proof of Theorem 2.4.1 we can observe case splitting by truth values of literals; e.g., in the splitting rule one set of clauses evaluates to true if a literal L is true in an interpretation Γ, another set of clauses evaluates to true if L is false in Γ. In the formalism of semantic trees we may describe Γ as a branch having two successors (at least) one of them verifying the set of clauses. Thus in an abstract sense the rules of Davis and Putnam may be interpreted as an efficient semantic tree procedure. The Davis–Putnam method is essentially "reductive", i.e., the satisfiability problem of a set of ground clauses $\mathcal{D}$ containing $n + 1$ different atoms is reduced to one ($\mathcal{D}^\sim$) or two problems ($\mathcal{D}_1, \mathcal{D}_2$) containing at most n different atoms. The reduction process can be represented most conveniently in the form of trees, where unary nodes belong to the rules 1), 2), 3) and binary nodes to the rule 4).

Definition 2.4.3. *A DP-tree $\mathcal{T}$ for a (finite) reduced set of ground clauses $\mathcal{D}$ is a tree with the following properties:*

1. *The nodes of $\mathcal{T}$ are (finite) sets of reduced ground clauses.*
2. *All nodes of $\mathcal{T}$ are of degree ≤ 2.*
3. *The root is $\mathcal{D}$.*

4. If $\mathcal{D}'$ is a node of degree 1 and $(\mathcal{D}', \mathcal{D}'')$ is an edge in $\mathcal{T}$ then $\mathcal{D}'' = \mathcal{D}'^\sim$ *(for some application of the rules 1), 2), 3)).*
5. If $\mathcal{D}'$ is a node of degree 2 and $(\mathcal{D}', \mathcal{M}), (\mathcal{D}', \mathcal{N})$ are edges in $\mathcal{T}$ (from left to right) then $\mathcal{M} = \mathcal{D}'_1, \mathcal{N} = \mathcal{D}'_2$ for some application of the splitting rule.

A DP-tree for a set of ground clauses $\mathcal{D}$ is called a DP–decision tree for C if either all leaf nodes are $\square$ or there exists a leaf node which is $\emptyset$.

Example 2.4.2.

$$\mathcal{D} = \{P \vee Q, \quad R \vee S, \quad \neg R \vee S, \quad R \vee \neg S, \quad \neg R \vee \neg S, \quad P \vee \neg Q \vee \neg P\}$$

P, Q, R, S represent some ground atoms different form each other. Applying the tautology rule we obtain

$$\mathcal{D}_1 = \mathcal{D} - \{P \vee \neg Q \vee \neg P\}.$$

Applying the pure literal rule to $\mathcal{D}_1$ we obtain

$$\mathcal{D}_2 = \{R \vee S, \quad \neg R \vee S, \quad R \vee \neg S, \quad \neg R \vee \neg S\}.$$

On $\mathcal{D}_2$ the splitting rule yields $\mathcal{D}_3 = \{R, \neg R\}, \mathcal{D}_4 = \{R, \neg R\}$. On $\mathcal{D}_3, \mathcal{D}_4$ the one literal rule gives $\{\square\}$.

Figure 2.11 shows the reduction tree corresponding to the reduction steps performed on $\mathcal{D}$.

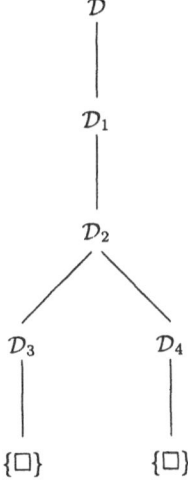

Fig. 2.11. A DP–decision tree for $\mathcal{D}$

The DP-tree of the last example is in fact a DP–decision tree; it proves that $\mathcal{D}$ is unsatisfiable.

Definition 2.4.4. *Let $\mathcal{T}$ be a DP–decision tree for a set of reduced ground clauses $\mathcal{D}$. We say that $\mathcal{T}$ proves the satisfiability of $\mathcal{D}$ if there is a leaf node which is $\emptyset$ and $\mathcal{T}$ proves the unsatisfiability of $\mathcal{D}$ if all leaf nodes contain $\square$.*

It is apparent from Definition 2.4.4 that finding a DP–decision tree for $\mathcal{D}$ algorithmically decides satisfiability. To facilitate further analysis we may assume that $\mathcal{D}$ does not contain tautologies. In fact, if $\mathcal{D}' = \mathcal{D}$ after tautology elimination, in the DP–trees for $\mathcal{D}'$ there can be no edges corresponding to the tautology rule (note that the rules 2, 3, and 4 cannot introduce tautologies). Thus tautology elimination can be considered as a "preprocessing" step. We call a set of reduced ground clauses *taut-reduced* if it does not contain tautologies.

We are in the position now to formulate correctness and completness of the Davis–Putnam method.

Theorem 2.4.2 (Correctness of the Davis–Putnam method). *Let $\mathcal{D}$ be a set of reduced and taut-reduced ground clauses and let $\mathcal{T}$ be a DP–decision tree such that $\mathcal{T}$ proves the satisfiability (unsatisfiability) of $\mathcal{D}$. Then $\mathcal{D}$ is indeed satisfiable (unsatisfiable).*

Proof. By an easy induction on the number of nodes in $\mathcal{T}$ and by use of Theorem 2.4.1. The details are left as an exercise. $\Diamond$

In a usual completeness result it is sufficient to prove the existence of the required deduction; in our case this corresponds to the proof that every reduced, taut-reduced set of ground clauses possesses a DP–decision tree. But such a result is not completely satisfying for getting a decision procedure. Thus we will prove the stronger result that every DP-tree can be extended into a decision tree. Saying that $\mathcal{T}'$ is an extension of $\mathcal{T}$ we mean that $\text{NOD}(\mathcal{T}) \subseteq \text{NOD}(\mathcal{T}')$ and $\text{E}(\mathcal{T}) \subseteq \text{E}(\mathcal{T}')$.

Theorem 2.4.3 (completeness of the Davis–Putnam method).
Let $\mathcal{D}$ be a (finite) set of reduced, taut-reduced set of ground clauses and let $\mathcal{T}$ be a DP-tree for $\mathcal{D}$. Then $\mathcal{T}$ can be extended to a DP–decision tree $\mathcal{T}'$.

Remark 2.4.1. Theorem 2.4.3 is in fact a strong completeness result. It shows that, no matter how we start using the Davis–Putnam rule, we always get a decision tree. That means, every DP-tree which cannot be extended properly is in fact a DP–decision tree.

Proof. We proceed by induction on $atn(\mathcal{D}) = $ number of different ground atoms occurring in $\mathcal{D}$ (we mean the cardinality of the set of ground atoms, not the number of ocurrences).

$atn(\mathcal{D}) = 0$: Then $\mathcal{D} = \emptyset$ or $\mathcal{D} = \{\square\}$. The only DP-tree for $\mathcal{D} = \emptyset$ is the tree with root $\emptyset$; but this tree is also a DP–decision tree. For $\mathcal{D} = \{\square\}$ the only DP-tree is the tree with root $\{\square\}$, which is a DP–decision tree too.

(IH) Suppose that for all reduced, taut-reduced sets of ground clauses $\mathcal{D}$ such that
$atn(\mathcal{D}) \leq n$ every DP-tree $\mathcal{T}$ can be extended to a DP–decision tree $\mathcal{T}'$.

Now let us assume that $\mathcal{D}$ is a reduced, taut-reduced set of ground clasuses such that $atn(\mathcal{D}) = n + 1$; moreover let $\mathcal{T}$ be a DP-tree for $\mathcal{D}$.

case a) $\mathcal{T}$ consists of the root $\mathcal{D}$ only.

As $atn(\mathcal{D}) = n + 1$ there must be a clause in $\mathcal{D}$ containing a literal L. We select an arbitrary literal L occurring in $\mathcal{D}$. If L^d does not occur in $\mathcal{D}$ then the pure literal rule is applicable. If L occurs as unit clause in $\mathcal{D}$ then the one literal rule is applicable. If L neither occurs as unit clause nor as "pure" literal then the splitting rule is applicable. Thus we conclude two facts

1. One of the rules 2, 3, 4 must be applicable,
2. If we reduce by an arbitrary rule to a set of clauses $\mathcal{D}^\sim$ (or to two sets $\mathcal{D}_1, \mathcal{D}_2$ via rule 4 then $atn(\mathcal{D}^\sim) \leq n$ $(atn(\mathcal{D}_1), atn(\mathcal{D}_2) \leq n$ respectively).

Property 2 directly follows from the definition of the rules 2, 3, 4.

Thus for every extension of $\mathcal{T}$ to $\mathcal{T}'$ such that either

$\text{NOD}(\text{T}') = \{\mathcal{D}, \mathcal{D}^\sim\}, \quad \text{E}(\text{T}') = \{(\mathcal{D}, \mathcal{D}^\sim)\}$ or
$\text{NOD}(\text{T}') = \{\mathcal{D}, \mathcal{D}_1, \mathcal{D}_2\}, \quad \text{E}(\text{T}') = \{(\mathcal{D}, \mathcal{D}_1), (\mathcal{D}, \mathcal{D}_2)\},$

we obtain $atn(\mathcal{D}^\sim) \leq n$ or $atn(\mathcal{D}_1), atn(\mathcal{D}_2) \leq n$ respectively.

Clearly every extension of $\mathcal{T}$ is defined by an extension of some $\mathcal{T}'$. So let $\mathcal{T}'$ be as above and let $\mathcal{D}^\sim$ be the successor node of $\mathcal{D}$ or $\mathcal{D}_1, \mathcal{D}_2$ be the successor nodes of $\mathcal{D}$ respectively.

By (IH) every DP-tree with root $\mathcal{D}^\sim(\mathcal{D}_1, \mathcal{D}_2$ respectively) can be extended to a DP–decision tree. We conclude that every DP-tree $\mathcal{T}'$ above can be extended to a DP–decision tree. In particular, $\mathcal{T}$ can be extended to a DP–decision tree.

case b) $\mathcal{T}$ contains edges.

Let $\mathcal{D}_1, \ldots, \mathcal{D}_n$ be all leaf nodes of $\mathcal{T}$. If all $\mathcal{D}_i$ contain $\square$ or there exists a $\mathcal{D}_i$ such that $\mathcal{D}_i = \emptyset$ then $\mathcal{T}$ is a DP–decision tree and we have achieved our goal.

Otherwise all $\mathcal{D}_i$ are sets of clauses such that $atn(\mathcal{D}_i) \leq n$; this can be obtained by an easy induction argument based on arguments like in case (a).

Let $\mathcal{T}_i$ be the trees $(\{\mathcal{D}_i\}, \emptyset)$ for $i = 1, \ldots, k$. By (IH) every $\mathcal{T}_i$ can be extended to a DP–decision tree $\mathcal{T}_i'$. Let $\mathcal{T}' = T$ after replacement of the nodes $\mathcal{D}_i$ by the trees $\mathcal{T}_i'$. Then (by definition of DP–decision trees) $\mathcal{T}'$ is a DP–decision tree (note that in $\mathcal{T}'$ either all leaf nodes contain $\square$, or there is one which is $\emptyset$). But $\mathcal{T}'$ is an extension of $\mathcal{T}$. This concludes the proof of case n + 1.   ◊

Theorem 2.4.3 suggests the following decision method for the satisfiability problem of (finite) sets of ground clauses $\mathcal{D}$:

DP 1) $\mathcal{D} \leftarrow \mathcal{D}$ after reduction of all clauses in $\mathcal{D}$;
DP 2) $\mathcal{D} \leftarrow \mathcal{D}$ after deletion of all tautologies;
DP 3) construct a DP–decision tree for $\mathcal{D}$.

Because in a taut-reduced, reduced set of ground clauses $\mathcal{D}$ every application of the rules 2, 3, 4 decreases the number of different atom symbols, the tree constructed in point 3 above is of depth $\leq atn(\mathcal{D})$. However the method, due to the splitting rule, may be exponential (i.e. the total number of atom occurrences in nodes of the tree may be exponential in the number of atom occurrences in $\mathcal{D}$).

Combining the propositional decision method DP 1–DP 3 with the production of the sets of H_n-instances we get the (Herbrand type) proof method of Davis and Putnam displayed in Figure 2.12.

```
begin {C is a finite set of clauses}
    if C does not contain function symbols
    then apply DP1) - DP3) to C'₀
        else begin
            n ← 0; contr ← FALSE;
            while ¬ contr do
                begin
                    perform DP1) -DP3) on C'ₙ;
                    if the DP-decision tree proves unsatisfiability
                    then contr ← TRUE
                    else contr ← FALSE:
                    n ← n + 1
                end
        end
end.
```

Fig. 2.12. The Davis–Putnam method

The algorithm in Figure 2.12 is nondeterministic, because the construction of the DP–decision tree in DP 3) is nondeterministic. If $\mathcal{C}$ does not contain function symbols then the method always terminates and yields a decision procedure. If $\mathcal{C}$ is satisfiable and contains function symbols the while loop in the algorithm is an endless loop. As clause logic is undecidable (it is a reduction class of predicate logic), we cannot expect termination in all cases. However we will define resolution methods in Chapter 5 which favor termination in a stronger way.

We have seen that the algorithm in Figure 2.11 yields a decision procedure for function-free sets of clauses. This is straightforward as for function-free sets of clauses $\mathcal{C}, H(\mathcal{C})$ – and thus $AT(\mathcal{C})$ – is finite. Using skolemization and reduction to clause form, we thus can solve the following decision problem:

Let $A : (\forall x_1) \ldots (\forall x_n)(\exists y_1) \ldots (\exists y_m)M$ be an arbitrary formula in predicate logic, where M neither contains quantifiers nor function symbols. Is A valid?

In 1928 [BS28] Bernays and Schönfinkel showed that this problem is decidable. As Herbrand's theorem was not available then (it was published 3 years later), the solution was not so easy as it is today. In deciding the problem above we simply perform the following steps:

1. Transfom $\neg A$ to clause form C.
2. Apply DP 1)–DP 3) to C_0'

By subjecting $\neg A$ to step 1) of our normal form reduction in Section 2.2 we obtain a form

$$F_1 : (\exists x_1) \ldots (\exists x_n)(\forall y_1) \ldots (\forall y_m)M'.$$

By iterated skolemization of F we get

$$F_2 : (\forall y_1) \ldots (\forall y_m)M'\{x_1 \leftarrow c_1, \ldots, x_n \leftarrow c_n\}$$

for new constant symbols $c_1, \ldots, c_n$. Clearly, the clause form of F_2 is function free. It is easily verified that also the validity problem for formulas of the type:

$$B : ((\exists^*)(\forall^*)M_1 \wedge \ldots \wedge (\exists^*)(\forall^*)M_n) \to (\forall^*)(\exists^*)M_0$$

is transformed (by the normal form reduction in Section 2.2) to a function-free clause set C, provided the M_i are all quantifier and function free. Note that not all prefix forms of $\neg B$ give function-free sets of clauses; but as the reductions in Section 2.2 avoid prefixing completely, we don't get such problems here.

Exercises

Exercise 2.4.1. Give a proof of Theorem 2.4.2.

Exercise 2.4.2. Let

$$C_n = \{P(a)\} \cup \bigcup_{i=0}^{n-1} \{\neg P(f^{(i)}(a)) \vee P(f^{(i+1)}(a))\} \cup \{\neg P(f^{(n)}(b))\}$$

be a sequence of sets of ground clauses ($f^0(t) = t$, $f^{i+1}(t) = f(f^i(t))$ for $t \in T$). Prove (using the rules of Davis and Putnam) that all C_n are satisfiable.

Exercise 2.4.3. Show that the satisfiability of a set of ground Horn clauses (see Definition 2.2.4) can be decided by the rules of Davis and Putnam without the splitting rule.

Let T be a DP–decision tree for a set of ground clauses C and N be the union of all nodes of T (note that the nodes are sets of ground clauses). Then we call $|N|$ (the number of clauses occurring in nodes of T) the size of T.

Exercise 2.4.4. Prove that for every set of ground Horn clauses there exists a DP–decision tree of polynomial size, i.e., there exists a polynomial p with the following property: For every set of ground Horn clauses $\mathcal{C}$ there exists a DP–decision tree $T(\mathcal{C})$ such that size($T(\mathcal{C})$) $\leq p(|\,\mathcal{C}\,|)$.

2.5 The Propositional Resolution Principle

The Davis–Putnam method defined in Section 2.4, although defined nondeterministically, is a decision method for the satisfiability problem of sets of ground clauses. Because the sets C'_n, defined by ground substitutions over H_n, are defined successively in the form of a level saturation method, a semi-decision procedure for the satisfiability problem of ground clause logic would not guarantee completeness. The reason is that we might fail to decide whether the "candidate" C'_n produced so far is indeed unsatisfiable. But let us assume (for theoretical reasons first) that we already have an unsatisfiable set of ground instances C' and our problem is to prove that C' is indeed unsatisfiable. To solve this problem we may resort to the common inference paradigm, i.e., we search for a calculus which refutes unsatisfiable sets of ground clauses. Such a calculus can be obtained by the propositional resolution principle.

To give a motivation we first consider two ground instances C'_1, C'_2 of clauses C_1, C_2. Suppose that C'_1, C'_2 are of the form

$$C'_1 = A \vee P \vee B, \quad C'_2 = D \vee \neg P \vee E$$

where P is a ground atom and A, B, D, E are ground clauses. Let Γ be an arbitrary interpretation of the set of clauses, containing C_1 and C_2, such that $v_\Gamma(C'_1 \wedge C'_2) = \mathbf{T}$. Then either $v_\Gamma(P) = \mathbf{T}$ or $v_\Gamma(P) = \mathbf{F}$. If $v_\Gamma(P) = \mathbf{T}$ then $v_\Gamma(C'_2) = v_\Gamma(D \vee E) = \mathbf{T}$ (because $v_\Gamma(C'_2)$ must be $\mathbf{T}$); if $v_\Gamma(P) = \mathbf{F}$ we conclude similarly that $v_\Gamma(C'_1) = v_\Gamma(A \vee B) = \mathbf{T}$. We conclude that either $v_\Gamma(A \vee B) = \mathbf{T}$ or $v_\Gamma(D \vee E) = \mathbf{T}$. This proves the validity of

$$C'_1 \wedge C'_2 \rightarrow A \vee B \vee D \vee E.$$

The above argument shows the correctness of the inference schema of *propositional resolution*:

$$\frac{A \vee L \vee B \quad D \vee L^d \vee E}{A \vee B \vee D \vee E}$$

where L is a metavariable for literals and A, B, D, E are metavariables for clauses. Although this rule is a propositional one it is also valid on general clause forms. Let

$$\begin{aligned} C_1 &= A \vee P \vee B; \\ C_2 &= D \vee \neg P \vee E \end{aligned}$$

be clauses. C_1 and C_2 are represented by the PL-formulas

$$F(\{C_1\}) \;\; = \;\; (\forall x_1)\ldots(\forall x_n)(A \vee P \vee B) \quad \text{and}$$
$$F(\{C_2\}) \;\; = \;\; (\forall y_1)\ldots(\forall y_m)(D \vee \neg P \vee E).$$

But (by the substitution principle) $F(\{C_1\}) \to C_1$, $F(\{C_2\}) \to C_2$ are both valid. The same argument as above shows that also

$$C_1 \wedge C_2 \to A \vee B \vee D \vee E$$

is valid.

It follows that $F(\{C_1\}) \wedge F(\{C_2\}) \to A \vee B \vee D \vee E$ is valid. Because the last formula is valid for all variable environments we conclude (applying a simple quantifier shifting) that

$$F(\{C_1, C_2\}) \to F(A \vee B \vee D \vee E)$$

is valid. Thus we have shown that the propositional resolution rule is correct in clause logic.

Definition 2.5.1 (propositional resolvent, p-resolvent). *Let C_1, C_2 be clauses such that $C_1 = A \vee L \vee B$ and $C_2 = D \vee L^d \vee E$ for some literal L and clauses A, B, D, E. Then $A \vee B \vee D \vee E$ is called* p-resolvent *of C_1 and C_2. (p-resolvents are the result of the propositional resolution rule).*

Using p-resolution without additional rules cannot result in a refutationally complete inference system on ground clause logic. Consider

$$C_1 = P(a) \vee P(a), \quad C_2 = \neg P(a) \vee \neg P(a).$$

There is one p-resolvent of C_1, C_2 (although obtained in two ways), namely $C_3 = P(a) \vee \neg P(a)$. Resolving C_3 with C_1 we get C_1 again and similarly for C_2. Although $\{C_1, C_2\}$ is clearly unsatisfiable we cannot deduce the empty clause. It is obvious that we also need a contraction rule. We define such a rule for (general) clause logic.

Definition 2.5.2 (p-reduct). *Let C be a clause. Then C' is called* p-reduct *of C if C' is C after omission of some multiple literals. Formally, if*

$$C = C_1 \vee L \vee C_2 \vee L \vee C_3$$

then $C, C_1 \vee C_2 \vee L \vee C_3$ and $C_1 \vee L \vee C_2 \vee C_3$ are p-reducts of C; every p-reduct of a p-reduct is also a p-reduct (of C).

Example 2.5.1. $C = P(x) \vee R(x) \vee P(x) \vee P(x)$.
The p-reducts of C are

$$C, \;\; R(x) \vee P(x) \vee P(x), \;\; P(x) \vee R(x) \vee P(x), \;\; R(x) \vee P(x) \;\; \text{and} \;\; P(x) \vee R(x).$$

Using p-resolution together with p-reduction it is easy to refute the set

$\{P(a) \lor P(a), \neg P(a) \lor \neg P(a)\}$: By p-reduction we obtain $P(a), \neg P(a)$ and by p-resolution $\square$. We will see that in combining p-resolution and p-reduction we get a refutationally complete inference principle for propositional clause logic. But we will show more: By Herbrand's theorem we know that unsatisfiable sets of clauses $\mathcal{C}$ possess finite, unsatisfiable sets $\mathcal{C}'$ of ground instances. Let $\mathcal{C}$ be an unsatisfiable set of clauses and $\mathcal{R}$ be a complete inference principle for propositional logic. If we first produce a finite, unsatisfiable set $\mathcal{C}'$ of ground instances by substitution and then apply $\mathcal{R}$ to $\mathcal{C}'$ we obtain a principle of inference which is complete for (general) clause logic. This motivates the following definitions:

Definition 2.5.3. *Let $\mathcal{C}$ be a set of clauses and C be a clause. A sequence $C_1, \ldots, C_n$ is called* PR-deduction *of C from $\mathcal{C}$ if the following conditions are fulfilled:*

1. *$C_n = C$*
2. *For all $i = 1, \ldots n$:*
   a) *either C_i is an instance of a clause in $\mathcal{C}$ or*
   b) *C_i is a p-resolvent of clauses D, E, where D, E are p-reducts of some clauses C_j, C_k for $j, k < i$.*

A PR-deduction of $\square$ from $\mathcal{C}$ is called a PR-refutation *of $\mathcal{C}$.*

Definition 2.5.4. *A PR-deduction is called a* GR-deduction *(ground resolution deduction) if, in the definition of PR-deductions, 2(a) is replaced by "C_i is a ground instance of a clause in $\mathcal{C}$".*

It is easy to verify that the calculi defined in Definitions 2.5.3 and 2.5.4 are sound.

Proposition 2.5.1 (soundness of PR-deduction). *Let $\mathcal{C}$ be a set of clauses and Γ be a PR-deduction of a clause C from $\mathcal{C}$. Then $F(\mathcal{C}) \to F(\{C\})$ is valid.*

Proof. By an easy induction on the length of the deduction Γ. Essentially we have to prove that instantiation, p-reducts and p-resolution are correct rules. For p-resolution this has been shown in the beginning of this section, for instantiation and p-reduct it is trivial. $\Diamond$

Corollary 2.5.1. *Let $\mathcal{C}$ be a set of clauses and Γ be a GR-deduction of a clause C from $\mathcal{C}$. Then $F(\mathcal{C}) \to F(\{C\})$ is valid.*

Proof. Every GR-deduction is a PR-deduction. $\Diamond$

An immediate consequence of Proposition 2.5.1 is the following: If $\mathcal{C}$ is a satisfiable set of clauses then $\square$ is not derivable. The requirement that, in case of satisfiability, $\square$ cannot be derived is in fact weaker than soundness as expressed in Proposition 2.5.1 (we call this weaker principle *refutational soundness*). Skolemization (interpreted as a rule) is refutationally sound, but not (strongly) sound.

The main result of this chapter will be the (refutational) completeness of PR- and GR-deductions.

Theorem 2.5.1 (completeness of GR-deduction). *If C is an unsatisfiable set of clauses then there exists a GR-refutation of C.*

Proof. Suppose that C is an unsatisfiable set of clauses. By Lemma 2.3.2 there exists a finite, closed semantic tree T for C.

We prove the existence of a GR-refutation by induction on $|\mathrm{NOD}(T)|$, the number of nodes of T.

$|\mathrm{NOD}(T)| = 1$: Because already the root falsifies a clause in C, $\square$ must be in C. But then $\square$ is a GR-refutation of C.

(IH) Suppose that for all C, such that C possesses a finite, closed semantic tree T with $|\mathrm{NOD}(T)| \leq n$, there exists a GR-refutation of C.

Now let C be an (unsatisfiable) set of clauses, such that T is a closed semantic tree with $|\mathrm{NOD}(T)| = n + 1$. Then there exists a node N in T that is not a failure node, yet both sons N_1, N_2 are failure nodes (otherwise T would be infinite). The situation is graphically represented in Figure 2.13.

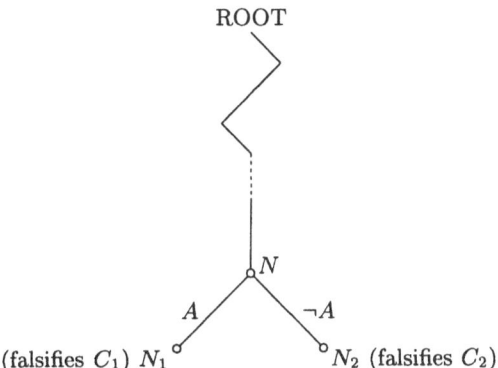

Fig. 2.13. An inference node in a semantic tree

Because N_1 and N_2 are failure nodes there are clauses $C_1, C_2 \in C$ of whose N_1 falsifies C_1 and N_2 falsifies C_2. That means there are ground instances C'_1, C'_2 of C_1, C_2 such that:

For all L in $C'_1 : L^d \in \gamma_{N_1}$,

For all L in $C'_2 : L^d \in \gamma_{N_2}$.

Moreover, $\neg A$ must occur in C_1' and A in C_2' (otherwise N_1, N_2 have ancestor nodes falsifying C_1' or C_2', which contradicts the property of being a failure node).

Let C_1'' be a reduced form of C_1' and C_2'' be a reduced form of C_2'. Then C_1'', C_2'' are specific p-reducts of C_1', C_2'. Still $\neg A$ is in C_1'', A in C_2'' and we have

$$C_1'' = D_1 \vee \neg A \vee D_2,$$
$$C_2'' = E_1 \vee A \vee E_2$$

for ground clauses D_1, D_2, E_1, E_2 which neither contain A nor $\neg A$ (note that C_1'' cannot contain A, because in this case it would be a tautology and thus cannot be falsified at all).

By the occurrence of A and $\neg A$ in C_1'', C_2'' there exists a p-resolvent C' : $D_1 \vee D_2 \vee E_1 \vee E_2$ of C_1'' and C_2'';

C' neither contains A nor $\neg A$, but $\mathrm{LIT}(C') \subseteq \mathrm{LIT}(C_1') \cup \mathrm{LIT}(C_2')$. It follows that for all L in C', $L^d \in \gamma_N$. Consequently N itself or an ancestor node of N is a failure node for $\mathcal{C} \cup \{C'\}$ (N falsifies C', but there may be an ancestor node of N which also falsifies C').

Thus let T_1 be a closed semantic tree for $\mathcal{C} \cup \{C'\}$ obtained from T by removing N_1 and N_2 and cutting the corresponding path at a failure node. Then, clearly,

$\mid \mathrm{NOD}(T_1) \mid < \mid \mathrm{NOD}(T) \mid$ and therefore
$\mid \mathrm{NOD}(T_1) \mid \leq n.$

By the induction hypothesis (IH) there exists a GR-refutation Π of $\mathcal{C} \cup \{C'\}$. The sequence C_1', C_2', C' is a GR-deduction of C' from $\mathcal{C}$; C' is obtained from C_1'' and C_2'' by p-resolution and C_1'', C_2'' are p-reducts of C_1', C_2'. By definition of GR-deductions, the sequence

$$C_1', C_2', C', \Pi$$

is a GR-refutation of $\mathcal{C}$.   ◊

Corollary 2.5.2. *Let $\mathcal{C}$ be an unsatisfiable set of clauses; then there exists a PR-refutation of $\mathcal{C}$.*

Proof. Trivial, as every GR-refutation is a PR-refutation.   ◊

The concept of PR-deduction does not represent "the" resolution principle (in the usual terminology) because it is not based on most general unification (to be introduced in the next chapter). The weak point in the concepts of PR- and GR-deductions is the unlimited substitution rule (Definition 2.5.3, item 2.a), making PR-(GR-) deductions inappropriate for modeling automated deduction. However we will use both concepts in analyzing the complexity of resolution refutations.

On sets of ground clauses GR-deductions are reduced to the application of p-reduction and p-resolution. It is easy to see that p-resolution alone suffices, if all clauses are kept in reduced form (and all p-resolvents are reduced immediately after computation). Thus Theorem 2.5.1 immediately gives the completeness of the propositional resolution principle. Propositional resolution can also be applied as a decision procedure on sets of ground clauses (like the Davis–Putnam method); but we do not go into detail here and refer to Chapter 5, which is devoted to resolution decision procedures in general (but see Exercise 2.5.1). Finally, we give an example of a GR-refutation.

Example 2.5.2. Let $C = \{C_1, C_2, C_3, C_4\}$ such that

$$
\begin{aligned}
C_1 &= P(x, f(y)) \vee P(x, f(x)), \\
C_2 &= \neg P(x, y) \vee P(y, x), \\
C_3 &= \neg P(x, y) \vee P(f(x), y), \\
C_4 &= \neg P(f(f(z)), z).
\end{aligned}
$$

The deduction in Figure 2.14 is represented in form of a tree; the nodes in the tree are labeled with the clauses actually appearing in the PR-refutation. The edges labeled by "sub" represent the application of a substitution, those labelled by "p-red" the application of a p-reduction; "res" represents the application of propositional resolution. The corresponding sequence, according to Definition 2.5.3, is

$$\Pi : P(x, f(x)) \vee P(x, f(x)), \; \neg P(x, f(x)) \vee P(f(x), x), \; P(f(x), x),$$
$$\neg P(f(x), x) \vee P(f(f(x)), x), \; P(f(f(x)), x), \neg P(f(f(x)), x), \; \square.$$

A special feature of PR-deductions is their "robustness" under substitution: Let t be an arbitrary term; then

$$\Pi[\tfrac{x}{t}] : P(t, f(t)) \vee P(t, f(t)), \; \neg P(t, f(t)) \vee P(f(t), t), \; P(f(t), t),$$
$$\neg P(f(t), t) \vee P(f(f(t)), t), \; P(f(f(t)), t), \; \neg P(f(f(t)), t), \; \square$$

is also a PR-refutation of C. If $t \in H(C)$ then Π_t^x is a GR-refutation of C.

Exercises

Let GCL be the class of all finite sets of reduced ground clauses (we call GCL ground clause logic).

Exercise 2.5.1. Let $C \in$ GCL and $\text{Res}(C)$ be the set of all p-resolvents from C in reduced form. We define $R(C) = C \cup \text{Res}(C)$ and for all $i \geq 1 : R^{i+1}(C) = R(R^i(C))$ and $R^*(C) = \bigcup_{i=1}^{\infty} R^i(C)$. Prove that R defines a decision algorithm for the satisfiability problem of GCL, i.e.:

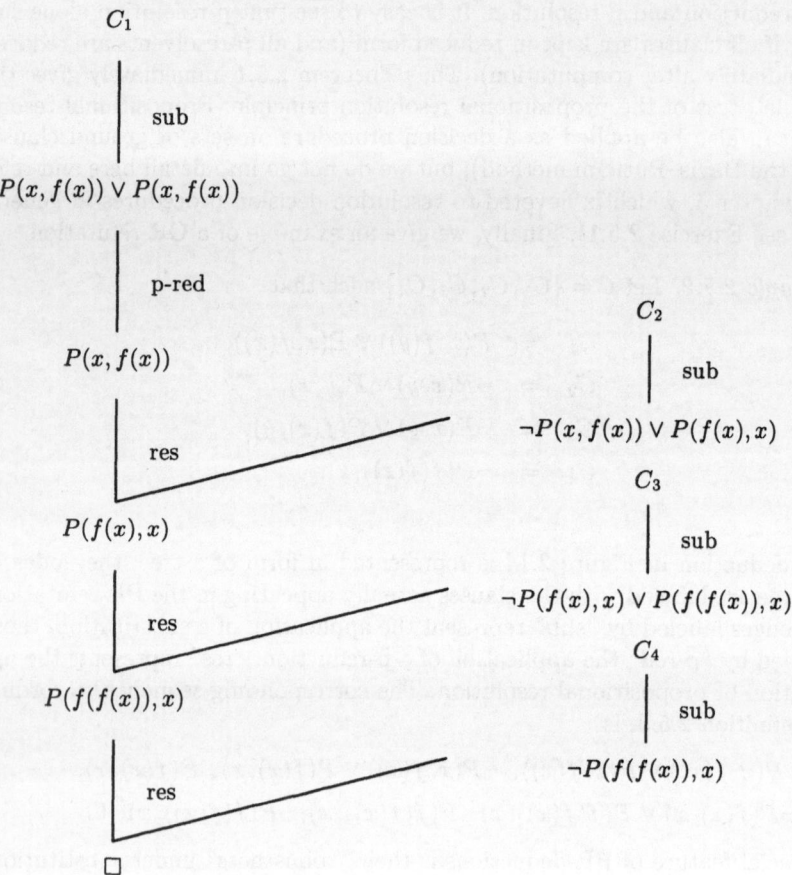

Fig. 2.14. A deduction tree

a) for all $\mathcal{C} \in \text{GCL} : R^*(\mathcal{C}) \in \text{GCL}$ (and thus $R^*(\mathcal{C})$ is finite).
b) $\square \in R^*(\mathcal{C})$ iff $\mathcal{C}$ is unsatisfiable.

We denote the set of all finite sets of Krom clauses (see Definition 2.2.4) by KL; the subclass of all set of reduced ground clauses in KL is written as GKL.

If $\mathcal{C}$ is a set of clauses we write $\text{occl}(\mathcal{C})$ for the number of all occurrences of literals in $\mathcal{C}$.

Exercise 2.5.2. Let R be defined as in Exercise 2.5.1. Prove that the satisfiability problem for GKL is polynomially decidable; that means prove the existence of a polynomial p such that

a) for all $C \in \text{GCL} : occl(R^*(C)) \leq p(occl(C))$ and
b) $\square \in R^*(C)$ iff C is unsatisfiable.

(Hint: A p-resolvent of two Krom clauses is also a Krom clause).

Exercise 2.5.3. Let C be a (finite) unsatisfiable set of Horn clauses. Show that there exists a GR-refutation of C without p-reduction, i.e., we use a concept of GR-deduction based on p-resolution only. (Hint: use a restriction of resolution where one of two clauses to be resolved must be a positive unit clause and show that this restriction is complete).

2.6 Substitution and Unification

In Example 2.5.2 it is clearly visible that the substitution rule always serves the purpose of making literals complementary and of preparing clauses for p-resolution. If two literals are made complementary, their atoms are made equal and we speak about unification. We have seen that, for the PR-refutation Π in Example 2.5.2, there are infinitely many "versions" Π_t^x.

Figure 2.15 shows a segment (a deduction appearing in Π_t^x) of Π_t^x. Let us

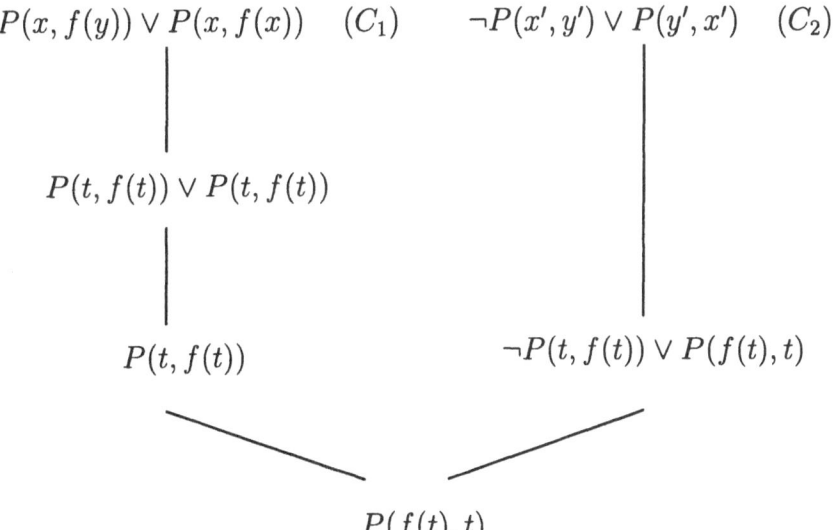

Fig. 2.15. Resolution and unification

consider the literals in C_1 and the literal $\neg P(x', y')$ in C_2 in Figure 2.15. The set of their atoms is $\{P(x, f(y)), P(x, f(x)), P(x', y')\}$. Replacing x by t, y by t, x' by t and y' by $f(t)$ we get the atom $P(t, f(t))$, which is cut out by p-resolution. Thus the substitution serves as a unifier. There are infinitely many terms t over C and infinitely many deductions corresponding to t. We will see that *one* unifying substitution will always do the job and that, even under this restriction, resolution remains complete. Precisely this fact makes resolution much more powerful than any ground deduction method.

Before starting the study of unification in more detail, we prepare the necessary machinery concerning substitutions.

For the formal definition of a substitution see Definition 2.1.10 (substitution) in Section 2.1.

Definition 2.6.1 (ground substitution). *We call a substitution σ based on a set of clauses C if all terms in $rg(\sigma)$ are contained in $T(C)$. A substitution is called a* ground substitution *(based on C) if all terms in $rg(\sigma)$ are contained in $H(C)$.*

By Definition 2.1.10 the action of a substitution canonically extends to all terms; there they represent specific endomorphisms of the term algebra. As, by this extension, σ is of the type $T \to T$, it is possible to concatenate substitutions. If σ, τ are two substitutions, $\sigma \circ \tau$ simply denotes the composition of the mappings σ and τ. For $t \in T$ we get $(\sigma \circ \tau)(t) = \sigma(\tau(t))$ – writing the application of the mappings in prefix notation. As it is usual in automated theorem proving to use the postfix notation for substitutions, we henceforth write $t\tau\sigma$ instead of $(\sigma \circ \tau)(t)$. Algebraically, substitutions define a monoid, where ϵ denotes the empty substitution:

$$\sigma(\tau\mu) = (\sigma\tau)\mu,$$
$$\sigma\epsilon = \epsilon\sigma = \sigma$$

for all substitutions σ, τ, μ.

We denote the set of substitutions by SUBST.

SUBST$_\Sigma$ is the subset of all $\sigma \in$ SUBST such that $rg(\sigma) \subseteq T_\Sigma$; similarly we define SUBST(C).

In clause logic not only terms, but also atoms and literals are subjected to substitutions. The extension of substitutions to atoms and literals is defined in Definition 2.1.10. To make the mathematical analysis smoother we define the concept of expression.

Definition 2.6.2 (expression). *An* expression *is a term, an atom or a literal.*

Let $W = \{E_1, \ldots, E_n\}$ be a set of expressions. By $W\sigma$ we denote the set $\{E_1\sigma, \ldots, E_n\sigma\}$.

Definition 2.6.3 (instance). *An expression E_1 is an* instance *of an expression E_2 if there exists a substitution σ such that $E_1 = E_2\sigma$. In this case we call σ a* match *and use the term "E_1 matches E_2".*

Substitutions and expressions may be compared with regard to their "generality", i.e., whether they can be obtained from other substitutions (expressions) by instantiation. We obtain a relation $\leq_s$:

Definition 2.6.4 (generality). *Let E_1 and E_2 be expressions, then we define that E_1 is more (or equally)* general *than E_2 (notation $E_1 \leq_s E_2$) if there exists a substitution σ such that $E_1\sigma = E_2$. Let σ, τ be two substitutions. We define $\sigma \leq_s \tau$ (σ is more general than τ) if there exist a substitution ϑ such that $\sigma\vartheta = \tau$.*

The relation $\leq_s$ defines a quasiordering on the set of all substitutions. There exists a (unique) minimal element with regard to $\leq_s$, namely ϵ. We write $E_1 =_s E_2$ for $E_1 \leq_s E_2$ and $E_2 \leq_s E_1$; clearly $=_s$ is an equivalence relation. Note that there are alternative definitions of substitution orderings appearing in the literature (for an alternative definition see Exercise 2.6.1).

Definition 2.6.5 (permutation). *A substitution σ is called a* permutation *if σ is one-one and $rg(\sigma) \subseteq V$. A permutation σ is called a* renaming *of a set of expressions E if $V(E) \cap rg(\sigma) = \emptyset$.*

We can go back now to the problem of unifying atoms, literals, and terms (i.e., expressions) by substitutions. Recall the set of atoms $\{P(x, f(y)), P(x, f(x)), P(x', y')\}$ appearing as a unification problem in Figure 2.15. Let t be an arbitrary term different from x. Then the substitution $\sigma_t = \{x \leftarrow t, y \leftarrow t, x' \leftarrow t, y' \leftarrow f(t)\}$ fulfils

$$P(x, f(y))\sigma_t = P(x, f(x))\sigma_t = P(x', y')\sigma_t = P(t, f(t)).$$

The same property holds for the substitution

$$\sigma = \{y \leftarrow x, x' \leftarrow x, y' \leftarrow f(x)\}.$$

This motivates the following definition.

Definition 2.6.6 (unifier). *Let W be a nonempty set of expressions. A substitution σ is called* unifier *of W if $|W\sigma| = 1$.*
If W consists of only one element (where there is actually nothing to unify) then, according to our definition, every substitution is a unifier of W.

Example 2.6.1. $W = \{P(x, f(y)), P(x, f(x)), P(x', y')\}$.
All substitutions σ, σ_t are unifiers of W (as T is infinite there are infinitely many unifiers with domain $\subseteq V(W)$). Moreover, we see that the unifier σ plays an exceptional role:
For

$$\sigma = \{y \leftarrow x, x' \leftarrow x, y' \leftarrow f(x)\} \text{ and}$$
$$\sigma_t = \{x \leftarrow t, y \leftarrow t, x' \leftarrow t, y' \leftarrow f(t)\} \quad \text{we have}$$
$$\sigma\{x \leftarrow t\} = \sigma_t, \quad \text{i.e.,} \quad \sigma \leq_s \sigma_t.$$

It is easy to verify that for all unifying substitutions ϑ (including those with $dom(\vartheta) - V(W) \neq \emptyset$) we obtain $\sigma \leq_s \vartheta$. σ is more general than all other unifiers of W, it is indeed "most" general. However, σ is not the only most general unifier; for the unifier $\lambda : \{y \leftarrow x', x \leftarrow x', y' \leftarrow f(x')\}$ we get

$$\lambda \leq_s \sigma, \sigma \leq_s \lambda, \text{ and } \lambda \leq_s \vartheta \text{ for all unifiers } \vartheta \text{ of } W.$$

More generally, if λ is a most general unifier of a set of expressions and μ is a permutation (see Definition 2.6.5) then $\lambda\mu$ is a most general unifier too.

Definition 2.6.7 (most general unifier). *A unifier σ of a set of expressions W is called* most general unifier *(henceforth abbreviated by m.g.u.) of W if for every unifier τ of W: $\sigma \leq_s \tau$.*

If W is a set of expressions, we write $UN(W)$ for the set of all unifiers ϑ of W fulfilling $dom(\vartheta) \subseteq V(W)$ and $rg(\vartheta) \subseteq T(W)$ ($T(W)$ is the set of all terms over the signature of W). $UN(W)$ essentially describes the set of unifiers over the "syntax" of W. It is easy to verify that W is unifiable iff W is unifiable by a substitution in UN(W) (Exercise 2.6.3).

While the set of all unifiers of a finite set of expressions W is always infinite (by replacing dummy variables outside W), $UN(W)$ may be finite. As an example consider $W = \{P(x), P(a)\}$, where $UN(W) = \{\{x \leftarrow a\}\}$.

Let $W = \{P(x, f(y)), P(x, f(x)), P(u, v)\}$ as in Example 2.6.1. Then $UN(W) = \{\sigma, \sigma_1, \sigma_2\} \cup \{\sigma_t \mid t \in T'\}$, where

$$\sigma = \{y \leftarrow x, \quad u \leftarrow x, v \leftarrow f(x)\},$$
$$\sigma_1 = \{x \leftarrow u, \quad y \leftarrow u, v \leftarrow f(u)\},$$
$$\sigma_2 = \{x \leftarrow y, \quad u \leftarrow y, v \leftarrow f(y)\},$$
$$\sigma_t = \{x \leftarrow t, \quad y \leftarrow t, u \leftarrow t, \quad v \leftarrow f(t)\}, \quad \text{and}$$
$$T' = T(W) - \{x, y, u\}.$$

The $\leq_s$ relations among the $\vartheta \in UN(W)$ are the following:

$\sigma \leq_s \sigma_1$ and $\sigma_1 \leq_s \sigma$ (we write $\sigma =_s \sigma_1$)
$\sigma =_s \sigma_2$ and (because $=_s$ is an equivalence relation)
$\sigma_1 =_s \sigma_2$.

For all $\vartheta \in UN(W) - \{\sigma, \sigma_1, \sigma_2\}$ we get $\sigma \leq_s \vartheta$ but not $\vartheta \leq_s \sigma$ (and therefore also $\sigma \neq_s \vartheta$).

$\sigma, \sigma_1, \sigma_2$ are all most general unifiers of W, but they only differ with regard to variable permutations within $V(W)$:

$$\sigma\{x \leftarrow u\} = \sigma_1, \quad \sigma_1\{u \leftarrow x\} = \sigma,$$
$$\sigma\{x \leftarrow y\} = \sigma_2, \quad \sigma_2\{y \leftarrow x\} = \sigma,$$
$$\sigma_1\{u \leftarrow y\} = \sigma_2, \quad \sigma_2\{y \leftarrow u\} = \sigma_1,$$

It is easy to see that a finite set W of expressions can have only a finite number of m.g.u.'s. First of all, if there is an m.g.u. of a set of expressions W it must be contained in $UN(W)$. But if $\sigma \in UN(W)$ there are only finitely many $\vartheta \in UN(W)$ such that $\sigma =_s \vartheta$ (see Exercise 2.6.4).

The question remains, whether there is always an m.g.u. of W in case $UN(W) \neq \emptyset$. We will give a positive answer and design an algorithm which always computes an m.g.u. if $UN(W) \neq \emptyset$ and stops if $UN(W) = \emptyset$. Before we define such a unification algorithm we show that the problem of unifying an arbitrary finite set of two or more expressions can be reduced to unifying a set $\{E_1, E_2\}$ consisting of two expressions only.

Let $W = \{E_1, \ldots, E_n\}$ be a set of expressions and $n \geq 2$. If W is a set of terms we define $W' = W$. If W contains atoms or literals we proceed in the following manner: Let $\{P_1, \ldots, P_m\}$ be the set of all predicate symbols appearing in W. For every P_i we introduce a new function symbol f_i (which is not contained in W) of the same arity. We reserve a one-place function symbol g to describe negation of atoms (g is different from $f_1, \ldots, f_m$). Then we translate

$$P_i(S_1^i, \ldots, S_{k_i}^i) \quad \text{into} \quad f_i(S_1^i, \ldots, S_{k_i}^i)$$

and

$$\neg P_i(S_1^i, \ldots, S_{k_i}^i) \quad \text{into} \quad g(f_i(S_1^i, \ldots, S_{k_i}^i)).$$

Let W' be the set of all translated expressions.

Clearly ϑ is a unifier of W iff ϑ is a unifier of W'. Thus let $W' = \{E_1', \ldots, E_n'\}$ be the set of terms corresponding to W and let f be an arbitrary n-ary function symbol.

Then $E_1'\sigma = \ldots = E_n'\sigma$ iff $f(E_1', \ldots, E_n')\sigma = f(E_1', \ldots, E_1')\sigma$, and W' is unifiable by unifier ϑ iff $\{f(E_1', \ldots, E_n'), f(E_1', \ldots, E_1')\}$ is unifiable by unifier ϑ.

This completes the reduction of the unification problem for W to a unification problem of $\{t_1, t_2\}$, t_1 and t_2 being terms.

Example 2.6.2. $W = \{P(x, f(y)), P(x, f(x)), P(u, v)\}$.
We transfer the unification problem to a unification problem for two terms.

$$W' = \{h(x, f(y)), h(x, f(x)), h(u, v)\} \text{ for } h \in FS_2.$$

Now let $i \in FS_3$. We define

$$W'' \quad = \quad \{i(h(x, f(y)), h(x, f(x)), h(u, v)), i(h(x, f(y)), h(x, f(y)), h(x, f(y)))\}.$$
Then
$$\sigma \quad = \quad \{y \leftarrow x, u \leftarrow x, v \leftarrow f(x)\} \text{ (an m.g.u. of } W)$$
$$\text{is also an m.g.u. of } W''.$$
Indeed
$$W''\sigma \quad = \quad \{i(h(x, f(x)), h(x, f(x)), h(x, f(x)))\}.$$

By the transformation described above it is justified to reduce unification to "binary" unification. We now investigate the syntactical structures within $\{E_1, E_2\}$ which characterize unifiability. Let

$$E = \{P(t_1, \ldots, t_n), P(s_1, \ldots, s_n)\}.$$

By elementary properties of substitutions we obtain

$$P(t_1, \ldots, t_n)\sigma = P(t_1\sigma, \ldots, t_n\sigma),$$
$$P(s_1, \ldots, s_n)\sigma = P(s_1\sigma, \ldots, s_n\sigma).$$

Then E is unifiable iff there exists a substitution σ such that

$$s_1\sigma = t_1\sigma, \ldots, s_n\sigma = t_n\sigma.$$

We observe that the unifiability of E is equivalent to the simultaneous unifiability of $(s_1, t_1), \ldots, (s_n, t_n)$. The s_i or the t_i may be again of the form $f(u_1, \ldots, u_m)$ and the property of decomposition holds recursively. This leads to the following definition:

Definition 2.6.8 (corresponding pairs). *Let E_1, E_2 be two expressions. The set of* corresponding pairs $\mathrm{CORR}(E_1, E_2)$ *is defined as follows:*

1. $(E_1, E_2) \in \mathrm{CORR}(E_1, E_2)$
2. *If $(\neg A, \neg B) \in \mathrm{CORR}(E_1, E_2)$ then $(A, B) \in \mathrm{CORR}(E_1, E_2)$*
3. *If $(F_1, F_2) \in \mathrm{CORR}(E_1, E_2)$ such that $F_1 = F(s_1, \ldots, s_n)$ and $F_2 = F(t_1, \ldots, t_n)$, where $F \in FS \cup PS$, then $(s_i, t_i) \in \mathrm{CORR}(E_1, E_2)$ for all $i = 1, \ldots, n$.*
4. *Nothing else is in $\mathrm{CORR}(E_1, E_2)$.*

A pair (F_1, F_2) is called *irreducible* if the leading symbols of F_1, F_2 are different.

There are two different types of irreducible pairs. Take for example the pairs $(x, f(y)), (x, f(x))$ and $(f(x), g(a))$. All these pairs are irreducible. But for $\sigma = \{x \leftarrow f(y)\}$ the pair $(x, f(y))\sigma = (f(y), f(y))$ is reducible and even identical; thus there exists a substitution which removes the irreducibility of $(x, f(y))$. We show now that no such substitutions exist for the pairs $(f(x), g(a))$: For arbitrary substitutions λ we have the property

$$(f(x), g(a))\lambda = (f(x\lambda), g(a))$$

and thus $(f(x), g(a))\lambda$ is irreducible for all λ.

Let us consider the pair $(x, f(x))$ and the substitution $\lambda = \{x \leftarrow f(y)\}$ then $(x, f(x))\lambda = (f(y), f(f(y)))$; $(f(y), f(f(y)))$ is reducible but reduction yields $(y, f(y))$. Because, for all substitutions $\lambda, x\lambda$ is properly contained in $f(x)\lambda$ the set $\{x, f(x)\}$ is not unifiable.

We call a pair of expressions (E_1, E_2) *unifiable* if the set $\{E_1, E_2\}$ is unifiable. In this terminology, $(x, f(y))$ is irreducible but unifiable, but $(x, f(x))$ and $(f(x), g(a))$ are irreducible and nonunifiable. It is easy to realize that (E_1, E_2) is only unifiable if all irreducible elements in $\mathrm{CORR}(E_1, E_2)$ are (separately) unifiable; note that this property is necessary but not sufficient.

Example 2.6.3.

$$
\begin{aligned}
E_1 &= P(x, f(x, y)), \\
E_2 &= P(f(u, u), v), \\
\mathrm{CORR}(E_1, E_2) &= \{(E_1, E_2), (x, f(u, u)), (f(x, y), v)\}.
\end{aligned}
$$

We eliminate the irreducible pair $(x, f(u, u))$ by applying the substitution

$$\sigma_1 = \{x \leftarrow f(u, u)\}.$$

Applying σ_1 we obtain $(E_1\sigma_1, E_2\sigma_1)$ and

$$
\mathrm{CORR}(E_1\sigma_1, E_2\sigma_1) = \{(P(f(u, u), f(f(u, u), y)), \quad P(f(u, u), v)),
$$
$$
(f(u, u), f(u, u)), (f(f(u, u), y), v), (u, u)\}.
$$

In $\mathrm{CORR}(E_1\sigma_1, E_2\sigma_1)$ the only irreducible pair is $(f(f(u, u), y), v)$ which can be eliminated by the substitution $\sigma_2 = \{v \leftarrow f(f(u, u), y)\}$.

Now it is easy to see that $E_1\sigma_1\sigma_2 = E_2\sigma_1\sigma_2$ and that $\mathrm{CORR}(E_1\sigma_1, \sigma_2, E_2\sigma_1\sigma_2)$ consists of identical pairs only. $\sigma_1\sigma_2$ is clearly a unifier of $\{E_1, E_2\}$ (it is even the m.g.u.). For the unification above it was essential that all irreducible pairs were unifiable.

For an algorithmic treatment of unification the reducible and identical corresponding pairs are irrelevant; it suffices to focus on irreducible pairs. We are led to the following definition:

Definition 2.6.9 (difference set). *The set of all irreducible pairs in* $\mathrm{CORR}(E_1, E_2)$ *is called the* difference set *of* (E_1, E_2) *and is denoted by* $\mathrm{DIFF}(E_1, E_2)$.

Example 2.6.4. Let $(E_1, E_2) = (P(x, f(x, y)), P(f(u, u), v))$ as in Example 2.6.3. Then

$$\mathrm{DIFF}(E_1, E_2) = \{(x, f(u, u)), (f(x, y), v)\}.$$

By application of $\sigma_1 = \{x \leftarrow f(u, u)\}$ we first obtain the pair $(f(u, u), f(u, u))$ which is in $\mathrm{CORR}(E_1\sigma_1, E_2\sigma_1)$, but not in $\mathrm{DIFF}(E_1\sigma_1, E_2\sigma_1)$. Thus we obtain $\mathrm{DIFF}(E_1\sigma_1, E_2\sigma_1) = \{(f(f(u, u), y), v)\}$.

We have already mentioned that $\{E_1, E_2\}$ is unifiable only if all corresponding pairs are unifiable. By definition of $\mathrm{DIFF}(E_1, E_2)$ we get the following necessary condition for unifiability: For all pairs $(s,t) \in \mathrm{DIFF}(E_1, E_2), (s,t)$ is unifiable. The following proposition shows that the unification problem for (single) pairs in $\mathrm{DIFF}(E_1, E_2)$ is very simple.

Proposition 2.6.1. *Let E_1, E_2 be expressions and (s,t) be a pair in $\mathrm{DIFF}(E_1, E_2)$. Then (s,t) is unifiable iff the following two conditions hold:*

a) $s \in V$ or $t \in V$,

b) If $s \in V (t \in V)$ then s does not occur in t (t does not occur in s).

Proof. (s,t) is unifiable $\Rightarrow$:

By definition of $\mathrm{DIFF}(E_1, E_2)$ the pair (s,t) is irreducible; because it is unifiable, s or t must be a variable (otherwise s and t are terms with different head symbols and thus are not unifiable). Suppose now without loss of generality that $s \in V$. If s occurs in t then $s\lambda$ (properly) occurs in $t\lambda$ for all $\lambda \in \mathrm{SUBST}$; in this case (s,t) is not unifiable. We have shown that a) and b) both hold.

a), b) $\Rightarrow$

Suppose without loss of generality that $s \in V$. Define $\lambda = \{s \leftarrow t\}$; then $s\lambda = t$ and, because s does not occur in t, $t\lambda = t$. It follows that λ is a unifier of (s,t). $\diamond$

The idea of the unification algorithm shown in Figure 2.16 is the following: construct the difference set D. If there are nonunifiable pairs in D then stop with failure; otherwise eliminate a pair (x,t) in D by the substitution $\{x \leftarrow t\}$ and construct the next difference set.

```
algorithm UAL {input is a pair of expressions (E₁, E₂)};
begin
    ϑ ← ε;
    while DIFF(E₁ϑ, E₂ϑ) ≠ ∅ do
        if DIFF(E₁ϑ, E₂ϑ) contains a nonunifiable pair
        then failure
        else
            select a unifiable pair (s,t) ∈ DIFF(E₁ϑ, E₂ϑ)
            if s ∈ V
            then α := s; β := t
            else α := t; β := s
            end if
            ϑ := ϑ{α ← β}
        end if
    end while
    {ϑ is m.g.u.}
end.
```

Fig. 2.16. Unification algorithm

Note that UAL is a nondeterministic algorithm, because the selection of a unifiable pair (s,t) is nondeterministic. UAL can be transformed into different deterministic (implementable) versions by choosing appropriate search strategies. But even if the pairs (s, t) are selected from left to right (according to their positions in (E_1, E_2)), both s and t may be variables and thus $\{s \leftarrow t\}$ and $\{t \leftarrow s\}$ can both be used in extending the substitution ϑ. UAL is more than a decision algorithm in the usual sense, because in case of a positive answer (termination without failure) it also provides an m.g.u. for $\{E_1, E_2\}$.

Theorem 2.6.1 (unification theorem). *UAL is a decision algorithm for the unifiability of two expressions. Particularly the following two properties hold:*

a) *If $\{E_1, E_2\}$ is not unifiable then UAL stops with failure.*
b) *If $\{E_1, E_2\}$ is unifiable then UAL stops and ϑ (the final substitution constructed by UAL) is a most general unifier of $\{E_1, E_2\}$.*

Proof.

a) If (E_1, E_2) is not unifiable then for all substitutions λ $E_1 \lambda \neq E_2 \lambda$. Thus for every ϑ defined in UAL we get $\mathrm{DIFF}(E_1 \vartheta, E_2 \vartheta) \neq \emptyset$. In order to prove termination we have to show that the while-loop is not an endless loop: In every execution of the while loop a new substitution ϑ is defined as $\vartheta = \vartheta'\{x \leftarrow t\}$, where ϑ' is the substitution defined during the execution before and (x, t) (or (t, x)) is a pair in $\mathrm{DIFF}(E_1 \vartheta', E_2 \vartheta')$ with $x \in V$. Because $x \notin V(t)$ (otherwise UAL terminates with failure before), the pair $(E_1 \vartheta, E_2 \vartheta)$ does not contain x anymore; we conclude

$$| V(\{(E_1 \vartheta', E_2 \vartheta')\}) | > | V(\{(E_1 \vartheta, E_2 \vartheta)\}) | .$$

It follows that the number of executions of the while-loop must be $\leq k$ for $k = | V(\{E_1, E_2\}) |$. We see that, whatever result is obtained, UAL must terminate. Because UAL terminates and (by nonunifiability) $\mathrm{DIFF}(E_1 \vartheta, E_2 \vartheta) \neq \emptyset$ for all ϑ, it must stop with failure.

b) In the k-th execution of the while loop (provided termination with failure does not take place) the k-th definition of ϑ via $\vartheta := \vartheta\{\alpha \leftarrow \beta\}$ is performed. We write ϑ_k for the value of ϑ defined in the k-th execution.

Suppose now that η is an arbitrary unifier of $\{E_1, E_2\}$. We will show by induction on k, that for all ϑ_k there exist substitutions λ_k such that $\vartheta_k \lambda_k = \eta$. We are now in a position to conclude our proof as follows:

Because UAL terminates (see part a of the proof), there exists a number m such that the m-th execution of the while-loop is the last one.
From $\vartheta_m \lambda_m = \eta$ we get $\vartheta_m \leq_s \eta$. Moreover ϑ_m must be a unifier: Because the m-th execution is the last one, either $\mathrm{DIFF}(E_1 \vartheta_m, E_2 \vartheta_m) = \emptyset$ or there is a nonunifiable pair $(s, t) \in \mathrm{DIFF}(E_1 \vartheta_m, E_2 \vartheta_m)$;

but the second alternative is impossible, as $\eta = \vartheta_m \lambda_m$ and λ_m is a unifier of $(E_1 \vartheta_m, E_2 \vartheta_m)$. Because ϑ_m is a unifier, η is an arbitrary unifier and $\vartheta_m \leq_s \eta$, ϑ_m is a m.g.u. of $\{E_1, E_2\}$ (note that m and ϑ_m depend on (E_1, E_2) only, but λ_m depends on η).

Therefore it remains to show that the following statement $A(k)$ holds for all $k \in N$:

$A(k)$: Let ϑ_k be the substitution defined in the k-th execution of the while-loop. Then there exists a substitution λ_k such that $\vartheta_k \lambda_k = \eta$.

We proceed by induction on k:

$A(0)$: $\vartheta_0 = \epsilon$.

We choose $\vartheta_0 = \eta$ and obtain $\vartheta_0 \lambda_0 = \eta$.

(IH) Suppose that $A(k)$ holds.

If ϑ_{k+1} is not defined by UAL (because it stops before) the antecedent of $A(k+1)$ is false and thus $A(k+1)$ is true. So we may assume that ϑ_{k+1} is defined by UAL.

Then $\vartheta_{k+1} = \vartheta_k \{x \leftarrow t\}$ where $x \in V$, $t \in T$ and $(x, t) \in$ DIFF$(E_1 \vartheta_k, E_2 \vartheta_k)$ or $(t, x) \in$ DIFF$(E_1 \vartheta_k, E_2 \vartheta_k)$.

By the induction hypothesis (IH) we know that there exists a λ_k such that $\vartheta_k \lambda_k = \eta$. Our aim is to find an appropriate substitution λ_{k+1} such that $\vartheta_{k+1} \lambda_{k+1} = \eta$.

Because λ_k is a unifier of $(E_1 \vartheta_k, E_2 \vartheta_k)$ it must unify the pair (x, t), i.e., $x\lambda_k = t\lambda_k$. Therefore λ_k must contain the element $x \leftarrow t\lambda_k$. We define

$$\lambda_{k+1} = \lambda_k - \{x \leftarrow t\lambda_k\}.$$

The substitution λ_{k+1} fulfils the property

$$(*) \quad \{x \leftarrow t\}\lambda_{k+1} = \lambda_{k+1} \cup \{x \leftarrow t\lambda_k\}.$$

To prove $(*)$ it is sufficient to show that

$$v\{x \leftarrow t\}\lambda_{k+1} = v(\lambda_{k+1} \cup \{x \leftarrow t\lambda_k\})$$

holds for all $v \in dom(\lambda_k)$ (note that $dom(\lambda_{k+1}) \subseteq dom(\lambda_k)$). If $v \neq x$ then $v\{x \leftarrow t\} = v$ and $v\{x \leftarrow t\}\lambda_{k+1} = v\lambda_{k+1}$. If $v = x$ then

$$x\{x \leftarrow t\}\lambda_{k+1} \quad = \quad t\lambda_{k+1} \text{ and}$$
$$x(\lambda_{k+1} \cup \{x \leftarrow t\lambda_k\}) \quad = \quad t\lambda_k.$$

By definition of UAL, ϑ_{k+1} is only defined it $x \notin V(t)$. But $x \notin V(t)$ implies $t\lambda_{k+1} = t\lambda_k$ and

$$x\{x \leftarrow t\}\lambda_{k+1} = x(\lambda_{k+1} \cup \{x \leftarrow t\lambda_k\}).$$

We see that $(*)$ holds.

It follows

$$\vartheta_k \lambda_k = \vartheta_k(\lambda_{k+1} \cup \{x \leftarrow t\lambda_k\}) = \vartheta_k\{x \leftarrow t\}\lambda_{k+1} = \vartheta_{k+1}\lambda_{k+1}.$$

This concludes the proof of $A(k+1)$. ◊

Example 2.6.5. Let $E_1 = P(x, h(x, y), z)$ and $E_2 = P(u, v, g(v))$. We compute an m.g.u. of $\{E_1, E_2\}$ by using UAL.

$\vartheta_o = \epsilon$; $\mathrm{DIFF}(E_1, E_2) = \{(x, u), (h(x, y), v), (z, g(v))\}$.

All pairs in $\mathrm{DIFF}(E_1, E_2)$ are unifiable. $\vartheta_1 = \vartheta_0\{z \leftarrow g(v)\} = \{z \leftarrow g(v)\}$. $\mathrm{DIFF}(E_1\vartheta_1, E_2\vartheta_1) = \{(x, u), (h(x, y), v)\}$. Again all pairs in the difference set are unifiable.

$\vartheta_2 = \vartheta_1\{x \leftarrow u\} = \{z \leftarrow g(v), \ x \leftarrow u\}$
$\mathrm{DIFF}(E_1\vartheta_2, E_2\vartheta_2) = \{(h(u, y), v)\}$.

As v does not occur in $h(u, y)$ we continue and obtain

$\vartheta_3 = \vartheta_2\{v \leftarrow h(u, y)\} = \{z \leftarrow g(h(u, y)), \ x \leftarrow u, \ v \leftarrow h(u, y)\}$

Now $\mathrm{DIFF}(E_1\vartheta_3, E_2\vartheta_3) = \emptyset$ and (due to Theorem 2.6.1) ϑ_3 is an m.g.u. of $\{E_1, E_2\}$. The expression $E_1\vartheta_3 (= E_2\vartheta_3)$ obtained by unification is $P(u, h(u, y), g(h(u, y)))$. If we replace E_2 by $E_2' = P(v, v, g(v))$ then we obtain

$\mathrm{DIFF}(E_1, E_2') = \{(x, v), (h(x, y), v), (z, g(v))\}$.

By defining $\vartheta_1' = \{x \leftarrow v\}$ we get

$\mathrm{DIFF}(E_1\vartheta_1', E_2\vartheta_1') = \{(h(v, y), v), (z, g(v))\}$.

Because the pair $(h(v, y), v)$ is not unifiable we stop with failure.

Example 2.6.5 shows that in computing $\mathrm{DIFF}(E_1\vartheta, E_2\vartheta)$ we do not need the expressions E_1, E_2 themselves, but merely $\mathrm{DIFF}(E_1, E_2)$. Our formulation is close to that of Martelli and Montanari [MM82] where the pairs are replaced by equations. UAL is an exponential algorithm because it computes the m.g.u. ϑ in an "explicit" form, i.e., ϑ is represented as a list $x_1 \leftarrow t_1, \ldots, x_n \leftarrow t_n$.

Example 2.6.6.

$$E_1^n = P(x_1, g(x_1, x_1), x_2, g(x_2, x_2), \ldots, x_n, g(x_n, x_n)),$$
$$E_2^n = P(g(y_1, y_1), y_2, g(y_2, y_2), \ldots, y_n, g(y_n, y_n), y_{n+1}).$$

We define the length of an expression as the number of occurrences of symbols (belonging to $V \cup CS \cup FS \cup PS$). Then length $(E_1^n) =$ length$(E_2^n) \leq \alpha n$ for all n and some constant α.

(E_1^n, E_2^n) is unifiable by m.g.u. $\sigma_n =$

$$\{y_{n+1} \leftarrow g(x_n, x_n)\}\{x_n \leftarrow g(y_n, y_n)\} \ldots \{y_2 \leftarrow g(x_1, x_1)\}\{x_1 \leftarrow g(y_1, y_1)\}$$

If we represent σ_n as a sequence

$$(x_1 \leftarrow s_1), \ldots, (x_n \leftarrow s_n), (y_2 \leftarrow t_2), \ldots, (y_{n+1} \leftarrow t_{n+1})$$

then the term t_{n+1} is already of exponential size. By defining

$w_1 = g(y_1, y_1)$
$w_{n+1} = g(w_n, w_n)$ for $n \geq 1$

we obtain $t_{n+1} = w_{2n}$. By

$$\text{length } (w_{k+1}) > 2 \text{ length } (w_k) \text{ and length } (w_1) = 3$$

we get $\text{length}(t_{n+1}) > 2^{2n}$. It follows that $\text{length}(t_{n+1})$ is exponential in n and $\text{length}(\sigma_n)$ (in explicit representation) is too. Particularly the unified atom $E_1^n \sigma_n$ is of exponential length with regard to $\text{length}(E_1^n) + \text{length}(E_2^n)$.

Martelli and Montanari [MM82] developed a unification algorithm which avoids the representation of substitutions in explicit form. Instead the substitutions are represented in a compositional form (note that the substitutions σ_n – represented as composition of "atomic" substitutions in Example 2.6.6 – are of linear length). The algorithm of Martelli and Montanari has an $O(n \log n)$-time worst case, which proves that unification is a polynomial-time decision problem. The problem was even shown linear-time decidable by Paterson and Wegman in [PW78]. However it should be noted that the computation of the unified expression itself is necessarily exponential unless we adopt a compositional notation: whatever algorithm is chosen to compute and represent an m.g.u. σ_n in Example 2.6.6, $E_1^n \sigma_n$ is of size exponential in the size of E_1^n. Because our approach is more logical and less directed to complexity theory, we may be content with UAL. Despite its exponential worst case UAL can be turned into a (practically) efficient decision algorithm by using clever selection methods for corresponding pairs.

Even without combination with the resolution principle, unification can be shown superior to deduction methods directly based on Herbrand's theorem.

Example 2.6.7. $C = \{P(x, f(x, b), u), \neg P(g(a), z, h(z))\}$

We show that C is unsatisfiable. Let $A_1 = P(x, f(x, b), u)$ and $A_2 = P(g(a), z, h(z))$. By Herbrand's theorem C is unsatisfiable iff there exists a finite unsatisfiable set of ground instances. By the rules of Davis and Putnam we conclude that this is only possible if A_1 and A_2 are unifiable, i.e., there exists a ground instance η such that $A_1 \eta = A_2 \eta$. On the other hand, C is unsatisfiable if A_1 and A_2 are unifiable (apply the unifier and then propositional resolution). It follows that C is unsatisfiable iff $\{A_1, A_2\}$ is unifiable. By applying UAL we find out that $\{A_1, A_2\}$ is indeed unifiable by m.g.u.

$$\sigma = \{x \leftarrow g(a), z \leftarrow f(g(a), b), u \leftarrow h(f(g(a), b))\}.$$

Thus we can show unsatisfiability by proving that A_1 and $\neg A_2$ are made complementary by σ. Indeed the unified atom is

$$P(g(a), f(g(a), b), h(f(g(a), b))).$$

σ is not only an m.g.u., it is also the only unifier of $\{A_1, A_2\}$ in $UN(\{A_1, A_2\})$.

While proving C unsatisfiable via unification is an easy task, it becomes unfeasible by the method of Davis and Putnam. The failure of the Davis–Putnam method mainly depends on the rapid growth of the Herbrand sets

H_i and of the corresponding set of ground instances C_i'. However as a propositional proof procedure the Davis–Putnam method is quite efficient. In fact the true barrier is the principle of ground saturation.

Let us abbreviate $\mid H_i(C) \mid$ by h_i; then in our example $h_0 = 2$, $h_1 = 10$, $h_2 = 122$, $h_3 = 15130$ and in general

$$h_{i+1} = h_i + (h_{i-1} + h_i + 2)(h_i - h_{i-1}) \text{ for } i \geq 1.$$

Because σ is the only unifier (if no dummy variables are substituted) and $u \leftarrow h(f(g(a), b)) \in \sigma$, $\tau(h(f(g(a), b))) = 3$, C_0', C_1' and C_2' are all satisfiable. Thus C_3' is computed in the Davis–Putnam method. But $\mid C_3' \mid = h_3^2 + h_3 > 2 \cdot 10^8$. Thus only one substitution is needed to refute C by using UAL, while in the Davis–Putnam method more than 200 million ground instances are produced.

There is a further advantage in using UAL instead of DP. Let us generalize the set of clauses in Example 2.6.7 to $C = \{L_1, L_2\}$. C is unsatisfiable iff $\{L_1, L_2^d \eta\}$ is unifiable (for a variable-renaming substitution η such that $V(L_1) \cap V(L_2 \eta) = \emptyset$). In order to decide satisfiability we only apply UAL to $(L_1, L_2^d \eta)$; note that UAL always terminates. On the other hand, the Davis–Putnam method never terminates if C is satisfiable. We conclude that UAL is not only faster but can actually do "much more" – at least on examples of that type. However, there are improved ground instance enumeration procedures using the unification principle (they generate ground instances of m.g.u.'s only); we just mention Plaisted's linking and hyperlinking method [LP92].

Exercises

Let $\leq_{s0}$ be the following relation on the set of all substitutions SUBST: $\sigma \leq_{s0} \tau$ iff there exists a $\vartheta \in$ SUBST such that for all $x \in dom(\sigma) : x\sigma\vartheta = x\tau$.

Exercise 2.6.1.

a) Show that $\sigma \leq_s \tau$ (see Definition 2.6.4) always implies $\sigma \leq_{s0} \tau$.
b) Define two substitutions σ, τ such that $\sigma \leq_{s0} \tau$ holds but $\sigma \leq_s \tau$ does not hold.

Exercise 2.6.2. Let E be an expression, σ be a substitution with $dom(\sigma) = V(E)$ and $E\sigma \leq_s E\tau$ for some substitution τ. Show that $\sigma \leq_{s0} \tau$ holds, but $\sigma \leq_s \tau$ need not be true.

Comments on Exercises 2.6.1, 2.6.2:

Exercises 2.6.1 and 2.6.2 show that, in some sense, the relation $\leq_{s0}$ is more natural than $\leq_s$ (in defining the concept "more general"). However, $\leq_s$ is more adequate for the analysis of most general unification. Note that the existence of a most general unifier (defined via $\leq_s$) is a stronger result than it would be for $\leq_{s0}$, indeed, if μ is m.g.u., we know that for all unifiers λ we have $\mu \leq_s \lambda$ ($\mu \leq_{s0} \lambda$ is an immediate consequence by Exercise 2.6.1).

Exercise 2.6.3. Show that a set of expressions W is unifiable iff $UN(W) \neq \emptyset$ (recall that $\vartheta \in UN(W)$ iff ϑ is a unifier, $dom(\vartheta) \subseteq V(W)$ and $rg(\vartheta) \subseteq T(W)$).

Exercise 2.6.4. Let W be a set of expressions and $\sigma \in UN(W)$. Show that there are only finitely many substitutions $\vartheta \in UN(W)$ such that $V(rg(\vartheta)) \subseteq V(W)$ and $\sigma =_s \vartheta$.

Exercise 2.6.5. Show that the most general unifiers defined by the algorithm UAL are always idempotent (i.e., if σ is an m.g.u. of a set of expressions then $\sigma\sigma = \sigma$). Does this property hold for all most general unifiers in $UN(W)$?

2.7 The General Resolution Principle

In Section 2.5 we have shown that propositional resolution is complete. A characteristic of PR-deductions is the use of unrestricted substitution in instantiating input clauses. On the other hand only unifying substitutions are relevant to PR-deductions. It would be an easy task to show the completeness of PR-deductions under the restriction that all instances are also unifiers (either for resolution or for unifying literals within a clause). Although such a restriction would be an obvious improvement, there may still be infinitely many unifiers of two literals. The real strength of the general resolution principle is based on the rigorous use of most general unification. That means unification (in every required form) appears as most general unification only and can be performed algorithmically (e.g., by UAL). Some additional techniques are necessary if substitution (which is a unary rule) is replaced by unification (which is binary).

Example 2.7.1. Let $\mathcal{C}$ be $\{C_1, C_2, C_3\}$ for

$C_1 = \neg P(x) \vee R(f(x))$, $C_2 = P(f(x)) \vee R(x)$ and $C_3 = \neg R(x)$.

$\Pi : \neg P(f(a)) \vee R(f(f(a))), P(f(a)) \vee R(a), R(f(f(a))) \vee R(a), \neg R(a),$
$\quad R(f(f(a))), \neg R(f(f(a))), \square$

is a GR-refutation of $\mathcal{C}$.

Although the substitutions $\vartheta_1 = \{x \leftarrow f(a)\}$ in C_1 and $\vartheta_2 = \{x \leftarrow a\}$ in C_2 lead to the complementary literals $\neg P(f(a))$ and $P(f(a)), \{P(x), P(f(x))\}$ was not unified in the sense of Section 2.6. The reason is that $\{P(x), P(f(x))\}$ is not unifiable by a substitution applied to both atoms simultaneously. But, as clauses represent closed, universal disjunctions, the names of variables are clearly irrelevant. By renaming C_2 to $C_2' : P(f(y)) \vee R(y)$ we can unify $\{P(x), P(f(y))\}$ by m.g.u. $\sigma = \{x \leftarrow f(y)\}$.

$$\Pi_1 : \neg P(f(y)) \vee R(f(f(y))), \quad P(f(y)) \vee R(y), \quad R(f(f(y))) \vee R(y)$$

is a PR-deduction from $\{C_1, C_2, C_3\}$ based on σ.

We see that some kind of renaming is necessary in order to combine unification and resolution. By extending Π_1 to Π, but under most general unification, we obtain

$$\Pi_1, \neg R(x), R(f(f(x))).$$

Again we are in the situation that $\{R(x), R(f(f(x)))\}$ is not unifiable.

This problem can easily be solved by replacing every clause by an appropriate renamed version. Thus we may resolve $\neg P(x) \vee R(f(x))$ and $P(f(x)) \vee R(x)$ by replacing them (internally) by variants $\neg P(y) \vee R(f(y))$ and $P(f(z)) \vee R(z)$ which are variable-disjoint, and apply most general unifiers and propositional resolution afterwards. Using this method we get $R(f(f(z))) \vee R(z)$ and all its variants as resolvents of $\neg P(x) \vee R(f(x))$ and $P(f(x)) \vee R(x)$. In the sense of this new renaming method we obtain a deduction:

$$\neg P(x) \vee R(f(x)), \quad P(f(x)) \vee R(x), \quad R(f(f(z))) \vee R(z), \neg R(x), \quad R(f(f(y))), \quad \square.$$

Note that we can use $\neg R(x)$ twice, first in the resolution with $R(f(f(z))) \vee R(z)$ and then in resolving with (its resolvent) $R(f(f(y)))$.

Example 2.7.1 gives a motivation for the following definitions:

Definition 2.7.1 (variant). *A clause C is called a* variant *of a clause D if there exists a variable permutation (see Definition 2.6.5) η such that $C\eta = D$; we write $C \sim_v D$.*

It is easy to verify that $\sim_v$ is an equivalence relation on the set of all clauses. By $\mathcal{C}/\sim_v$ we denote the equivalence classes under $\sim_v$ defined by the set of clauses $\mathcal{C}$.

Definition 2.7.2 (binary resolvent). *Let C and D be two clauses and C_1, D_1 be variants of C, D such that $V(C_1) \cap V(D_1) = \emptyset$. Suppose that C_1 and D_1 are of the form $C_1 : A \vee L \vee B$, $D_1 : E \vee M \vee F$ such that $\{L^d, M\}$ is unifiable by m.g.u. σ. Then the clause $(A \vee B \vee E \vee F)\sigma$ is called* binary resolvent *of C and D. The atom formula of L is the atom "resolved upon".*

Because there are infinitely many variants C_1, D_1 of C, D, two clauses may have infinitely many binary resolvents. The following simple proposition shows that this infinity is "inessential":

Proposition 2.7.1. *Let C, D be two clauses and $Res(\{C, D\})$ be the set of all binary resolvents of C and D. Then $Res(\{C, D\})/\sim_v$ is finite.*

Proof. Let C_1, D_1 be two variants of C, D such that $V(C_1) \cap V(D_1) = \emptyset$. Because there are maximally $\mid C_1 \mid \cdot \mid D_1 \mid$ pairs of complementary literals, there are only finitely many "direct" binary resovents of C_1, D_1 (i.e., resolvents obtained without renaming C_1, D_1 again). Suppose now that $C_1 = C\eta_1$ and $D_1 = D\lambda_1$ for renaming substitutions η_1, λ_1 and $C = A \vee L \vee B$, $D = E \vee M \vee F$. The binary resolvent, obtained resolving upon $L\eta_1$ and $M\lambda_1$, is

$$R_1 = (A\eta_1 \vee B\eta_1 \vee E\lambda_1 \vee F\lambda_1)\sigma_1$$

where σ_1 is an m.g.u. of $\{L^d\eta_1, M\lambda_1\}$.

Taking another pair of variants $C_2 = C\eta_2$ and $D_2 = D\lambda_2$ and resolving upon $L\eta_2$ and $M\lambda_2$ we get

$$R_2 = (A\eta_2 \vee B\eta_2 \vee E\lambda_2 \vee F\lambda_2)\sigma_2$$

where σ_2 is an m.g.u. of $\{L^d\eta_2, M\lambda_2\}$.

Let $W_1 = \{L^d\eta_1, M\lambda_1\}$ and $W_2 = \{L^d\eta_2, M\lambda_2\}$ then $W_1 \sim_v W_2$ and there exists an η such that $W_1\eta = W_2$.

Let us denote $A\eta_1 \vee B\eta_1 \vee E\lambda_1 \vee F\lambda_1$ by G_1, $A\eta_2 \vee B\eta_2 \vee E\lambda_2 \vee F\lambda_2$ by G_2.

Then $G_1 \sim_v G_2$ because $V(A\eta_1 \vee B\eta_1) \cap V(E\lambda_1 \vee F\lambda_1) = \emptyset$, and $V(A\eta_2 \vee B\eta_2) \cap (E\lambda_2 \vee F\lambda_2) = \emptyset$ and $\eta_1, \lambda_1, \eta_2, \lambda_2$ are all permutations (and thus are invertible).

Let ϑ be the renaming substitution with $G_1\vartheta = G_2$. Because ϑ is a renaming and σ_1, σ_2 are m.g.u.'s of expressions which are variants of each other there exists a renaming ρ such that $\vartheta\sigma_2 = \sigma_1\rho$. But then we obtain

$$R_1\rho = G_1\sigma_1\rho = G_1\vartheta\sigma_2 = G_2\sigma_2 = R_2.$$

Similarly we can construct a renaming τ such that $R_2\tau = R_1$. It follows that two binary resovents defined by resolving different variants, but upon variants of the same literals, are $\sim_v$ equivalent. Particularly R_1 and R_2 are in the same $\sim_v$ equivalence class. $\Diamond$

The resolution rule defined in Definition 2.7.2 is called "binary" resolution, because only a pair of literals is made complementary by most general unification. This binary unification principle, however, is too weak to result in a complete inference principle.

Example 2.7.2. Let $\mathcal{C}$ be the unsatisfiable set of clauses

$$\{P(x) \vee P(y), \neg P(x) \vee \neg P(y)\}.$$

With regard to $\sim_v$ the only resolvent of C_1 and C_2 is $C_3 : \neg P(x) \vee P(y)$. Note that $P(y) \vee \neg P(y)$ is a P-resolvent, but not a binary resolvent of C_1 and C_2. But (again under $\sim_v$) C_3 with C_1 reproduces C_1 and C_3 with C_2 reproduces C_2 again. We conclude that a deduction principle based only on the binary resolution rule (in Definition 2.7.2) is not complete. There is of course the following PR-refutation of $\mathcal{C}$:

$P(x) \vee P(x), \ \neg P(x) \vee \neg P(x), \ \Box.$

Note that $P(x) \vee P(x)$ defines the p-reduct $P(x)$. The substitution $\lambda : \{y \leftarrow x\}$ applied to C_1 works as an internal m.g.u. to unify literals within C_1. We recognize the necessity to unify inside clauses in order to make p-reduction possible.

Note that renaming substitutions (being permutations) cannot lead to unification within clauses. Moreover $C \sim_v D$ implies $F(\{C\}) \sim F(\{D\})$. This property does not hold for unifying substitutions, e.g., $F(\{P(x,y) \vee P(y,x)\})$ is not logically equivalent to $F(\{P(x,x) \vee P(x,x)\})$ (we performed the internal m.g.u. $\sigma = \{y \leftarrow x\}$). However, $F(\{P(x) \vee P(y)\})$ and $F(\{P(x)\})$ are equivalent due to the property of condensing, which will be discussed in Section 3.2.

Definition 2.7.3 (G-instance). *Let C be a clause and A be a nonempty set of literals occurring in C. If A is unifiable by m.g.u. μ then $C\mu$ is called a G-instance (general instance) of C. Every G-instance of a G-instance is also a G-instance (of C).*

Note that, under $\sim_v$, there are only finitely many G-instances of a clause C, while C may possess infinitely many instances which are not variants of each other. If A consists of only one element then $C\mu$ is a G-instance for every permutation μ; but, in this case, $C\mu \sim_v C$ and, under renaming, C itself is the only G-instance with respect to A. The following concept combines G-instantiation and p-reduction:

Definition 2.7.4 (factor). *A p-reduct of a G-instance of a clause C is called a factor of C. A factor C' of C is called nontrivial if the number of literals in C' is strictly smaller than the number of literals in C.*

Note that the concepts of factor and p-reduct coincide in case of ground clauses.

Example 2.7.3. $C = P(x) \vee R(x) \vee P(z) \vee R(y)$.

C_1 : $P(x) \vee R(x) \vee P(x) \vee R(y)$ is a G-instance of C by setting A in Definition 2.7.3 to $\{P(x), P(z)\}$.

C_2 : $P(x) \vee R(x) \vee P(x) \vee R(x)$ is G-instance of C_1 (via $A = \{R(x), R(y)\}$) and therefore G-instance of C. $P(x) \vee R(x) \vee R(y)$ and $P(x) \vee R(x)$ are factors of C.

We are now in possession of all technical concepts to define the general principle of R-deduction.

Definition 2.7.5 (resolvent). *Let C, D be clauses and C', D' be factors of C and D. If E is a binary resolvent of C' and D' then E is called a resolvent of C and D.*

Note that the definition of resolvent in Definition 2.7.5 differs from the concepts introduced in [Rob65], [CL73] and [Lov78]. A detailed analysis of the differences is given in Section 2.8.

Definition 2.7.6 (R-deduction). *Let C be a set of clauses and let C be a clause. A sequence $C_1, \ldots, C_n$ is called an R-deduction (resolution deduction) of C from C if it fulfils the following conditions:*

1. *$C_n = C$,*
2. *For all $i = 1, \ldots, n$:*
   a) *C_i is a variant of a clause in C or*
   b) *C_i is a resolvent of clauses C_j, C_k for $j, k < i$.*

An R-deduction of $\square$ from C is called an R-refutation of C.

If $\Gamma = C_1, \ldots, C_n$ is an R-(PR-, GR-) deduction we define $l(\Gamma) = n$ as the *length* of Γ.

Note that we did not list factors explicitly in the definition of an R-deduction. The reason can be found in the fact that a clause, being a closed $\forall$-quantified form semantically, describes all of its (finitely many) factors implicitly.

Unlike the concepts of PR- and GR- deductions, R-deductions are locally finite. Every clause defines only a finite number of factors and resolvents (modulo renaming). Therefore there are only finitely many R-deductions of length $\leq n$ under $\sim_v$ for every $n \in \mathbb{N}$. This is a very important feature and one of the reasons for the efficiency of resolution. In particular, the branching rate of the search space is finite, whereas in classical systems it is infinite. Moreover, R-deduction is not only locally finite (as also a depth-bounded ground term saturation would be) but only produces "relevant" instances by using substitution as a binary principle. In all traditional calculi for predicate logic (Hilbert-type, natural deduction, sequent calculus, etc.) the substitution rule is unary and unrestricted; in the sequent calculus we have the $\forall$-introduction left (see [Tak75]):

$$\frac{A(t), \Gamma \vdash \Delta}{(\forall x) A(x), \Gamma \vdash \Delta}$$

for any term t which does not contain variables bound in $A(x)$. In the calculus of natural deduction [Pra71] there exists the rule $\forall$-elimination:

$$\frac{(\forall x) A(x)}{A(t)}$$

Again t may not contain variables which are bound in $A(x)$. This restriction (required for soundness) is very weak, e.g., t may be an arbitrary ground term. Although the method of Davis and Putnam essentially differs from classical logical calculi, it also (potentially) produces all ground instances of clauses. It is the thoroughgoing use of most general unification (as the *only* substitution concept) that separates classical logical calculi from "computational" calculi. Inference systems based on the unification principle are commonly subsumed under the name of *computational logic*. It is this principle (rather than just the rule of p-resolution) which represents the real power of the resolution calculus. We will see in Chapter 5 that the local finiteness of resolution combined with some proof-theoretic refinements yields a powerful basis for decision procedures in clausal predicate logic.

In Section 2.5 we have shown refutational completeness of GR- and of PR-deduction. The proof of completeness of R-deduction is essentially based on the technique of "lifting". Lifting is the replacement of an R-deduction by a more general one. That means every kind of instantiation can somehow be replaced by most general unification.

Example 2.7.4. $C = \{P(x), \neg P(y) \vee R(f(y)), \neg R(z)\}$.

Let t be an arbitrary ground term. Then the set of clauses
$C_t = \{P(t), \neg P(t) \vee R(f(t)), \neg R(f(t))\}$ is unsatisfiable. The following deduction Γ_t is a R-refutation of C_t

$$\Gamma_t = P(t),\ \neg P(t) \vee R(f(t)),\ R(f(t)),\ \neg R(f(t)),\ \square.$$

Γ_t can be lifted to an R-refutation Γ of C for:

$$\Gamma = P(x),\ \neg P(y) \vee R(f(y)),\ R(f(y)),\ \neg R(z),\ \square.$$

Γ is more general than Γ_t in the sense:

$$\Gamma \lambda_t = \Gamma_t \text{ for } \lambda_t = \{x \leftarrow t, y \leftarrow t, z \leftarrow f(t)\}.$$

All refutations Γ_t of C_t are instances of Γ. Γ is obtained from Γ_t by "lifting".

Definition 2.7.7. *Let* $\Gamma : C_1, \ldots, C_n$ *and* $\Delta : D_1, \ldots, D_n$ *be R-deductions.* Γ *is called* more general *than* Δ *if there are substitutions* $\vartheta_1, \ldots, \vartheta_n$ *such that* $(C_1\vartheta_1, \ldots, C_n\vartheta_n) = \Delta$; *we write* $\Gamma \leq_s \Delta$. *In particular we define* $C \leq_s D$ *for clauses* C, D *if there exists a substitution* λ *such that* $C\lambda = D$.

Note that $C \leq_s D$ implies the validity of $F(\{C\}) \rightarrow F(\{D\})$ but not vice versa. Moreover $C \leq_s D$ is not identical with the (stronger) subsumption relation to be defined in Section 4.2. The following two lemmas show that the lifting property holds for factoring and for resolution.

Lemma 2.7.1. *Let C, D be clauses such that $C \leq_s D$ holds. If F is a factor of D then there exists a factor E of C with $E \leq_s F$ (factors of instances are instances of factors).*

Proof. By induction on the degree of the factor F of D. If X' is a factor of a clause X we define the degree of X' as the number $|X| - |X'|$ (or we say that X' is a k-factor for $|X| - |X'| = k$). 0-factors are the trivial factors. We start with factors of degree 0:

By Definition 2.7.4 $F = D\sigma$ for a G-instance $D\sigma$ of D. Moreover $D = C\lambda$ for a substitution λ. We simply set $E = C$, then E is a trivial factor of C and $E\lambda\sigma = F$.

(IH) suppose that the theorem holds for all factors F of D of degree k.

Let F be a factor of degree $k + 1$ of D. By Definition 2.7.4 there exists a k-factor F' of D such that F is a 1-factor and even a p-reduct of F'. Therefore F' must be of the form

$F' = F_1 \vee L \vee F_2 \vee L \vee F_3$ and
$F = F_1 \vee L \vee F_2 \vee F_3$, or
$F = F_1 \vee F_2 \vee L \vee F_3$ respectively.

By (IH) there exists a factor E' of C such that $E' \leq_s F'$. Therefore

$E' = E_1 \vee M_1 \vee E_2 \vee M_2 \vee E_3$ and
$E_1\lambda = F_1, \quad M_1\lambda = M_2\lambda = L, \quad E_2\lambda = F_2$ and
$E_3\lambda = F_3$

for an appropriate substitution λ. Note that M_1 and M_2 may be different literals. But, by $M_1\lambda = M_2\lambda$, λ is a unifier of $\{M_1, M_2\}$. E' is a p-reduct of a G-instance $C\tau$ of C.
If $M_1 = M_2$ then we define $E = E_1 \vee M_1 \vee E_2 \vee E_3$. Then E is a p-reduct of E' and thus a factor of C fulfilling $E\lambda = F$; so we have $E \leq_s F$.

If $M_1 \neq M_2$ then there are literals K_1, K_2 in C such that $K_1\tau = M_1$ and $K_2\tau = M_2$. Clearly $\{K_1, K_2\}$ is not unified by τ. Because λ is a unifier of $\{M_1, M_2\}$ there exists an m.g.u. σ such that $\sigma \leq_s \lambda$. By Definition 2.7.3 $C\tau\sigma$ is a G-instance of C. Moreover there exists a literal L_0 such that $M_1\sigma = M_2\sigma = L_0$ and $L_0\vartheta = L$ for some substitution ϑ.

By defining a p-reduct E'' from $C\tau\sigma$ – contracting analogous literals as for E' – we obtain

$$E'' = E_1' \vee L_0 \vee E_2' \vee L_0 \vee E_3'.$$

By Definition 2.7.4 E'' is a factor of C. Because $E' \leq_s F'$ and σ is m.g.u. of $\{M_1, M_2\}$ (while λ is an arbitrary unifier $\{M_1, M_2\}$) we also obtain $E'' \leq_s F'$. Then

$E = E_1' \vee L_0 \vee E_2' \vee E_3'$

is a p-reduct of E'' and $E \leq_s F$. But E is also a factor of C. This concludes the proof of case $k + 1$.   ◊

Lemma 2.7.2 (Lifting Lemma). *Let C, D, C', D' be clauses such that $C \leq_s C'$ and $D \leq_s D'$. If E' is a binary resolvent of C' and D' then there exists a binary resolvent E of C and D with $E \leq_s E'$.*

Proof. Let $C' = C_1' \lor L' \lor C_2'$ and $D' = D_1' \lor M' \lor D_2'$. By Definition 2.7.2 we may assume without loss of generality that $V(C') \cap V(D') = \emptyset$ (otherwise we consider variants of C', D'). Then the binary resolvent is of the form:

$$E' = (C_1' \lor C_2' \lor D_1' \lor D_2')\sigma$$

where σ is m.g.u. of $\{L'^d, M'\}$.

To simplify the proof we may assume $V(C) \cap V(D) = \emptyset$. By $C \leq_s C'$ and $D \leq_s D'$ there are substitutions λ, μ such that

$$C = C_1 \lor L \lor C_2, \quad D = D_1 \lor M \lor D_2 \quad \text{and} \quad C_1\lambda = C_1', \quad L\lambda = L',$$
$$C_2\lambda = C_2', \quad D_1\mu = D_1', \quad M\mu = M', \quad D_2\mu = D_2'.$$

Because $V(C) \cap V(D) = \emptyset$ there exists a substitution η such that

$$dom(\eta) \subseteq dom(\lambda) \cup dom(\mu), \quad C\lambda = C\eta \text{ and}$$

$$D\mu = D\eta, \quad C\eta = C' \text{ and } D\eta = D'.$$

Because σ is an m.g.u. of $\{L'^d, M'\}$ and $L' = L\eta$, $M' = M\eta$, σ is m.g.u. of $\{L, M\}\eta$. Consequently $\eta\sigma$ is a unifier of $\{L^d, M\}$. By the Unification Theorem 2.6.1 there exists an m.g.u. τ of $\{L, M\}$. As $\eta\sigma$ is an arbitrary unifier of $\{L, M\}$ there exists a φ such that $\tau\varphi = \eta\sigma$. But

$$E = (C_1 \lor C_2 \lor D_1 \lor D_2)\tau$$

is a binary resolvent of C and D and we obtain:

$$E\varphi = (C_1 \lor C_2 \lor D_1 \lor D_2)\tau\varphi = (C_1 \lor C_2 \lor D_1 \lor D_2)\eta\sigma = (C_1' \lor C_2' \lor D_1' \lor D_2')\sigma = E'.$$

It follows that $E \leq_s E'$.   ◇

Theorem 2.7.1 (Lifting Theorem). *Let C be a set of clauses and C' be a set of instances of clauses in C. Let Δ be an R-deduction from C'. Then there exists an R-deduction Γ from C such that $\Gamma \leq_s \Delta$.*

Proof. By induction on $l(\Delta)$, the length of the deduction Δ.
Induction basis $l(\Delta) = 1$:
Δ must be of the form D' for some variant D' of a clause D in C'. By definition of C' there must be a $C \in C$ and a substitution λ such that $C\lambda = D$. But then there exists a λ' such that $C\lambda' = D'$.
Therefore $\Gamma : C$ is an R-deduction from C and $\Gamma \leq_s \Delta$.

(IH) Suppose that for every set of instances C' of C and for every R-deduction Δ from C' with $l(\Delta) \leq n$ there exists an R-deduction Γ from C such that $\Gamma \leq_s \Delta$.

Let Δ be an R-deduction of length $n+1$ from C'. Then $\Delta = \Delta', C$ for some R-deduction of length n from C'. For C there are two possibilities:

a) C is a variant of a clause in C' or
b) C is binary resolvent of clauses D', E' such that D', E' are factors of clauses D, E in Δ'.

case a) Let $C = D\eta$ for a renaming η and a $D \in C'$. As C' is a set of instances from C there exists a clause $E \in C$ such that $E \leq_s D$. Because η is a renaming (and thus admits an inverse) we also obtain $E \leq_s C$. By (IH) there exists an R-deduction Γ' from C such that $\Gamma' \leq_s \Delta'$. By definition of $\leq_s, \Gamma' \leq_s \Delta'$ and $E \leq_s C$ implies

$$\Gamma', E \leq_s \Delta', C.$$

Defining $\Gamma = \Gamma', E$ we obtain $\Gamma \leq_s \Delta$. Because Γ' is an R-deduction from C and $E \in C, \Gamma$ is an R-deduction from C.

case b) $\Delta' = \Delta_1, D, \Delta_2, E, \Delta_3$ for appropriate clause sequences Δ_1, Δ_2, and Δ_3.

By (IH) there exists an R-deduction Γ' such that $\Gamma' \leq_s \Delta'$. Then Γ' must be of the form $\Gamma_1, F, \Gamma_2, G, \Gamma_3$ such that $F \leq_s D$ and $G \leq_s E$. By assumption, D' and E' are factors of D and E. Thus by Lemma 2.7.1 we obtain factors F' of F and G' of G such that $F' \leq_s D'$ and $G' \leq_s E'$.

By the assumption b) C is binary resovent of D' and E'. By Lemma 2.7.2 there exists a binary resolvent H of F' and G' such that $H \leq_s C$. By definition of resolution, H is a resolvent of F and G.

Thus we define $\Gamma : \Gamma', H$.

Then Γ is an R-deduction from C and $\Gamma \leq_s \Delta$.

This concludes the proof of case $n+1$.   ◇

The lifting theorem easily yields the completeness of R-deduction.

Theorem 2.7.2 (completeness of R-deduction). *If C is an unsatisfiable set of clauses then there exists an R-refutation of C.*

Proof. From Theorem 2.5.1 we know that GR-deduction is complete. Thus let Π be a GR-refutation of C and let C' be the set of all ground instances from C appearing in Π. Because, on ground clauses, p-reduction coincides with factoring, Π is also an R-refutation of C'. By Theorem 2.7.1 there exists an R-deduction Γ from C such that $\Gamma \leq_s \Pi$. But a refutation can only be an instance of a deduction, which is a refutation too. Therefore Γ is an R-refutation of C.   ◇

In most of the literature on automated theorem proving the lifting theorem is called "lifting lemma". We call it a theorem because the $\leq_s$-ordering of deductions according to the $\leq_s$-ordering of clauses is interesting from a proof-theoretic point of view as a result on its own. Note that in the proof of Theorem 2.7.2 no semantical concepts are involved (semantical concepts are only contained in the completeness proof of GR-deductions); instead we proved the (relative) completeness of R-deductions relative to that of GR-deductions. The lifting theorem is the key not only to the proof of completeness of R-deductions, but to all completeness proofs of resolution refinements and of deletion methods.

Exercises

Exercise 2.7.1. Let $\Gamma : C_1, \ldots, C_n$ and $\Delta : D_1, \ldots, D_n$ be R-deductions such that $\Gamma \leq_s \Delta$ (see definition 2.7.7). Then there exists an R-deduction Γ' and a (single) substitution λ such that $\Gamma' = C_1', \ldots, C_n'$ (for some clauses C_i') and

$$(C_1', \ldots, C_n')\lambda = C_1'\lambda, \ldots, C_n'\lambda = \Delta \text{ (i.e. } \Gamma\lambda = \Delta).$$

Exercise 2.7.2.
Let

$$\begin{aligned}
\Gamma_n \;=\; & P_1(x),\; \neg P_1(x) \vee P_2(f(x)),\; P_2(f(x)), \ldots, \\
& P_i(f^{i-1}(x)),\; \neg P_i(x) \vee P_{i+1}(f(x)),\; P_{i+1}(f^i(x)), \ldots, \\
& P_n(f^{n-1}(x)),\; \neg P_n(x) \vee P_{n+1}(f(x)),\; P_{n+1}(f^n(x)), \\
& \neg P_{n+1}(x), \square
\end{aligned}$$

be an R-refutation of

$$C_n = \{P_1(x)\} \cup \{\neg P_i(x) \vee P_{i+1}(f(x)) | 1 \leq i \leq n\} \cup \{\neg P_{n+1}(x)\}$$

for $n \neq 1$.
Define a sequence of GR-refutations Δ_n (for $n \neq 1$) such that $\Gamma_n \leq_s \Delta_n$ (investigate the substitutions that have to be applied to Γ_n).

Exercise 2.7.3. Show that there exists a set of clauses C and an R-refutation Γ of C such that for no GR-refutation Δ of C $\Gamma \leq_s \Delta$ (i.e., the inverse operation to lifting is generally impossible).

2.8 A Comparison of Different Resolution Concepts

In the numerous publications on automated theorem proving several different basic concepts of resolution appear. Sometimes it is important to know which concept is actually used, particularly in theoretical investigations. The purpose of this section is to give a comparison of some widely used resolution concepts.

The first (and somehow trivial) difference can be found in the data structures for the representation of clauses. While we defined clauses as disjunctions, there are also the sequent notation (particularly in logic programming

[Llo87] and the representation as sets of literals [Rob65], [CL73], [Lov78]. If the differences would be only notational a comparison were pointless and boring; but this is not the case. Indeed Chang & Lee's and Loveland's concept differ from Robinson's, and our concept, defined in Section 2.7, . does not completely coincide with any of those.

First of all we choose the disjunction notation as basic and interpret the other concepts in this formalism. We only have to take care that the logical content of the rules is preserved. So we may consider the set notation as a normal form under the equivalence relation of commutativity, associativity, and idempotence for the connective "∨". In Section 3.2 we will present a flexible method to handle clause normalization which provides a general framework for all clausal data structures used in automated deduction. To facilitate the description of Robinson's and Chang & Lee's concepts we introduce some additional notations and definitions.

Let C be a clause. We write $C \setminus L$ for C after omission of all occurrences of the literal L (if L does not appear in C then $C \setminus L = C$).

Definition 2.8.1 (S-factor). *Let A be a subset of LIT(C) for a clause C and σ be an m.g.u. of A such that $A\sigma = \{L\}$. Then $(C\sigma \setminus L) \vee L$ is called an S-factor of C ("S" stands for "simple"). An S-factor is called trivial if $\sigma = \epsilon$.*

Example 2.8.1. Let C be the clause $P(x) \vee P(y) \vee R(u) \vee R(v)$. Then $R(u) \vee R(v) \vee P(x)$ and $P(x) \vee P(y) \vee R(u)$ are both S-factors; $P(y) \vee R(u) \vee R(v) \vee P(x)$ is a trivial S-factor. Note that $R(u) \vee R(v) \vee P(x)$ is not a factor in the sense of Definition 2.7.4.

$P(x) \vee R(u)$ is a factor (obtained by p-reduction of the G-instance $P(x) \vee P(x) \vee R(u) \vee R(u)$), but it is not an S-factor.

In a simple factor the literal resulting from unification and p-reduction is always the last one. If we resolve only on the rightmost literals of simple factors we can be sure that these are the literals that result from factoring. We thus obtain a closer connection between factoring and the resolution cut rule.

Definition 2.8.2 (Robinson resolvent). *Let C, D be clauses, C', D' variable disjoint variants of C, D and C_1, D_1 be S-factors of C', D' such that $C_1 = A \vee L$ and $D_1 = B \vee M$. Suppose that $\{L, M^d\}$ is unifiable by m.g.u. σ. Then the clause $E: (A \vee B)\sigma$ is called a Robinson resolvent of C and D.*

If C_1, D_1 are trivial S-factors then E is called a binary Robinson resolvent.

If C_1 and D_1 are reduced clauses then $(A \vee B)\sigma$ in Definition 2.8.2 is also binary resovent of C_1 and D_1. According to Robinson's original definition, clauses are thought of as sets, and variables are renamed in some standard way. Up to renaming, Robinson's concept is the following:

Let C_1, D_1 be variable-disjoint variants of C and D and $A \subseteq C_1, B \subseteq D_1$. If there exists a most general unifier σ of $A \cup B^d$ then the clause $(C_1 - A)\sigma \cup (D_1 - B)\sigma$ is called a resolvent of C and D.

It is easy to see that the set of literals in a Robinson resolvent (definition 2.8.2) forms a resolvent in the set notation as defined above.

Example 2.8.2. Let

$$C = P(x) \vee P(y) \vee R(u) \vee R(v) \quad \text{and}$$
$$D = \neg Q(x_1) \vee \neg P(x_1) \vee \neg R(x_1) \vee \neg P(x_2).$$

Then

$$C_1 \quad : \quad R(u) \vee R(v) \vee P(x),$$
$$D_1 \quad : \quad \neg Q(x_1) \vee \neg R(x_1) \vee \neg P(x_1).$$

are S-factors of C, D. $\sigma = \{x_1 \leftarrow x\}$ is an m.g.u. of $\{P(x), P(x_1)\}$. Consequently

$$E : R(u) \vee R(v) \vee \neg Q(x) \vee \neg R(x) \text{ is Robinson-resolvent of } C \text{ and } D.$$

In the set notation we obtain the clause $\{R(u), R(v), \neg Q(x), \neg R(x)\}$.

Note that $C_2 : P(x) \vee R(u) \vee R(v)$ and $D_2 : \neg Q(x_1) \vee \neg P(x_1) \vee \neg R(x_1)$ are factors of C and D and E is also a resolvent of C and D. The clause $C_3 : P(x) \vee R(u)$ is a factor of C and $F : R(u) \vee \neg Q(x) \vee \neg R(x)$ is a binary resolvent of C_3 and D_2 and thus a resolvent of C and D. But F is not a Robinson resolvent of C and D.

We show now that Robinson resolution yields a refinement of R-deduction (in the sense of Section 3.1).

Definition 2.8.3 (Robinson deduction). *Let C be a set of clauses and C be a clause. A sequence $C_1, \ldots, C_n$ is called Robinson deduction of C from C if it fulfills the following conditions:*

(1) $C_n = C$ and
(2) For all $i = 1, \ldots, n$:
    (2a) C_i is a variant of a clause in C or
    (2b) C_i is a Robinson resolvent of C_j, C_k for $j, k < i$.

Proposition 2.8.1. *Every Robinson deduction is an R-deduction.*

Proof. By Definitions 2.7.6 and 2.8.3 it suffices to prove that every Robinson resolvent is a resolvent (in the sense of Definition 2.7.5). Let C, D be variable-disjoint clauses and $C_1\sigma \vee L$, $D_1\mu \vee M$ be S-factors of C and D. Suppose that $\{L, M^d\}$ is unifiable by m.g.u. ϑ. Then

$$E : (C_1\sigma \vee D_1\mu)\vartheta$$

is a Robinson-resolvent of C and D.

We have to show that E is also a resolvent of C and D. The S-factor $C_1\sigma \vee L$ is a permutation variant of a "real" factor $A_1\sigma \vee L \vee A_2\sigma$ of C with $A_1\sigma \vee A_2\sigma = C_1\sigma$.

Similarly $D_1\mu \vee M$ is a permutation variant of a factor $B_1\mu \vee M \vee B_2\mu$ of D with $B_1\mu \vee B_2\mu = D_1\mu$.

$$F : A_1\sigma\vartheta \vee A_2\sigma\vartheta \vee B_1\mu\vartheta \vee B_2\mu\vartheta$$

is binary resolvent of $A_1\sigma \vee L \vee A_2\sigma$ and $B_1\mu \vee M \vee B_2\mu$ and thus resolvent of C and D. But $F = (A_1\sigma \vee A_2\sigma \vee B_1\mu \vee B_2\mu)\vartheta = (C_1\sigma \vee D_1\mu)\vartheta = E$. ◊

Now we give some definitions as a preparation for the definition of Chang & Lee's resolution concept.

Definition 2.8.4 (standard factor). *Let C be a clause and $\mathcal{A} \subseteq LIT(C)$ such that $\mathcal{A}$ is unifiable by an m.g.u. σ and $\mathcal{A}\sigma = \{L\}$. Every reduced p-reduct of $C\sigma$ is called* standard factor *of C.*

Standard factors are factors in the sense of Definition 2.7.4, while S-factors are not. But every S-factor is a permutation variant of some standard factor.

Example 2.8.3. $C = P(x) \vee P(y) \vee R(u) \vee R(v)$.

$P(x) \vee R(u) \vee R(v)$ and $P(x) \vee P(y) \vee R(u)$ are standard factors of C; the corresponding S-factors are $R(u) \vee R(v) \vee P(x)$ and $P(x) \vee P(y) \vee R(u)$. The factor $P(x) \vee R(u)$ is not a standard factor.

As in the case of S-factors, we define a standard factor as trivial if the factoring substitution is ϵ.

Definition 2.8.5 (Chang–Lee resolvent). *Let C_1 and C_2 be clauses, C_1' and C_2' be variable-disjoint variants of C_1, C_2 and $D_1 : A_1 \vee L_1 \vee B_1$, $D_2 : A_2 \vee L_2 \vee B_2$ be standard factors of C_1' and C_2'. If σ is an m.g.u. of $\{L_1, L_2^d\}$ then the clause*

$$E : (D_1\sigma \setminus L_1\sigma) \vee (D_2\sigma \setminus L_2\sigma)$$

is called a Chang–Lee resolvent *of C_1 and C_2. If D_1, D_2 are trivial (standard) factors then E is called a* binary Chang–Lee resolvent *of C_1 and C_2.*

Example 2.8.4. Let $C = P(x) \vee P(y) \vee R(u) \vee R(v)$ and $D = \neg R(a)$. Then $P(x) \vee R(u) \vee R(v)$ is a standard factor of C.

$E = P(x) \vee R(v)$ is a Chang–Lee resolvent of C and D.

E is not a Robinson resolvent, because the corresponding S-factor is $R(u) \vee R(v) \vee P(x)$, which does not define a Robinson resolvent with D.

$P(x) \vee R(u)$ is a factor of C, which is neither an S-factor nor a standard factor.

Therefore $P(x)$ is a resolvent, but neither a Robinson nor a Chang–Lee resolvent of C and D.

In comparing Definitions 2.8.5 and 2.7.5, we see that in a Chang–Lee resolvent the m.g.u. corresponding to the binary resolvent is applied before cutting out the literal. Combined with the reduction rule (removal of multiple literals) this can give some side-effects in binary resolution.

Proposition 2.8.2. *There are binary Chang–Lee resolvents which are not binary resolvents (in the sense of Definition 2.7.2).*

Proof. Let $C = P(a) \vee P(y)$ and $D = \neg P(a)$. Selecting $P(y)$ from C and $\neg P(a)$ from D we obtain an m.g.u. $\sigma = \{y \leftarrow a\}$ and the following binary Chang–Lee resolvent of C and D:

$$((P(a) \vee P(y))\sigma \setminus P(y)\sigma) \vee (\neg P(a)\sigma \setminus \neg P(a)\sigma) =$$
$$((P(a) \vee P(a)) \setminus P(a)) \vee \square = \square \vee \square = \square.$$

$\square$ is neither a binary resolvent nor a binary Robinson resolvent of C and D. $\square$ is resolvent of C and D, but it is only definable by the factor $P(a)$ of C.
                                                                                    $\Diamond$

The proof of Proposition 2.8.2 shows that binary Chang–Lee resolution "automatically" performs some factoring tests; we might speak about unintentional factoring. This kind of factoring can create side effects in the lifting of deductions [Lei89].

Example 2.8.5. We define instances $C\lambda, D\mu$ of C, D such that λ and μ do not unify literals in C and D and there exists a binary Chang–Lee resolvent E of $C\lambda, D\mu$ with the following property: There exists no binary Chang–Lee resolvent F of C and D with $F \leq_s E$.

   In some sense the lifting property does not hold for binary Chang–Lee resolvents. Let $C = P(x) \vee P(y)$ and $D = \neg P(a)$, $\lambda = \{x \leftarrow a\}$ and $\mu = \epsilon$. Then λ and μ fulfill the condition stated above. The instance $C\lambda : P(a) \vee P(y)$ is not a G-instance of C. In the proof of Proposition 2.8.2 we have shown that $\square$ is a binary Chang–Lee resolvent of $C\lambda$ and $D\mu$. On the other hand, $P(x)$ is the only binary Chang–Lee resolvent of C and D (under renaming). Clearly $\square$ is not an instance of $P(x)$ and $(\forall x)P(x)$ does not even imply $\square$.

   In [Lei89] it is shown that Robinson resolution (in contrast to Chang–Lee resolution) fulfills the "binary lifting" property.

If we do not care about factoring and focus on the "full" resolution concepts, Chang–Lee-resolution yields a refinement of R-deduction. This gives us the means to describe both Robinson and Chang–Lee resolution within the formalism developed in Section 2.7.

Definition 2.8.6 (Chang–Lee deduction). *Let C be a set of clauses and C be a clause. A sequence $C_1, \ldots, C_n$ is called a* Chang–Lee deduction *of C from $\mathcal{C}$ if it fulfills the following conditions:*

(1) $C_n = C$ and
(2) For all $i = 1, \ldots, n$:
   (2a) C_i is a variant of a clause in $\mathcal{C}$ or
   (2b) C_i is a Chang–Lee resolvent of clauses C_j, C_k for $j, k < i$.

Proposition 2.8.3. *Every Chang–Lee deduction is an R-deduction.*

Proof. According to Definitions 2.7.6 and 2.8.6 it is enough to show that every Chang–Lee resolvent is a resolvent in the sense of Definition 2.7.5. The proof of this property is not completely trivial as, by Proposition 2.8.2, binary Chang–Lee resolvents need not be binary resolvents.

Let C_1, C_2 be clauses and C'_1, C'_2 variable-disjoint variants of C_1 and C_2. Furthermore let D_1, D_2 be standard factors of C'_1, C'_2 and

$$D_1 = A_1 \vee L_1 \vee B_1, \quad D_2 = A_2 \vee L_2 \vee B_2.$$

We assume that $\{L_1, L_2^d\}$ is unifiable by m.g.u. σ. Then

$$G : (D_1\sigma \setminus L_1\sigma) \vee (D_2\sigma \setminus L_2\sigma)$$

is a Chang–Lee resolvent of C_1 and C_2. We have to show that G is also resolvent of C_1 and C_2.

If $D_1\sigma \setminus L_1\sigma = (A_1 \vee B_1)\sigma$ and $D_2\sigma \setminus L_2\sigma = (A_2 \vee B_2)\sigma$, then G is binary resolvent of D_1 and D_2 and thus Robinson resolvent of C_1 and C_2. If $LIT(D_1\sigma \setminus L_1\sigma) \subseteq LIT(A_1 \vee B_1)\sigma$ or $LIT(D_2\sigma \setminus L_2\sigma) \subseteq LIT(A_2 \vee B_2)\sigma$ then more literals are cut out by binary Chang–Lee resolution than by binary resolution. In this case we proceed as follows:
Let

$$
\begin{aligned}
\mathcal{A}_1 &= \{K \mid K \in LIT(C'_1) \ \ K\eta_1 = L_1\} \quad \text{and}\\
\mathcal{A}_2 &= \{K \mid K \in LIT(C'_2) \ \ K\eta_2 = L_2\},
\end{aligned}
$$

where η_1, η_2 are the factoring substitutions applied to C'_1, C'_2 in order to get D_1, D_2.
Moreover we define

$$
\begin{aligned}
\mathcal{B}_1 &= \{K \mid K \in LIT(C'_1) \ \ K\eta_1\sigma = L_1\sigma\}\\
\mathcal{B}_2 &= \{K \mid K \in LIT(C'_2) \ \ K\eta_2\sigma = L_2\sigma\}.
\end{aligned}
$$

Clearly $\mathcal{A}_1 \subseteq \mathcal{B}_1$ and $\mathcal{A}_2 \subseteq \mathcal{B}_2$.

The sets $\mathcal{B}_1, \mathcal{B}_2$ are unifiable by substitutions $\eta_1\sigma$ and $\eta_2\sigma$ respectively, but $\eta_1\sigma$ and $\eta_2\sigma$ need not be m.g.u.'s of $\mathcal{B}_1, \mathcal{B}_2$.

Let λ_1 be m.g.u. of $\mathcal{B}_1$, λ_2 be m.g.u. of $\mathcal{B}_2$. Then $\lambda_1 \leq_s \eta_1\sigma$ and $\lambda_2 \leq_s \eta_2\sigma$. Moreover there exist literals M_1, M_2 such that

$$\mathcal{B}_1\lambda_1 = \{M_1\}, \mathcal{B}_2\lambda_2 = \{M_2\} \text{ and } M_1 \leq_s L_1\sigma, M_2 \leq_s L_2\sigma.$$

The substitution $(\eta_1 \cup \eta_2)\sigma$ is m.g.u. of $\mathcal{B}_1 \cup \mathcal{B}_2^d$. As $M_1 \leq_s L_1\sigma$, $M_2 \leq_s L_2\sigma$ and $V(M_1) \cap V(M_2) = \emptyset$ (note that λ_1, λ_2 are substitutions defining G-instances), the m.g.u. τ of $\{M_1, M_2^d\}$ fulfills $\tau \leq_s \sigma$.

By definition of λ_1, λ_2 and τ, $(\lambda_1 \cup \lambda_2)\tau$ is m.g.u. of $\mathcal{B}_1 \cup \mathcal{B}_2^d$. As also $(\eta_1 \cup \eta_2)\sigma$ is an m.g.u. of $\mathcal{B}_1 \cup \mathcal{B}_2^d$ we must have $(\lambda_1 \cup \lambda_2)\tau =_s (\eta_1 \cup \eta_2)\sigma$ and $C_1'\lambda_1$ and $C_2'\lambda_2$ are G-instances of C_1', C_2'.

Let $E_1 = F_1 \vee M_1 \vee G_1$, $E_2 = F_2 \vee M_2 \vee G_2$ p-reducts of $C_1'\lambda_1$ and $C_2'\lambda_2$ such that $M_1 \notin LIT(F_1 \vee G_1)$, $M_2 \notin LIT(F_2 \vee G_2)$.

By definition of $\mathcal{B}_1, \mathcal{B}_2$ we must have

$$(F_1 \vee G_1)\tau \leq_s (D_1\sigma \setminus L_1\sigma) \vee (D_2\sigma \setminus L_2\sigma)$$

and by $(\lambda_1 \cup \lambda_2)\tau =_s (\eta_1 \cup \eta_2)\sigma$ even

$$(F_1 \vee G_1)\tau =_s (D_1\sigma \setminus L_1\sigma) \vee (D_2\sigma \setminus L_2\sigma).$$

But $(F_1 \vee G_1)\tau$ is a resolvent of C_1, C_2. As every variant of a resolvent is a resolvent too, G is resolvent of C_1 and C_2. ◊

It remains to mention Loveland's concept of resolution. In his definition [Lov78] binary resolution is called just resolution and factoring is treated as a separate rule (i.e., factors appear explicitly in resolution deductions). Because Loveland's definition of resolution coincides with Chang–Lee's binary resolution there are Loveland resolvents which are not binary resolvents (see Proposition 2.8.2). As factors, in the sense of Definition 2.7.4, are derivable in Loveland's resolution concept, every clause derivable in an R-deduction is also derivable by Loveland's resolution. Our decision, not to add factors explicitly, was motivated by the subsumption principle to be defined in Section 4.2.

In Robinson's book [Rob79] resolution is defined as a method of extending clausal sequents (representing sets of clauses) by resolvents; in this version of resolution elimination of tautologies (see Section 4.4) and (forward) subsumption (see Section 4.2) are built in a priori.

Clearly the differences between the concepts are less important than their common features, like the use of most general unification. However, the differences are worth mentioning and discussing, as they can be of some importance in the "proof technology" of completeness and lifting proofs. Moreover Robinson's concept defines a valuable refinement a priori (which is relevant to implementations of resolution) by concentrating factoring and cut onto the same group of literals.

3. Refinements of Resolution

3.1 A Formal Concept of Refinement

In Chapter 2 we have proved the correctness and completeness of the resolution principle. It might seem that these are already the key results of resolution theory. Although the (general) resolution principle is clearly superior to the PR-deduction principle and to methods based on Herbrand's theorem, there are several reasons for further refining the deduction principle. First of all the high number of resolvents derivable from a set of clauses is a serious obstacle to practical applications. Thus it is significant that Robinson presented a paper on hyperresolution [Rob65a], a refinement to be described later, in the same year as his landmark paper on resolution [Rob65] was published. So one motivation for restricting the R-deduction principle is efficiency.

Another important appplication of refinements is the construction of resolution decision methods for decidable clause classes, a problem area which will be presented in Chapter 5. In defining resolution decision procedures, the key idea is to find complete refinements, which – on the class of clauses under consideration – produce finitely many clauses only. In both applications we have to take care that the completeness of the deduction principle is preserved. However, we do not need completeness on whole clause logic in general; there are relevant refinements (e.g., unit-resolution on Horn logic) which are complete on subclasses of clause logic only.

Formally a refinement is a subset of the set of all R-deductions. But this property (being a subset) is far too general and includes many pathological cases. In the following we will develop an abstract concept of refinement which covers all refinements defined in this book and – at the same time · fulfils some reasonable properties of effectiveness.

Let $\mathcal{C}$ be a finite set of clauses and $\Omega(\mathcal{C})$ be the set of all R-deductions from $\mathcal{C}$; $\Omega = \bigcup_{\mathcal{C} \in \mathbf{CL}} \Omega(\mathcal{C})$ is the set of all R-deductions (**CL** denotes the set of all finite sets of clauses). Based on this notation, a *refinement* is some subset Ψ of Ω. Some refinements can be extracted from the basic concepts in a straightforward manner. Let Γ_1, Γ_2 be two deductions in Ω; if $\Gamma_1 =_s \Gamma_2$ (for $=_s$ see Definition 2.7.7) we can safely omit Γ_1 (or Γ_2 respectively) without losing the completeness of the deduction principle. In defining Ψ as a maximal subset of Ω such that $\Gamma_1, \Gamma_2 \in \Psi$ implies $\Gamma_1 \neq_s \Gamma_2$, we get a restriction that preserves completeness. Another restriction is to consider deductions $C_1, \ldots, C_n$ only, where C_i, C_j are variable-disjoint for $i \neq j$. Combining both restrictions we get a complete refinement Ψ, which can be specified recursively.

The following example shows that the "subset definition" of refinements is too general.

Example 3.1.1.

$$\text{Let } \Psi(\mathcal{C}) = \{\Gamma\} \text{ for some R-refutation } \Gamma \text{ of } \mathcal{C} \text{ if } \mathcal{C} \text{ is unsatisfiable,}$$
$$= \{C\} \text{ if } \mathcal{C} \text{ is satisfiable, for some } C \in \mathcal{C},$$
$$\text{and } \Psi = \bigcup_{\mathcal{C} \in \mathbf{CL}} \Psi(\mathcal{C}).$$

Clearly Ψ is a refinement (by $\Psi \subseteq \Omega$) and $\Psi(\mathcal{C})$ contains a refutation for every unsatisfiable $\mathcal{C}$. We can make Ψ a function in the strict sense in specifying that Γ must be the first refutation of $\mathcal{C}$ in Ω and C be the first clause in $\mathcal{C}$, where "first" refers to some fixed ordering of clauses and deductions. A pathological property of Ψ is the absence of a decision procedure for the predicate π defined by : $\pi(\mathcal{C}) = \mathbf{T}$ iff $\Psi(\mathcal{C}) \subseteq \mathcal{C}$.

Note that a decision procedure for π would directly yield another for the satisfiability of clause logic. But clause logic is a reduction class of predicate logic [Lew79] and thus its satisfiability problem is undecidable. Although $\Psi(\mathcal{C})$ is finite for all $\mathcal{C} \in \mathbf{CL}$, the function $\lambda \mathcal{C}[\Psi(\mathcal{C})]$ is not computable; because we can decide whether a given deduction is a refutation, an algorithm computing Ψ would yield a decision procedure for satisfiability immediately.

Definition 3.1.1 (refinement). *Let Ψ be a mapping from the set of all finite sets of clauses $\mathbf{CL}$ to the set Ω of all R-deductions. Ψ is called a resolution refinement if the following conditions are fulfilled:*

1. *For all $\mathcal{C} \in \mathbf{CL} : \Psi(\mathcal{C}) \subseteq \Omega(\mathcal{C})$.*
2. *$\{\Pi \mid \Pi \in \Psi(\mathcal{C})\}$ is decidable for every $\mathcal{C} \in \mathbf{CL}$.*
3. *There exists an algorithm α constructing $\Psi(\mathcal{C})$ for every $\mathcal{C} \in \mathbf{CL}$ (note that $\Psi(\mathcal{C})$ may be infinite).*
4. *$\Psi(\mathcal{C}) \subseteq \Psi(\mathcal{D})$ for $\mathcal{C} \subseteq \mathcal{D}$.*

Although Definition 3.1.1 is still very abstract, "refinements" of the type indicated in Example 3.1.1 are excluded now. Definition 3.1.1(2) indicates that refinements must be specified by "syntactic" properties. Definition 3.1.1(3) expresses the fact that there exists a (globally defined) recursive enumeration of $\Psi(\mathcal{C})$ for all $\mathcal{C} \in \mathbf{CL}$. Definition 3.1.1(4) excludes the case that the mere presence of more clauses makes some deductions impossible.

Definition 3.1.2. *A resolution refinement is called* complete *if, for every unsatisfiable set of clauses $\mathcal{C}$, $\Psi(\mathcal{C})$ contains an R-refutation of $\mathcal{C}$.*

Example 3.1.2. Let $\Psi(\mathcal{C}) = \{(C, D, E) \mid C, D \in \mathcal{C}, E$ is resolvent of C and $D\}$. Clearly Ψ fulfils 1–4 of Definition 3.1.1 and thus is a resolution refinement. However Ψ is not complete, as there exist unsatisfiable sets of clauses $\mathcal{C}$ such that $\Psi(\mathcal{C})$ does not contain a refutation of $\mathcal{C}$.

However, Ψ is complete on the class of all $\mathcal{C} \in \mathbf{CL}$ with $\mid \mathrm{LIT}(C) \mid= 1$ for all $C \in \mathcal{C}$ (Ψ is complete on unit clause logic).

The refinement Ψ of Example 3.1.2 may be further restricted to Ψ' such that for $\Gamma_1, \Gamma_2 \in \Psi' : \Gamma_1 \neq_s \Gamma_2$. Then $\Psi'(\mathcal{C})$ is finite for every $\mathcal{C} \in \mathbf{CL}$. The following proposition, which is easily proved, shows that such a property cannot hold for complete refinements.

Proposition 3.1.1. *Let Ψ be a complete refinement. Then there must exist (finite) sets of clauses $\mathcal{C}$ such that $\Psi(\mathcal{C})$ is infinite.*

Proof. Assume that Ψ is complete and that $\Psi(\mathcal{C})$ is finite for every $\mathcal{C} \in \mathbf{CL}$. By Definition 3.1.1(3) there exists an algorithm α with $\alpha(\mathcal{C}) = \Psi(\mathcal{C})$ (where $\alpha(\mathcal{C})$ denotes the result of the computation of α on $\mathcal{C}$). If $\Psi(\mathcal{C})$ is finite for all $\mathcal{C} \in \mathbf{CL}$ then α always terminates. Because Ψ is complete, $\Psi(\mathcal{C})$ contains a refutation iff $\mathcal{C}$ is unsatisfiable. Thus producing $\Psi(\mathcal{C})$ first and then testing whether $\Psi(\mathcal{C})$ contains a refutation gives a decision procedure for the satisfiablility problem of clause logic; but clause logic is undecidable (in fact it is a reduction class of predicate logic). We conclude that there must exist $\mathcal{C} \in \mathbf{CL}$ such that $\Psi(\mathcal{C})$ is infinite. ◇

Proposition 3.1.1 shows that the halting problem for complete refinements is always undecidable. However, we will specify clause classes $\mathbf{CL}'$ and complete refinements Ψ in Chapter 5, where Ψ always terminates on $\mathbf{CL}'$. Following the arguments in the proof of Proposition 3.1.1, we see that such a termination property yields a decision procedure for $\mathbf{CL}'$; consequently such an effect can appear only on decidable classes $\mathbf{CL}'$.

In general an application of a complete resolution theorem prover Ψ on a set of clauses $\mathcal{C}$ yields one of the following three outcomes:

a) Derivation of □.
b) $\Psi(\mathcal{C})$ is finite (Ψ terminates on $\mathcal{C}$) and $\Psi(\mathcal{C})$ does not contain a refutation: Because Ψ is complete we know that $\mathcal{C}$ is satisfiable.
c) $\Psi(\mathcal{C})$ is infinite and does not contain a refutation of $\mathcal{C}$: Ψ does not terminate on $\mathcal{C}$ and (by completeness) $\mathcal{C}$ is satisfiable. This case corresponds to entering an endless-loop in programming.

Many refinements such as ordering resolution (Section 3.3), semantic resolution (Section 3.6), and lock resolution (Section 3.4) can be specified by means of set operators. In these cases the set of all deductions, specified by a refinement Ψ, can be described by the set of all derivable clauses.

Let $\mathcal{R}es(C, D)$ be the set of all resolvents of the clauses C and D. We have seen in Section 2.7 that, although $\mathcal{R}es(C, D)$ is infinite, $\mathcal{R}es(C, D) \mid_{\sim v}$ is finite. In using resolution operators it is convenient to avoid different variants

of clauses in the set of derived resolvents. Because it is inconvenient to work
with equivalence classes $C \mid_{\sim v}$, we select one clause from every equivalence
class. By $\varphi(C)$ we denote the clauses which are the representatives of the
classes $C \mid_{\sim v}$. For example we can take $P(x_1) \vee Q(x_2)$ as a representative of
$\{P(x) \vee Q(y) \mid x, y \in V, x \neq y\}$. Generally we can use a standard method to
rename variables in clauses: replace the first variable (from the left) by x_1,
the second by x_2 etc ... . Such a method of standardization is similar to that
of Robinson [Rob65].

The standard renaming operator φ on clauses can be extended to sets
of clauses by $\varphi(C) = \{\varphi(C) \mid C \in C\}$. It is easy to see that φ fulfils the
properties: $\varphi(C) \subseteq \varphi(D)$ for $C \subseteq D, \varphi(\varphi(C)) = \varphi(C)$.

Definition 3.1.3. *Let C be a set of clauses. We define*

$$\begin{aligned}
\mathcal{R}es(C) &= \bigcup\{\mathcal{R}es(C, D) \mid C, D \in C\} \text{ and} \\
S_\emptyset^0(C) &= \varphi(C), \\
S_\emptyset^{i+1}(C) &= S_\emptyset^i(C) \cup \varphi(\mathcal{R}es(S_\emptyset^i(C))) \text{ for } i \in \mathbb{N}, \\
R_\emptyset(C) &= \bigcup_{i=0}^\infty S_\emptyset^i(C).
\end{aligned}$$

$R_\emptyset$ is called the operator of unrestricted resolution.

$R_\emptyset(C)$ is the set of all clauses (in variable standard form) derivable via
R-deductions from C. By using the operator φ we can guarantee, for each i,
the finiteness of the sets $S^i(C)$, i.e., of the set of all clauses derivable within
deduction depth $\leq i$. The finiteness of $R_\emptyset(C)$ itself (like that of $\Psi(C)$ for some
refinement Ψ) implies "termination of resolution" on C. The term termination
is justified, as $R_\emptyset$ can be considered as an algorithm defined on **CL**; the
computation can be performed according to the recursive Definition 3.1.2. If
$R_\emptyset(C)$ is finite then this computation terminates, otherwise it does not.

Definition 3.1.4 (refinement operator). *A resolution refinement operator R_x is a mapping from* **CL** *to the set of all sets of clauses fulfilling the
following conditions:*
*There exists a mapping $\rho_x : $ **CL** $\to$ **CL** *such that*

*a) $\rho_x(C)$ is a finite subset of $R_\emptyset(C)$ for all $C \in$ **CL**,*
b) there exists an algorithm α computing $\rho_x(C)$ on **CL**,

and R_x is defined (via ρ_x) in the following way:

$$\begin{aligned}
S_x^0(C) &= \varphi(C), \\
S_x^{i+1}(C) &= S_x^i(C) \cup \rho_x(S_x^i(C)) \quad for \quad i \in \mathbb{N}, \\
R_x(C) &= \bigcup_{i=0}^\infty S_x^i(C).
\end{aligned}$$

In Definition 3.1.4(a) we stated the condition $\rho_x(\mathcal{C}) \subseteq R_{\emptyset}(\mathcal{C})$ instead of $\rho_x(\mathcal{C}) \subseteq \varphi(Res(\mathcal{C}))$ only. This more general definition is motivated by the refinement of hyperresolution, where R-deductions of some specific type are contracted into single inference steps. By Definition 3.1.4(b) and by the definition of R_x as Kleene closure, the operator R_x is computable on **CL**. Clearly $R_x(\mathcal{C}) \subseteq R_{\emptyset}(\mathcal{C})$ holds for every refinement operator R_x. In Section 3.2 we will extend the concept of refinement to arbitrary normal forms for clauses (i.e., the sets $R_{\emptyset}(\mathcal{C})$ are replaced by the sets of all normalized clauses derivable by resolution). Definition 3.1.4 only covers refinements without (backward) deletion methods, which means produced clauses cannot be removed afterwards. In Section 4.2 we will define resolution operators and describe subsumption and other deletion methods, which are not refinements (in the formal sense).

Definition 3.1.5. *A resolution refinement operator R_x is called* complete *if for all unsatisfiable sets of clauses $\mathcal{C}$ in* **CL** *we have* $\square \in R_x(\mathcal{C})$.

It is quite easy to prove a result analogous to Proposition 3.1.1, namely that for every complete resolution refinement operator R_x there must exist a set of clauses $\mathcal{C} \in$ **CL** such that $R_x(\mathcal{C})$ is infinite.

Let Ψ be a set of R-deductions. By $\text{Der}(\Psi)$ we denote the set of all clauses in variable standard form derived by deductions in Ψ. Clearly $\text{Der}(\Omega(\mathcal{C})) = R_{\emptyset}(\mathcal{C})$ for all $\mathcal{C} \in$ **CL**.

In case of so-called linear refinements (see Section 2.5) the deduction sets Ψ cannot simply be described by refinement operators on sets of clauses. But, on the other hand, refinement operators always define refinements in the sense of Definition 3.1.1.

Proposition 3.1.2. *Let R_x be a resolution refinement operator; then there exists a refinement Ψ with $\text{Der}(\Psi(\mathcal{C})) = R_x(\mathcal{C})$ for all $\mathcal{C} \in$ **CL**.*

Proof. Let $\mathcal{C}$ be a finite set of clauses. Then every $C \in \mathcal{C}$ is also an R-deduction of length 1 and $\mathcal{C}$ can be considered as a set of deductions too.

We define Ψ via level saturation (like refinement operators).

(I) Let $\Psi^0(\mathcal{C}) = \phi(\mathcal{C})$ for $\mathcal{C} \in$ **CL**.
By Definition 3.1.4(b) for the operator ρ_x, ρ_x is computable; by Definition 3.1.4(a) $\rho_x(\mathcal{C})$ is finite and $\rho_x(\mathcal{C}) \subseteq R_{\emptyset}(\mathcal{C})$. Thus there exists an algorithm α (independent of $\mathcal{C}$) which produces a set of deductions $\Pi(\mathcal{C}) = \bigcup_{i=1}^{n} \Delta_i$ (from $\mathcal{C}$) such that $\text{Der}(\Pi(\mathcal{C})) = \rho_x(\mathcal{C})$ and Π is computable. We continue in defining Ψ:
Suppose that $\Psi^i(\mathcal{C})$ has already been constructed and let
$$\Psi^i(\mathcal{C}) = \{\psi_1, \ldots, \psi_{k(i)}\}.$$
By concatenating all deductions in $\Psi^i(\mathcal{C})$ we obtain a deduction
$$\chi(\Psi^i(\mathcal{C})) = \psi_1, \ldots, \psi_{k(i)}.$$
We then define:
(II) $\Psi^{i+1}(\mathcal{C}) = \Psi^i(\mathcal{C}) \cup \{\chi(\Psi^i(\mathcal{C}))\psi | \psi \in \Pi(\text{Der}(\Psi^i(\mathcal{C})))\},$

(III) $\Psi(\mathcal{C}) = \bigcup_{i=0}^{\infty} \Psi^i(\mathcal{C})$.

By definition $\Psi(\mathcal{C})$ is a set of deductions from $\mathcal{C}$. In (II) $\chi(\Psi^i(\mathcal{C}))$ is a deduction from $\mathcal{C}$ and ψ are deductions from $\mathrm{Der}(\Psi^i(\mathcal{C}))$. Because all clauses in $\mathrm{Der}(\Psi^i(\mathcal{C}))$ possess a deduction based on the clauses in $\chi(\Psi^i(\mathcal{C}))$, $\chi(\Psi^i(\mathcal{C}))\psi$ is a deduction from $\mathcal{C}$ for all deductions ψ from $\mathrm{Der}(\Psi^i(\mathcal{C}))$.

In order to show $\mathrm{Der}(\Psi(\mathcal{C})) = R_x(\mathcal{C})$ it is enough to prove $\mathrm{Der}(\Psi^i(\mathcal{C})) = S_x^i(\mathcal{C})$ for $i \in \mathbb{N}, \mathcal{C} \in \mathbf{CL}$ (note that $R_x(\mathcal{C}) = \bigcup_{i=0}^{\infty} S_x^i(\mathcal{C})$).

We proceed by induction on i:

$i = 0$: $\mathrm{Der}(\Psi^0(\mathcal{C})) = \mathrm{Der}(\phi(\mathcal{C})) = S_x^0(\mathcal{C})$.

(IH) Suppose that the assertion holds for i.

Case $i + 1$:

We first prove $S_x^{i+1}(\mathcal{C}) \subseteq \mathrm{Der}(\Psi^{i+1}(\mathcal{C}))$, which can be reduced to:

$$S_x^i(\mathcal{C}) \subseteq \mathrm{Der}(\Psi^{i+1}(\mathcal{C})) \text{ and } \rho_x(S_x^i(\mathcal{C})) \subseteq \mathrm{Der}(\Psi^{i+1}(\mathcal{C})).$$

By (IH) $S_x^i(\mathcal{C}) = \mathrm{Der}(\Psi^i(\mathcal{C}))$.

Because $\mathrm{Der}(\Psi^i(\mathcal{C})) \subseteq \mathrm{Der}(\Psi^{i+1}(\mathcal{C}))$ holds trivially, we obtain

$$S_x^i(\mathcal{C}) \subseteq \mathrm{Der}(\Psi^{i+1}(\mathcal{C})).$$

By definition of Π we have $\mathrm{Der}(\Pi(S_x^i(\mathcal{C}))) = \rho_x(S_x^i(\mathcal{C}))$. But for every $\psi \in \Pi(S_x^i(\mathcal{C}))$, $\chi(\Psi^i(\mathcal{C}))\psi$ is a deduction from $\mathcal{C}$ and

$$\mathrm{Der}(\chi(\Psi^i(\mathcal{C}))\psi) = \mathrm{Der}(\psi).$$

Therefore $\rho_x(S_x^i(\mathcal{C})) \subseteq \mathrm{Der}(\Psi^{i+1}(\mathcal{C}))$.

Now we show the other direction, $\mathrm{Der}(\Psi^{i+1}(\mathcal{C})) \subseteq S_x^{i+1}(\mathcal{C})$.

By (IH) we have $\mathrm{Der}(\Psi^i(\mathcal{C})) = S_x^i(\mathcal{C})$ and by definition of S_x^{i+1}, $S_x^i(\mathcal{C}) \subseteq S_x^{i+1}(\mathcal{C})$. Thus we get $\mathrm{Der}(\Psi^i(\mathcal{C})) \subseteq S_x^{i+1}(\mathcal{C})$.

If $\psi \in \Pi(\mathrm{Der}(\Psi^i(\mathcal{C})))$ then

$$\mathrm{Der}(\chi(\Psi^i(\mathcal{C}))\psi) = \mathrm{Der}(\psi).$$

By (IH) $S_x^i(\mathcal{C}) = \mathrm{Der}(\Psi^i(\mathcal{C}))$ and thus $\mathrm{Der}(\psi) \in \mathrm{Der}(\Pi(S_x^i(\mathcal{C})))$.

By definition of Π we obtain $\mathrm{Der}(\psi) \in S_x^{i+1}(\mathcal{C})$. Finally we get

$$\{\chi(\Psi^i(\mathcal{C}))\psi \mid \psi \in \Pi(\mathrm{Der}(\Psi^i(\mathcal{C})))\} \subseteq S_x^{i+1}(\mathcal{C}).$$

$\Diamond$

In many practical cases it is easier to define a refinement Ψ, corresponding to an operator R_x, than in the proof of Proposition 3.1.2. In most cases we have

$$\rho_x(\mathcal{C}) = \bigcup\{\rho_x(\{C_1, C_2\})/C_1, C_2 \in \mathcal{C}\}.$$

Let ψ_1, ψ_2 be deductions in Ψ such that

$$Der(\psi_1) = C_1, Der(\psi_2) = C_2.$$

If $C \in \rho_x(\{C_1, C_2\})$ then we add the deduction ψ_1, ψ_2, C to Ψ. Clearly we obtain a Ψ such that $\mathrm{Der}(\Psi) = R_x$.

3.2 Normalization of Clauses

In Robinson's famous paper on resolution [Rob65] a clause is defined as a set of literals subjected to some standard renaming of variables. Formally such a representation can be obtained by factoring under the equivalence relation defined by associativity, commutativity, and idempotence of $\vee$ and $\sim_v$ equivalence. Chang & Lee [CL73] and Loveland [Lov78] define clauses as sets of literals but do not keep clauses in variable standard form. In Section 3.1 we have seen that, without variable standard form, the set of resolvents $\mathrm{Res}(\mathcal{C})$ from a finite set of clauses $\mathcal{C}$ is infinite. The reader might now ask why we did not define clauses as sets of literals in advance. There are two main reasons for treating normalization explicitly:

1) There are resolution methods (particularly in Horn logic) based on clauses as lists or sequences rather than as sets.
2) Sometimes one would like to have stronger normalization principles than usually afforded by "set"-normalization and renaming.

In resolution decision theory (to be presented in Chapter 5) we need strongly nonredundant clause representations. Such representations can be obtained by the principle of condensing.

Example 3.2.1. $C = P(x) \vee P(a)$.
The variable standard form (i.e. naming the first variable x_1, the second x_2, etc.) of C is: $P(x_1) \vee P(a)$.
Clearly we have for all variables $w \in V : P(w) \vee P(a) \sim_v P(x_1) \vee P(a)$. If we denote by N_v the operator of variable normalization then $N_v(P(w) \vee P(a)) = P(x_1) \vee P(a)$ for all $w \in V$. But still we have $N_v(P(x) \vee P(a)) \neq N_v(P(a) \vee P(x))$.

Let N_o be the operator of lexicographic ordering of atoms, where positive literals are ordered before negative ones. Then $N_o(P(x) \vee P(a)) = P(a) \vee P(x)$. Moreover

$$N_v \circ N_o(P(x) \vee P(a)) = N_v \circ N_o(P(a) \vee P(y)) = P(a) \vee P(x_1) \text{ and}$$

$$N_o \circ N_v(P(x) \vee P(a)) = P(a) \vee P(x_1).$$

Let us define N as $N_v \circ N_o$.
N does not remove multiple occurrences of literals and $N(P(a) \vee P(a)) = P(a) \vee P(a)$.

By adding the operator N_r, i.e. the operator producing reduced clause forms (see Section 2.5), we obtain $N_r \circ N_v \circ N_o$ which essentially corresponds to Robinson's clause notation.

Let $N' = N_r \circ N_v \circ N_o$. It is easy to verify that $\mathrm{LIT}(C) = \mathrm{LIT}(D)$ implies $N(C) = N(D)$.

Even under the (relatively strong) normalization operator N' of Example 3.2.1 we get $N'(P(x) \vee P(a)) \neq N'(P(a))$. But $F(\{P(x) \vee P(a)\}) \sim$

$F(\{P(a)\})$ and $P(a)$ is a factor of $P(x) \vee P(a)$ which implies $P(x) \vee P(a)$. Indeed the factor $P(a)$ is also a subclause of $P(x) \vee P(a)$. Thus in some sense $P(x)$ is redundant in C and we may normalize $P(x) \vee P(a)$ to $P(a)$. This is the principle of condensing. Note that

$$N'(P(x_1) \vee \ldots \vee P(x_n)) = P(x_1) \vee \ldots \vee P(x_n)$$

for all $n \in \mathbb{N}$, while condensing gives $P(x_1)$ always ($P(x_1)$ is a factor which is a subclause).

Definition 3.2.1. *Let N_v be the operator of variable standardization. N_v is defined as $N_v(C) = C\eta\{v_1 \leftarrow x_1, \ldots, v_m \leftarrow x_m\}$, where $C\eta$ is a variant of C with $V(C\eta) \cap \bigcup_{i=1}^{\infty}\{x_i\} = \emptyset$ and v_i is the i-th variable appearing in $C\eta$ (from the left). Let N_o be an ordering for literals (positive before negative literals and lexicographic ordering among the groups) and N_r be the reduction operator (delete multiple occurrences of literals).*
The operator N_s, defined as $N_s = N_r \circ N_v \circ N_o$ is called the operator of *standard normalization.*

Definition 3.2.2 (condensation). *A clause C is called condensed if there is no literal L in C such that $C\backslash L$ is a factor of C.*
We define

$$\gamma(C) = \begin{cases} C & \text{if } C \text{ is condensed} \\ \gamma(C \setminus L) & \text{if } L \text{ is the first literal in } C \\ & \text{such that } C \setminus L \text{ is a factor of } C. \end{cases}$$

$N_c(C)$, *defined as $N_s \circ \gamma(C)$, is called condensed normalization (for short N_c-normalization) of C.*
Note that γ is well defined, as a clause contains finitely many literals only. The original concept of condensation was introduced by W.H. Joyner [Joy76]; in his (slightly different) concept there may be different condensed forms of a clause. However, the different "condensations" are all variants of each other. Thus the clause in condensed normalization (Definition 3.2.2) can be considered as a representative of this $\sim_v$ equivalence class.

Example 3.2.2. $C = P(x) \vee R(b) \vee P(a) \vee R(z)$.
Because $R(b) \vee P(a) \vee R(z)$ is a factor of C we obtain

$$\gamma(C) = \gamma(R(b) \vee P(a) \vee R(z)).$$

Note that $P(a) \vee R(z)$ and $R(b) \vee R(z)$ are not factors, but $R(b) \vee P(a)$ is a factor of $R(b) \vee P(a) \vee R(z)$.
Consequently $\gamma(R(b) \vee P(a) \vee R(z)) = \gamma(R(b) \vee P(a))$.
$R(b) \vee P(a)$ does not contain factors at all, thus $\gamma(R(b) \vee P(a)) = R(b) \vee P(a)$.

We conclude $\gamma(C) = R(b) \vee P(a)$ and $N_c(C) = P(a) \vee R(b)$.

Condensation is powerful, particularly if clauses contain variants of groups of literals within themselves. For example let $C = C_1 \lor C_2$ such that there exists a variable renaming η with $\eta(C_1) = C_2$ and $\eta(C_2) = C_2$. Then $\eta(C_1 \lor C_2) = C_2 \lor C_2$ and C_2 is a factor of C. Consequently $N_c(C_1 \lor C_2) = N_c(C_2)$. N_c is a strong normalization operator which removes "redundancy" within clauses and keeps clauses short. As already mentioned there are sequences of clauses C_n such that $N_s(C_n)$ are all C_n, but $N_c(C_n) = C$ for all n. This property is of particular importance to termination theory. Note that we did not remove all kinds of redundancy in sets of clauses by N_c, as N_c works on representation of single clauses only. Redundancy relative to other clauses will be removed by subsumption, a principle to be introduced in Chapter 4.

Proposition 3.2.1. *Condensing normalization is correct; i.e., for every clause C: $F(\{N_c(C)\}) \sim F(\{C\})$.*

Proof. It is enough to show that $F(\{\gamma(C)\}) \sim F(\{C\})$, because $F(\{N_s(C)\}) \sim F(\{C\})$ is trivial.

By definition of γ it suffices to show $F(\{C \setminus L\}) \sim F(\{(C)\})$ if $C \setminus L$ is a factor of C. $F(\{C \setminus L\}) \to F(\{C\})$ is valid for all clauses C and literals L. Because for every factor D of C $F(\{C\}) \to F(\{D\})$ holds and $C \setminus L$ is a factor of C we obtain $F(\{C\}) \to F(\{C \setminus L\})$.   ◊

Condensing is a principle which allows the replacement of clauses by factors without semantic change. It remains to show that condensed normalization can be built into resolution without loss of completeness.

Let N be a normalization operator which transforms clauses into some logically equivalent clauses in an algorithmic way. We can define a resolution principle according to N, where only clauses in N-normal form can be derived.

Definition 3.2.3 (N-resolvent). *Let N be a normalization operator for clauses and C_1, C_2 be clauses in N-normal form. Let C be a resolvent of C_1, C_2, then $N(C)$ is called N-resolvent of C_1 and C_2.*

Example 3.2.3.
$N = N_c$.
$C_1 = P(f(x_1)) \lor R(x_1)$, $C_2 = P(x_1) \lor \neg R(x_2)$.

C_1 and C_2 are in condensed normal form. After renaming C_2 by $\eta = \{x_1 \leftarrow x, x_2 \leftarrow y\}$ resolve (m.g.u.$= \{y \leftarrow x_1\}$) and obtain a resolvent C: $P(f(x_1)) \lor P(x)$. $N_c(C) = P(f(x_1))$ and thus $P(f(x_1))$ is N-resolvent of C_1 and C_2. The principle of N_s-resolution essentially coincides with Chang & Lee's resolution principle and thus is complete [CL73]. We do not show the completeness of N_c-resolution here, but delay this result. Indeed the completeness of resolution under N_c-normalization will follow from the completeness of A-ordering refinements, a result which will be proved in Section 3.3.

By replacing φ in Definition 3.1.2 by a normalization operator N we obtain the concept of N-resolution operators.

Definition 3.2.4. *Let N be a normalization operator for clauses. We define $\rho_N(\mathcal{C}) = N(Res(\mathcal{C}))$ and*

$$S_N^0(\mathcal{C}) = N(\mathcal{C}),$$

$$S_N^{i+1}(\mathcal{C}) = S_N^i(\mathcal{C}) \cup \rho_N(S_N^i(\mathcal{C})),$$

$$R_N(\mathcal{C}) = \bigcup_{i=0}^{\infty} S_N^i(\mathcal{C}).$$

R_N is called the operator of N-resolution.

Definition 3.2.5 (NR-deduction). *Let C be a set of N-clauses. An* NR-deduction *(N-resolution deduction) of C from $\mathcal{C}$ is a sequence of clauses $C_1, \ldots, C_n$ with the following properties:*

a) *All C_i are in N-normal form;*
b) *$C_n = C$*
c) *For every i either $C_i \in \mathcal{C}$ or $C_i \in \rho_N(\{C_j, C_k\})$ for some $j, k < i$.*

An NR-deduction of $\square$ is also called an NR-refutation of $\mathcal{C}$.

The concept of refinement defined in Section 3.1 can be generalized to N-resolution in an obvious way. In particular, stronger normalizations decrease the number of possible deductions.

Example 3.2.4. We present a resolution refutation in different normal forms.

$$\mathcal{C} = \{P(x) \vee R(x, y), \neg P(x) \vee R(x, f(y)), \neg R(a, f(z))\}.$$

$$\Gamma = P(x) \vee R(x, y), \neg P(x) \vee R(x, f(y)),$$
$$R(u, v) \vee R(u, f(w)), \neg R(a, f(z)), R(a, x), \square.$$

Γ is an R-refutation of $\mathcal{C}$.

$$\Gamma' = P(x_1) \vee R(x_1, x_2), R(x_1, f(x_2)) \vee \neg P(x_1), R(x_1, f(x_2)) \vee R(x_1, x_3),$$
$$\neg R(a, f(x_1)), R(a, x_1), \square.$$

Γ' is an $N_s R$-refutation of $\mathcal{C}$.

Γ' cannot be transformed into an $N_c R$-refutation "directly". The reason is that the clause $R(x_1, f(x_2)) \vee R(x_1, x_3)$ must be factored (!) in order to obtain an N_c-nomal form. Therefore we obtain the shorter $N_c R$-refutation

$$\Gamma'' = P(x_1) \vee R(x_1, x_2), R(x_1, f(x_2)) \vee \neg P(x_1), R(x_1, f(x_2)), \neg R(a, x_1), \square.$$

Normalization of clauses is not only of theoretical, but also of practical importance. In every implementation of a resolution theorem prover the programmer will take care of storage and keep clauses in a compact form. Moreover in stronger normal forms more clauses are equal and thus can be deleted. N_v-normalization is even necessary to keep the set of resolvents of a finite set of clauses finite. The gain of efficiency is represented mathematically by the fact that $\mid S_{N_1}^i(\mathcal{C}) \mid \leq \mid S_{N_2}^i(\mathcal{C}) \mid$ (in Definition 3.2.4) if N_1 is a stronger normalization principle than N_2. It may even be the case that $R_{N_1}(\mathcal{C})$ is finite, but $R_{N_2}(\mathcal{C})$ is infinite. Just take

$$\mathcal{C} = \{P(x_1) \vee P(x_2) \vee R(x_3), \ \neg P(x_1) \vee \neg P(x_2) \vee R(x_3)\},$$

for $N_1 = N_c$ and $N_2 = N_s$.

Exercises

Exercise 3.2.1. a) Show that $F(\{C\}) \sim F(\{D\})$ does not imply $N_c(C) = N_c(D)$ for all clauses C, D.
b) Find sufficiently strong syntactical criteria for clauses such that the implication in a) is valid (is it valid for ground clauses C, D?).

3.3 Refinements Based on Atom Orderings

Atom ordering refinements (or shortly A-ordering refinements) restrict the set of resolvents of two clauses, but do not express conditions on the form of deductions. The basic idea is to avoid resolvents which are too large with respect to an ordering defined on atom formulas. The use of ordering principles for refinements of resolution dates back to the very beginning of resolution theory. In 1967 [Sla67] Slagle defined the refinement of semantic clash resolution containing a principle of atom ordering. A more general concept of ordering has been developed by Kowalski and Hayes [KH69] and incorporated into a semantic-tree based resolution principle. In such a principle, resolution is restricted by the condition that some semantic tree must be reduced by addition of new resolvents. The definition given below is closely related to Joyner's [Joy76], but it is slightly more general.

Definition 3.3.1 (A-ordering). *An A-ordering (atom ordering) $<_A$ is a binary relation on the set of all atom formulas such that the following properties hold:*

(A1) $<_A$ is irreflexive,
(A2) $<_A$ is transitive,
(A3) for all $A, B \in AT$ and $\vartheta \in SUBST$: $A <_A B$ implies $A\vartheta <_A B\vartheta$.

(A1) and (A2) are fulfilled by every strict (partial) ordering principle. Property (A3) is important to ground lifting and will play a role in the completeness proof.

$A <_A B$ always implies the nonunifiability of A and B: Suppose to the contrary that $A <_A B$ and $A\vartheta = B\vartheta$ hold simultaneously; then by (A3) we obtain $A\vartheta <_A B\vartheta$ and $<_A$ cannot be irreflexive, which contradicts (A1).

According to this property, $<_A$-orderings cannot be total on the set of all atoms (e.g. the atoms $P(x)$ and $P(y)$ cannot be in an $<_A$-ordering relation). Nevertheless the partial ordering $<_A$ can be made total on the set of all ground atoms.

$<_A$-orderings are essentially based on the term complexity of atoms. One such complexity measure is the term depth τ, which has already been defined in Section 2.1. But mere term depth is a very rough measure of atom complexity. As an example consider $A_1 : P(x)$ and $A_2 : P(f(a))$; clearly $\tau(A_1) < \tau(A_2)$, but $\tau(A_1\vartheta) > \tau(A_2\vartheta)$ for infinitely many instances $A_1\vartheta$. Because we always deal with substitution instances in computational logic we have to take into account the term depth of the instances too. A finer complexity measure results when also the depths of variable occurrences are considered as well.

Definition 3.3.2. *The maximal depth of a variable x within an expression E (denoted by $\tau_{\max}(x, E)$) is defined as follows:*
For $t \in T$ we set

$$\tau_{\max}(x, t) = \begin{cases} 0 & \text{if } x \notin V(t) \text{ or } x = t, \\ 1 + \max\{\tau_{\max}(x, t_i) | i = 1, \ldots, n\} & \text{if } x \in V(t) \text{ and} \\ & t = f(t_1, \ldots, t_n), \ f \in FS. \end{cases}$$

We extend $\tau_{\max}$ to atoms, literals, and clauses:

$$\tau_{\max}(x, P(t_1, \ldots, t_n)) = \max\{(x, t_i) \mid i = 1, \ldots, n\}$$
$$\text{for atom formulas } P(t_1, \ldots, t_n)$$

$$\tau_{\max}(x, L) = \tau_{\max}(x, at(L)) \text{ for literals } L,$$

$$\tau_{\max}(x, L_1 \vee \ldots \vee L_n) = \max\{\tau_{\max}(x, L_i) \mid i = 1, \ldots, n\} \text{ for clauses.}$$

Example 3.3.1. Let A, B be arbitrary atoms. We define $A <_d B$ iff

1) $\tau(A) < \tau(B)$ and
2) for all $x \in V(A) : \tau_{\max}(x, A) < \tau_{\max}(x, B)$
   (including the property $V(A) \subseteq V(B)$).

Irreflexivity and transitivity of $<_d$ easily follow from 1) and 2).

If $\tau_{\max}(x, A) < \tau_{\max}(x, B)$ for all $x \in V(A)$ and $\tau(A) < \tau(B)$ then for all substitutions ϑ and $y \in V(A\vartheta)$:

$$\tau_{\max}(y, A\vartheta) < \tau_{\max}(y, B\vartheta)$$

and

$$\tau(A\vartheta) < \tau(B\vartheta).$$

Note that condition 1 would not suffice to define an $<_A$-ordering. As an example take the atoms $P(x), P(f(a))$ fulfilling $\tau(P(x)) < \tau(P(f(a)))$; but

$P(x)$ and $P(f(a))$ are unifiable and thus cannot be in an $<_A$-ordering relation. Condition 2 does not suffice either, because, e.g.,

$$\tau_{\max}(x, Q(f(a), a)) < \tau_{\max}(x, Q(f(x), b))$$

but

$$\tau_{\max}(x, Q(f(a), a))\vartheta = \tau_{\max}(x, Q(f(x), b))\vartheta = 0.$$

for $\vartheta = \{x \leftarrow a\}$. For $<_d$ we have

$$\begin{aligned} P(x, x) &<_d Q(f(x), y) \text{ and}\\ P(x, y) &<_d R(g(x, y)), \end{aligned}$$

but not

$$\begin{aligned} P(x, f(a)) &<_d Q(x, f(x)) \quad (\text{ 1}) \text{ is violated})\\ P(x, a) &<_d P(f(a), x) \quad (\text{ 2}) \text{ is violated}) \end{aligned}$$

The ordering $<_A$ is an ordering for atoms, not for literals (i.e., the sign does not influence the ordering relation). We thus extend $<_A$-orderings by

$$L <_A M \text{ iff } at(L) <_A at(M)$$

for any atom ordering $<_A$ and literals L, M.

Definition 3.3.3 (resolved atom). *Let $C : C_1 \vee L \vee C_2$ and $D = D_1 \vee M \vee D_2$ be variable disjoint clauses and $(C_1 \vee C_2 \vee D_1 \vee D_2)\sigma$ be a binary resolvent of C and D (under m.g.u. σ). Then $at(L)\sigma$ is called the resolved atom of the resolution. If E is a binary resolvent of renamed factors C', D' of two clauses C and D then the resolved atom of the binary resolvent is also called the resolved atom of the (this) resolution of C and D.*
Note that the resolved atom generally does not coincide with the atom as it is contained in the clause. The relevance of resolved atoms is based on the fact that a clause is a universal form, whose only relevant instances are those obtained by most general unifications.

For the concept of $<_A$-resolution we choose the strong N_c-normal form for clauses (we could in fact define it relatively to arbitrary clause normalizations). The reason for this choice can be found in resolution decision theory to be presented in Chapter 5.

Definition 3.3.4 ($<_A$-resolvent). *Let C be a set of N_c-clauses and $<_A$ be an atom ordering. Let C be an (ordinary) resolvent of two clauses $C_1, C_2 \in C$. Then (the condensation of) C is an $<_A$-resolvent of C_1 and C_2 (i.e., $N_c(C) \in \rho_{<_A}(C)$) iff there is no literal L in C such that $B <_A L$, where B is the resolved atom of the resolution of C_1 and C_2.*

Example 3.3.2. Let $\mathcal{C} = \{C_1, C_2\}$ for

$$C_1 \;=\; Q(f(x_1), x_1) \vee \neg R(f(x_1)) \quad \text{and}$$

$$C_2 \;=\; R(f(x_1)) \vee \neg Q(x_1, x_2).$$

C_1 and C_2 are N_c-clauses.
For resolution we choose the variants

$$C_1' \;=\; Q(f(x), x) \vee \neg R(f(x)) \quad \text{and}$$

$$C_2' \;=\; R(f(y)) \vee \neg Q(y, z).$$

$C : Q(f(x), x) \vee \neg Q(x, z)$ is a resolvent obtained via m.g.u. $\{y \leftarrow x\}$ and resolved atom $R(f(x))$.

$R(f(x)) \not<_d L$ for $L = Q(f(x), x)$ or $L = \neg Q(x, z)$. Thus C is "admissible" and the normalized resolvent $Q(f(x_1), x_1) \vee \neg Q(x_1, x_2)$ is in $\rho_{<_d}(\mathcal{C})$.

$D : \neg R(f(x)) \vee R(f(f(x)))$ is another resolvent obtained via m.g.u. $\{y \leftarrow f(x), z \leftarrow x\}$ and resolved atom $Q(f(x), x)$.

But $Q(f(x), x) <_d R(f(f(x)))$ and therefore D is not admissible. We conclude that the N_c-resolvent $R(f(f(x_1))) \vee \neg R(f(x_1))$ is not in $\rho_{<_d}(\mathcal{C})$.

Example 3.3.2 also shows that $<_A$-ordering resolution is an *a posteriori* ordering refinement. In the original clause forms C_1, C_2 (before application of the m.g.u.'s) there exists no $<_d$-ordering relation among the literals. Particularly $\neg Q(x_1, x_2) \not<_d R(f(x_1))$, but $Q(f(x), x) <_d R(f(f(x)))$ by most general unification. We thus realize that an *a priori* $<_d$-ordering of literals in clauses (a literal L in C cannot be resolved if there is an M in C such that $L <_d M$) would not have blocked the second resolution in Example 3.3.2. On the other hand, a resolvent defined via the a priori ordering principle (in case of the existence of a maximal literal) is also admissible via the a-posteriori one: Suppose that $C_1 = E_1 \vee L_1$ and $C_2 = E_2 \vee L_2$ such that $L <_A L_i$ for all L in E_i, $i = 1, 2$. Then by (A3) $L\vartheta <_A L_i \vartheta$ for all L in E_i, $\vartheta \in \text{SUBST}$; thus if σ is m.g.u. of $\{L_1, L_2^d\}$ and $(E_1 \vee E_2)\sigma$ is the resolvent then $N_c((E_1 \vee E_2)\sigma) \in \rho_{<_A}(\{C_1, C_2\})$.

Definition 3.3.5 ($R_{<_A}$-deduction). *Let $\mathcal{C}$ be a set of N_c-clauses and $<_A$ be an atom ordering. An $R_{<_A}$-deduction of an N_c-clause C from $\mathcal{C}$ is a sequence $C_1, \ldots, C_n$ such that*

a) $C_n = C$ *and*
b) For all $i = 1, \ldots, n$: Either $C_i \in \mathcal{C}$ or $C_i \in \rho_{<_A}(\{C_j, C_k\})$ for some $j, k < i$.

Clearly every $R_{<_A}$-deduction is also an $N_c R$-deduction. The set of all $R_{<_A}$-deducible clauses can be described by a resolution refinement operator of the type described in Section 3.1.

Definition 3.3.6. *Let C be an arbitrary set of N_c-clauses. We define*

$$S^0_{<_A}(C) = C,$$

$$S^{i+1}_{<_A}(C) = S^i_{<_A}(C) \cup \rho_{<_A}(S^i_{<_A}(C)),$$

$$R_{<_A}(C) = \bigcup_{i=0}^{\infty} S^i_{<_A}(C).$$

It is easy to verify that the set of all clauses derivable via $R_{<_A}$-deduction from a set of N_c-clauses C coincides with $R_{<_A}(C)$.

The principle of $R_{<_A}$-deduction is complete for arbitrary atom orderings $<_A$. The proof of completeness itself proceeds in two steps as usual: First show the completeness of $R_{<_A}$-ground deductions and lift to general $R_{<_A}$-deductions afterwards.

Lemma 3.3.1. *Let $\mathcal{D}$ be a finite, unsatisfiable set of ground N_c-clauses and $<_A$ be an atom ordering. Then there exists an $R_{<_A}$-refutation of $\mathcal{D}$.*

Proof. Let $\mathcal{A}$ be the set of all atom formulas occurring in $\mathcal{D}$. First we order the atoms in $\mathcal{A}$ according to $<_A$. Because $<_A$ need not be total on ground atoms (on all atoms it *cannot* be total) we complete it to a total, strict ordering $<$ in an arbitrary manner. Thus $A <_A B$ implies $A < B$ for all $A, B \in \mathcal{A}$.

Let $\mathcal{A} = \{A_1, \ldots, A_n\}$ such that $A_i < A_j$ for $i < j \le n$. On the basis of $<$ we define a semantic tree T of the form outlined in Figure 3.1. We say that T fulfills the order condition with respect to $<_A$.

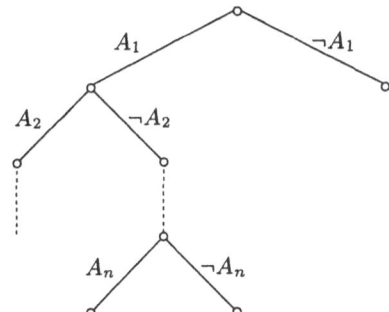

Fig. 3.1. An ordered semantic tree defined by $A_i < A_j$ for $i < j$

The tree T coincides with $\hat{T}(\mathcal{D})$ defined in Section 2.3, the atom ordering being based on $<$. By Theorem 2.3.2, T is closed because $\mathcal{D}$ is unsatisfiable. We conclude that on every branch there is a failure node. As T is closed there must exist a node N in T having two sons N_1, N_2 such that both N_1 and N_2 are failure nodes; we call N an inference node. Let $A_i, \neg A_i$ be the literals corresponding to the edges (N, N_1) and (N, N_2). Because N_1 and N_2 are failure nodes they falsify some clauses in $\mathcal{D}$; so let D_1 be a clause falsified by N_1 and D_2 falsified by N_2. Because N itself is not a failure node, $\neg A_i$ must occur in D_1 and A_i in D_2. Therefore $E : (D_1 \setminus \neg A_i) \vee (D_2 \setminus A_i)$ is a resolvent of D_1, D_2. Note that D_1, D_2 are in N_c-normal form and thus contain every literal at most once. Consequently E neither contains A_i nor $\neg A_i$ and N falsifies E.

We show that E is an $R_{<_A}$-resolvent of D_1 and D_2. E is obtained from D_1, D_2 by a resolution based on the resolved atom A_i. Because E is falsified by N, every literal L in E appears in the dualized form L^d on the branch ending in N. By construction of T every L must be an A_j or $\neg A_j$ for some $j < i$. According to the ordering of the atoms we obtain $A_j < A_i$ and therefore $L < A_i$. We conclude that $N_c(E) \in \rho_{<_A}(\{D_1, D_2\})$.

It is an easy task now to prove the existence of an $R_{<_A}$-refutation of $\mathcal{D}$. For this purpose we proceed by induction on the number of nodes K in a minimal closed semantic tree $T'(\mathcal{D})$ for $\mathcal{D}$ (i.e., every leaf is a failure node), which fulfills the ordering condition.

$k = 1$: $\square$ must be contained in $\mathcal{D}$; $\square$ is an $R_{<_A}$-refutation of $\mathcal{D}$.

(IH): Suppose that the assertion holds for all sets of clauses $\mathcal{D}$ such that the number of nodes k in $T'(\mathcal{D})$ is $\leq n$.

$k = n + 1$: $T'(\mathcal{D})$ must possess the properties indicated above, i.e., there exists an inference node N such that N falsifies an $R_{<_A}$-resolvent E of clauses $D_1, D_2 \in \mathcal{D}$. Let $\mathcal{D}' = \mathcal{D} \cup \{E\}$; because N falsifies E, N itself or an ancestor of N is a failure node.

Therefore a minimal closed semantic tree $T'(\mathcal{D}')$ for $\mathcal{D}'$ must be smaller than $T'(\mathcal{D})$; moreover $T'(\mathcal{D}')$ fulfills the ordering condition. By (IH) there exists an $R_{<_A}$-refutation Γ of $\mathcal{D}'$. By definition of $R_{<_A}$-deductions, D_1, D_2, Γ is an $R_{<_A}$-refutation of $\mathcal{D}$. $\diamond$

Because of the normalization operator N_c, lifting of $R_{<_A}$-ground deductions is impossible using $\leq_s$ only. As an example take

$$C = P(f(y)) \vee P(x) \vee R(x) \vee R(y).$$

Then $N_c(C) = P(f(x_1)) \vee P(x_2) \vee R(x_1) \vee R(x_2)$ (note that C is already condensed).

The clause

$$D : P(f(a)) \vee P(f(a)) \vee R(a) \vee R(f(a))$$

is a ground instance of $N_c(C)$; its normalized form is $P(f(a)) \vee R(a) \vee R(f(a))$. Therefore $N_c(C) \leq_s N_c(D)$ is not valid, but there exists a substitution ϑ such that $N_c(C\vartheta) \leq_s N_c(D)$. This gives a motivation for the following technical notion:

Definition 3.3.7. *Let C, D be clauses in N_c-normal form. Then $C \leq_{sc} D$ if there exists a substitution ϑ such that $N_c(C\vartheta) = D$.*

It is easy to see that $C \leq_{sc} D$ implies the validity of $F(\{C\}) \to F(\{D\})$. Thus $C \leq_{sc} \square$ is only possible for $C = \square$. Note that $F(\{C\}) \to F(\{D\})$ does not imply $C \leq_{sc} D$ (Exercise 3.3.1).

The $\leq_{sc}$ relation can be extended to $N_c R$-deductions, where it serves as a tool to describe properties of lifting.

Definition 3.3.8. *Let $\Gamma = C_1, \ldots, C_n$ and $\Delta = D_1, \ldots, D_n$ be two $N_c R$-deductions. Then $\Gamma \leq_{sc} \Delta$ if for all $i = 1, \ldots, n : C_i \leq_{sc} D_i$.*

If Γ and Δ are $N_c R$-deductions, $\Gamma \leq_{sc} \Delta$, and Δ is a refutation then Γ is a refutation too. The proof of the following completeness theorem is based on $\leq_{sc}$ lifting.

Theorem 3.3.1 (Completeness of $R_{<_A}$-deduction). *Let $\mathcal{C}$ be an unsatisfiable set of N_c-clauses and $<_A$ be an atom ordering. Then there exists an $R_{<_A}$-refutation of $\mathcal{C}$.*

Proof. Because $\mathcal{C}$ is unsatisfiable there exists, by Herbrand's theorem, a finite unsatisfiable set of ground clauses $\mathcal{C}'$ of $\mathcal{C}$. Let $\mathcal{D} = N_c(\mathcal{C}')$; then $\mathcal{D}$ is a finite, unsatisfiable set of N_c-ground clauses. By Lemma 3.3.1 there exists an $R_{<_A}$-refutation Δ of $\mathcal{D}$. We will show the existence of an $R_{<_A}$-refutation Γ of $\mathcal{C}$ such that $\Gamma \leq_{sc} \Delta$. In fact we show the more general result:

(*) Let Δ be an $R_{<_A}$-deduction from $\mathcal{D}$. Then there exists an $R_{<_A}$-deduction Γ from $\mathcal{C}$ with $\Gamma \leq_{sc} \Delta$.

If Δ is an $R_{<_A}$-refutation and Γ is an $R_{<_A}$-deduction fulfilling $\Gamma \leq_{sc} \Delta$ then Γ is an $R_{<_A}$-refutation too. Thus it remains to prove (*). For this purpose we use induction on $l(\Delta)$, the length of the ground deduction Δ.

$l(\Delta) = 1$:

In this case $\Delta = D$ for a clause $D \in \mathcal{D}$. By definition of $\mathcal{D}$ there exists a $C' \in \mathcal{C}'$ such that $N_c(C') = D$. Moreover $C' = C\vartheta$ for some $C \in \mathcal{C}$ and $\vartheta \in \mathrm{SUBST}$. Therefore $N_c(C\vartheta) = D$, i.e., $C \leq_{sc} D$ and $\Gamma : C$ is the required deduction.

(IH): Suppose that (*) holds for all $R_{<_A}$-deductions Δ from $\mathcal{D}$ such that $l(\Delta) \leq n$.

$l(\Delta) = n + 1$: Let $\Delta = D_1, \ldots, D_{n+1}$ be an $R_{<_A}$-deduction from $\mathcal{D}$. Then either

a) $D_{n+1} \in \mathcal{D}$ or
b) $D_{n+1} \in \rho_{<_A}(\{D_i, D_j\})$ for $i, j \leq n$.

Let us denote $D_1, \ldots, D_n$ by Δ_n; then Δ_n is an $R_{<_A}$-deduction from $\mathcal{D}$ and – by (IH) – there exists an $R_{<_A}$-deduction Γ_n from $\mathcal{C}$ such that $\Gamma_n \leq_{sc} \Delta_n$.

case a) : Let C be a clause in $\mathcal{C}$ such that $N_c(C\vartheta) = D_{n+1}$ for a ground substitution ϑ. Then $C \leq_{sc} D_{n+1}$ and Γ_n, C is the required deduction Γ.

case b) : By $i, j \leq n$ and $\Gamma_n \leq_{sc} \Delta_n$ there are (by (IH)) clauses C_i, C_j in Γ_n such that $C_i \leq_{sc} D_i$ and $C_j \leq_{sc} D_j$. Our aim is the construction of an $R_{<_A}$-resolvent C of C_i, C_j such that $C \leq_{sc} D_{n+1}$. If there is such a resolvent then, clearly, Γ_n, C is the required $R_{<_A}$-deduction.

Let $D_i = A_1 \vee L_1 \vee B_1$ and $D_j = A_2 \vee L_2 \vee B_2$ such that $L_1^d = L_2$. Then D_{n+1}, being the N_c-normal form of the resolvent, is a permutation variant of the clause $A_1 \vee B_1 \vee A_2 \vee B_2$. By $C_i \leq_{sc} D_i$ and $C_j \leq_{sc} D_i$ there are substitutions ϑ_i, ϑ_j such that $N_c(C_i\vartheta_i) = D_i$ and $N_c(C_j\vartheta_j) = D_j$.

Let E_i, E_j be the reduced forms of $C_i\vartheta_i$ and of $C_j\vartheta$, respectively. Then D_i, D_j are merely permutation variants of E_i, E_j. Moreover there are (renamed) factors $C_i\sigma_i, C_j\sigma_j$ of C_i and C_j respectively such that $C_i\sigma_i \leq_s E_i$, $C_j\sigma_j \leq_s E_j$.

Such factors always exist, no matter whether C_i and C_j are in condensed form or not (note that condensations are obtained by factoring). Note that condensed forms of instances $C\eta$ cannot contain more literals than the (condensed) clause C itself.

The clause $E : (E_i \setminus L_1) \vee (E_j \setminus L_2)$ is a resolvent of E_i, E_j fulfilling $N_c(E) = D_{n+1}$. By definition of $C_i\sigma_i$ and $C_j\sigma_j$ there must exist literals M_1 in $C_i\sigma_i$ and M_2 in $C_j\sigma_j$ such that $M_1 \leq_s L_1$, $M_2 \leq_2 L_2$ and

$$C = (C_i\sigma_i \setminus M_1)\sigma \vee (C_j\sigma_j \setminus M_2)\sigma \leq_s E$$

for an m.g.u. σ of $\{M_1, M_2^d\}$ (note that $L_1 = L_2^d$). Clearly C is a resolvent of C_i, C_j such that $C \leq_{sc} D_{n+1}$. It remains to show that $N_c(C)$ is in $\rho_{<_A}(\{C_i, C_j\})$. Suppose (without loss of generality) that M_1 is a positive literal. Then $M_1\sigma$ is the resolved atom of the resolution.

We have to show that there is no literal L in C fulfilling $M_1\sigma <_A L$.

For this purpose we assume the contrary and derive a contradiction:

Let L be in C such that $M_1\sigma <_A L$. As $<_A$ is irreflexive we obtain $L \neq M_1\sigma$ and there exists a literal K in $(C_i\sigma_i \setminus M_1) \vee (C_j\sigma_j \setminus M_2)$ such that $L = K\sigma$; thus also

$$M_1\sigma <_A K\sigma.$$

By $M_1\sigma \leq_s L_1$ there exists a substitution ϑ such that $M_1\sigma\vartheta = L_1$, $N_c(C\vartheta) = D_{n+1}$ and $K\sigma\vartheta$ is a literal in D_i or in D_j.

By the property (A3) of A-orderings

$$M_1\sigma <_A K\sigma \text{ implies } M_1\sigma\vartheta <_A K\sigma\vartheta.$$

Thus there exists a literal K' in D_{n+1} such that $L_1 <_A K'$. On the other hand, D_{n+1} is an $R_{<_A}$-resolvent of D_i and D_j based on the resolved atom L_1; consequently such a literal K' cannot exist and we obtain a contradiction. We conclude that C cannot contain a literal L such that $M_1\sigma <_A L$ and therefore $C \in \rho_{<_A}(\{C_i, C_j\})$. $\Diamond$

Corollary 3.3.1. *If C is an unsatisfiable set of N_c-clauses and $<_A$ is an atom ordering then $\square \in R_{<_A}(C)$.*

Proof. Trivial consequence of Theorem 3.3.1.

In the proof of Theorem 3.3.1 we used Lemma 3.3.1, a result which is based on finite semantic trees. Although we used infinite semantic trees to prove Herbrand's theorem itself, we don't need them any longer (once we can refer to Herbrand's theorem). For this reason our result is somewhat more general than Joyner's [Joy76], who demands that the extension of the atom orderings to ground atoms must be of order type less than or equal to ω (i.e., the ordering is linear and there exists no element which is larger than infinitely many other elements). In our terminology the ordering $P(t) < Q(t)$ (for all terms t and two one-place predicate symbols P and Q) is an atom ordering, although there are infinitely many atoms A such that $A < Q(a)$. Such limit points in the ordering are harmless, because in a finite set of atoms we easily obtain the restriction of $<$ to this finite set and can construct an appropriate finite semantic tree.

The ordering property (A3) is of central importance to lifting and has been used in the proof of Theorem 3.3.1. We show now that (A3) cannot be replaced by the following, somewhat weaker condition: $A <_A B$ implies "$A\vartheta \leq_A B\vartheta$" for $\vartheta \in$ SUBST, where "$A\vartheta \leq_A B\vartheta$" is shorthand for not $(B\vartheta <_A A\vartheta)$.

Example 3.3.3. Let $<_v$ be a binary relation on the set of atoms defined by $A <_v B$ iff $V(A) \subset V(B)$.

Clearly $<_v$ is irreflexive and transitive and $V(A) \subset V(B)$ implies $V(A\vartheta) \subseteq V(B\vartheta)$ for all substitutions ϑ. Thus $A <_v B$ implies not $(B\vartheta <_v A\vartheta)$. However, $<_v$ is incompatible with ground lifting. Consider the following $R_{<_v}$-ground deduction

$$\Gamma' : P(a,a) \vee Q(a), \neg Q(a), P(a,a).$$

Note that, as $<_v$ is always false on the ground level (for two ground atoms A, B we always have $V(A) = V(B) = \emptyset$), every ground $N_c R$-deduction is a ground $R_{<_v}$-deduction too. Suppose that $P(x_1, x_2) \vee Q(x_1)$ and $\neg Q(x_1)$ are the N_c-clauses having $P(a,a) \vee Q(a)$ and $\neg Q(a)$ as instances.

The deduction

$$\Gamma : P(x_1, x_2) \vee Q(x_1), \neg Q(x_1), P(x_1, x_2)$$

is the (only possible) lifted $N_c R$-deduction corresponding to Γ'.

However, Γ is not an $R_{<_v}$-deduction: Take the renamed versions $P(x,y) \vee Q(x)$ and $\neg Q(z)$ and resolve via m.g.u. $\sigma = \{z \leftarrow x\}$. Then $Q(x)$ is the resolved atom and $P(x,y)$ is the resolvent. But $Q(x) <_v P(x,y)$ and, according to the definition of A-ordering resolvents, $P(x,y)$ (or its normal form $P(x_1, x_2)$) is not admissible. We see that Γ is not an $R_{<_v}$-deduction.

It is not only the lifting property which does not hold for $R_{<_v}$-deduction, but the $R_{<_v}$ refinement is incomplete too. Simply take $\mathcal{C} = \{P(x_1) \vee Q(x_2), \neg P(a), \neg Q(a)\}$. There are two resolvents, $Q(x_2)$ and $P(x_1)$. But $P(a) <_v Q(x_2)$ and $Q(a) <_v P(x_1)$ and $R_{<_v}(\mathcal{C}) = \mathcal{C}$. As $\mathcal{C}$ is obviously unsatisfiable, $R_{<_v}$-deduction is incomplete. However, we obtain a refutation if we use $<_v$ as an a priori ordering principle (clearly neither $P(x_1) <_v Q(x_2)$ nor $Q(x_2) <_v P(x_1)$). A careful analysis of the use of A-ordering properties in the completeness proof (Theorem 3.3.1) shows that the transitivity of $<_A$ is not really needed. In fact it suffices to guarantee that the relation $<_A$ is cycle-free. A general analysis of ordering refinements (including the more general concept of Π-ordering) can be found in [FLTZ93], Chapter 4.

Exercises

Exercise 3.3.1.
Let C, D be two arbitrary clauses such that $|C| = |D|$ and D is not a tautology. Show that the validity of $F(\{C\}) \rightarrow F(\{D\})$ does not (in general) imply $C \leq_{sc} D$.

3.4 Lock Resolution

While A-orderings are based on syntactical (term) properties of atom formulas, locking (or indexing) is a quite different order type. In locking every (occurrence of a) literal gets a number in advance and inherits this number in resolution deductions, independent of the term structure of the literals. Two syntactically identical literals may be labeled by different numbers and thus can be distinguished.

Example 3.4.1.
Let $\mathcal{C} = \{C_1, C_2, C_3\}$ be a set of clauses with

$$C_1 = \neg P(x, y) \vee P(y, x), \ C_2 = P(u, a) \text{ and } C_3 = \neg P(v, b).$$

The set of all resolvents, normalized under N_c, is

$$N_c(Res(\mathcal{C})) = \{P(a, x_1), \ \neg P(b, x_1), \ P(x_1, x_2) \vee \neg P(x_1, x_2)\}.$$

Note that by resolving C_1 with itself we obtain the tautology $P(x_1, x_2) \vee \neg P(x_1, x_2)$.

None of the resolvents in $N_c(Res(\mathcal{C}))$ can be excluded by an A-ordering refinement (although the third resolvent, being a tautology, can be deleted – a principle which will be discussed in Chapter 4). In fact $R_{<_A}(\mathcal{C}) = N_c(Res(\mathcal{C}))$ for every A-ordering $<_A$ (note that the resolved atoms are always unifiable with all atoms occurring in the resolvent). Indexing of literals changes this situation:

Let $C' = \{\neg P(\overset{1}{x}, y) \lor P(\overset{2}{y}, x), P(\overset{3}{u}, a), \neg P(\overset{4}{v}, b)\}$ be an indexed version of C.

Under the restriction that only literals with lowest index may be resolved (the others are "locked"), the only resolvent we can obtain is $P(a, u)$ (indices are inherited). Figure 3.2 shows a complete lock refutation of C (with respect to the indexed version C').

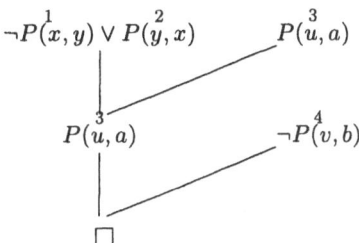

Fig. 3.2. A lock refutation

As numbers are inherited, the order type of a literal within a clause depends on its history, i.e., it is dependent on the deduction of the clause. The A-ordering relation, on the other hand, depends on the syntactical structure of atoms only, no matter where they come from. Because literals with different indices must be considered as different objects, we need new concepts of literals and clauses.

Definition 3.4.1. *A pair (L, i) where L is a literal and i is a natural number, is called an* indexed literal. *Indexed clauses are defined by: Indexed literals and $\square$ are indexed clauses; if A, B are indexed clauses then also $A \lor B$.*

Although identical literals having different indices are different (as indexed literals) we can perform some weak normalization within clauses, by which different indices must belong to different literals. Moreover we can delete multiple occurrences of indexed literals within clauses. The following principle of "merging low" yields such a normalization:

1) Order all literals in a clause according to their indices (i.e., (L, i) left of (M, j) for $i < j$).
2) Delete multiple occurrences of indexed literals (keep the leftmost occurrence only).
3) Delete every (L, i) from C for which there exists an (L, j) in C with $j < i$.

A clause obtained after all possible transformations of type 1–3 is said to be in *lock normal form*. If l is the locking and C is an indexed clause (with respect to l) then the normal form of C is denoted by $N_l(C)$. Instead of $N_v \circ N_l$ we write N_{vl} (the operator N_v can be extended to indexed clauses in a straightforward manner).

Example 3.4.2. $C : (P(x), 1) \vee (P(x), 4) \vee (Q(x), 3) \vee (Q(x), 2)$ is an indexed clause.

Its lock normal form is $N_l(C) = (P(x), 1) \vee (Q(x), 2)$.

Additional variable normalization yields $N_{vl}(C) = (P(x_1), 1) \vee (Q(x_1), 2)$.

Although, formally, indexed literals are tuples of the form (L, i) we will denote them (on the object level) by L^i.

Substitutions can be extended to indexed clauses simply by defining:

$$(L, i)\vartheta = (L\vartheta, i) \text{ and (as always)}$$
$$(A \vee B)\vartheta = A\vartheta \vee B\vartheta, \quad \square\vartheta = \square.$$

The following definition restricts the factoring principle (originally defined in Section 2.7) according to the principle of index minimality.

Definition 3.4.2. *Let C be an indexed N_l-clause and (L, i) be a literal of lowest index in C. Let $(L_1, i_1), \ldots, (L_n, i_n)$ be arbitrary indexed literals in C such that the set of (ordinary) literals $\{L, L_1, \ldots, L_n\}$ is unifiable by m.g.u. σ. Then $N_l(C\sigma)$ is called a* lock factor *of C.*

Example 3.4.3. $C = P(\overset{1}{x}) \vee P(\overset{2}{y}) \vee Q(\overset{2}{x}) \vee Q(\overset{3}{a})$.
$P(\overset{1}{x}) \vee Q(\overset{2}{x}) \vee Q(\overset{3}{a})$ is a lock factor of C, but $P(\overset{1}{a}) \vee P(\overset{2}{y}) \vee Q(\overset{2}{a})$ is not.

The principle of lock resolution results from the combination of binary resolution on minimal literals and lock factoring.

Definition 3.4.3. *Let C_1, C_2 be two indexed clauses in lock- and variable normal form C (i.e. $N_{vl}(C_i) = C_i$ for $i = 1, 2$). Let C_1', C_2' be two variable-disjoint variants of C_1, C_2 and D_1, D_2 be lock factors of C_1', C_2'. Suppose that (L_1, i) is a literal having minimal index in D_1 and similarly (L_2, j) in D_2 such that $\{L_1, L_2^d\}$ is unifiable by an m.g.u. σ. Then the clause $N_{vl}((D_1 \setminus (L_1, i))\sigma \vee (D_2 \setminus (L_2, j))\sigma)$ is called* lock resolvent *of C_1, C_2. If C is a (finite) set of indexed clauses then the set of all lock resolvents definable by clauses $C_1, C_2 \in C$ is denoted by $\rho_l(C)$ (where l denotes the specific locking of C).*

There are in fact different concepts of lock resolution in the literature on automated theorem proving. Our concept of lock factoring is related to order factoring and can be found in [CL73]. In Loveland's book [Lov78] a very restricted form of lock factoring is defined, where it is sufficient to factor literals having the same index. Lock resolution, even under this restricted concept of factoring, is complete.

Note that lock resolution is related to a priori ordering refinements, as the order type of literals in a clause can be derived from its uninstantiated form. As an example consider $L_1 = P(\overset{1}{x}, y)$, $L_2 = P(\overset{2}{x}, z)$. For all substitutions ϑ_1, ϑ_2 $P(\overset{1}{x}, y)\vartheta_1$ is of smaller index than $P(\overset{2}{x}, z)\vartheta_2$. Consider, on the other hand, the A-ordering $<_d$ defined in Section 3.3. With regard to $<_d$ $P(x, y)$ and $P(x, z)$ are incomparable, but we obtain

$$P(x,y)\vartheta_1 \quad <_d \quad P(x,z)\vartheta_1 \text{ for } \vartheta_1 = \{y \leftarrow x, z \leftarrow f(x)\} \text{ and}$$
$$P(x,z)\vartheta_2 \quad <_d \quad P(x,y)\vartheta_2 \text{ for } \vartheta_2 = \{y \leftarrow f(x), z \leftarrow x\}.$$

We see that A-ordering relations within a clause, unlike locking relations, are highly substitution-dependent.

Definition 3.4.4 (lock deduction). *Let C be a set of indexed N_{vl}-clauses for some locking l. A sequence $C_1, \ldots, C_n$ of indexed N_{vl}-clauses is called an R_l-deduction of C from $\mathcal{C}$ if the following conditions are fulfilled:*

1) *$C_n = C$ and*
2) *For every $i \in \{1, \ldots, n\}$:*
   2a) *$C_i \in \mathcal{C}$ or*
   2b) *There exist j, k such that $j, k < i$ and C_i is lock-resolvent of C_j, C_k.*

A lock deduction of $\square$ from $\mathcal{C}$ is called a lock refutation of $\mathcal{C}$.

Note that, by the definition above, the indices of lock-deduced clauses are always the same as in the original set of clauses (indices are inherited). Thus only a fixed set of indices appear in all lock deductions from $\mathcal{C}$.

The set of all lock-derivable clauses can be expressed by a set operator R_l; we only have to take into account that R_l works on indexed clauses. For every locking l and every set of indexed N_{vl}-clauses $\mathcal{C}$ we define $\rho_l(\mathcal{C}) =$ The set of all l-lock resolvents definable by (indexed) clauses in $\mathcal{C}$.

Definition 3.4.5. *Let $\mathcal{C}$ be a set of indexed N_v-clauses and l be a locking for $\mathcal{C}$. We define*

$$S_l^0(\mathcal{C}) \quad = \quad \mathcal{C},$$
$$S_l^{i+1}(\mathcal{C}) \quad = \quad S_l^i(\mathcal{C}) \cup \rho_l(S_l^i(\mathcal{C})),$$
$$R_l(\mathcal{C}) \quad = \quad \bigcup_{i=0}^{\infty} S_l^i(\mathcal{C}).$$

The proof of completeness of lock deduction is of the same structure as the proof for the completeness of $R_{<_A}$-deduction. We first show the completeness of the ground deduction principle, define an appropriate substitution ordering and then perform lifting.

Lemma 3.4.1. *Let $\mathcal{D}$ be a finite, unsatisfiable set of indexed ground N_{vl}-clauses and l be a locking for $\mathcal{D}$. Then there exists an R_l-refutation of $\mathcal{C}$.*

Proof. We apply a proof technique different from the one in Lemma 3.3.1. Instead of semantic trees, we use the excess literal parameter introduced by Anderson and Bledsoe [AB70].

Let $e(\mathcal{D}) = (\Sigma_{D \in \mathcal{D}} | D |) - | \mathcal{D} |$, i.e., the total number of literals occurring in clauses of $\mathcal{D}$ minus the number of clauses in $\mathcal{D}$. We proceed by induction on $e(\mathcal{D})$.

$e(\mathcal{D}) = 0$: In this case $\mathcal{D}$ is a set of unit (indexed) clauses. A refutation of $\mathcal{D}$ necessarily consists in resolving two complementary literals to $\square$. But such a refutation is also a lock refutation (independent of the specific locking).

(IH) Suppose that for all finite, unsatisfiable sets of ground N_{vl}-clauses $\mathcal{D}$ fulfilling $e(\mathcal{D}) \le n$, $\mathcal{D}$ is lock refutable.

$e(\mathcal{D}) = n + 1$: By $e(\mathcal{D}) > 0$ there exists an indexed clause $D \in \mathcal{D}$ such that $\mid D \mid \ge 2$. Let r be the highest index occurring in $\mathcal{D}$ and (L, r) be a corresponding literal. Then D is of the form

$$D_1 \vee (L, r) \vee D_2.$$

Because of the validity of $(D_1 \vee D_2) \to D$ and of $(L, r) \to D$, the sets of clauses

$$\mathcal{D}_1 : (\mathcal{D} - \{D\}) \cup \{D_1 \vee D_2\}$$

and

$$\mathcal{D}_2 : (\mathcal{D} - \{D\}) \cup \{(L, r)\}$$

are both unsatisfiable. Both $\mathcal{D}_1$ and $\mathcal{D}_2$ contain the same number of clauses as $\mathcal{D}$; moreover $e(\mathcal{D}_1), e(\mathcal{D}_2)$ are strictly smaller than $e(\mathcal{D})$. Therefore (IH) applies and there are lock refutations Γ_1 of $\mathcal{D}_1$ and Γ_2 of $\mathcal{D}_2$.

Let Γ_1' be the sequence of clauses obtained from Γ_1 after replacement of $D_1 \vee D_2$ by D everywhere in Γ_1 and by appending (L, r) (at the rightmost position) to all resolvents defined via $D_1 \vee D_2$ in Γ_1. By N_{vl}-normalization multiple occurrences of (L, r) are removed; if E is N_{vl}-normal form then either $E = N_{vl}(E) = N_{vl}(E \vee (L, r))$ (if $(L, j) \in E$ for some $j < r$) or $E \vee (L, r)$ is in N_{vl}-normal form because (L, r) is of maximal index. Thus Γ_1' is an R_l-deduction.

Because r is the highest index no lock resolution can be blocked by introducing (L, r) into the clauses in Γ_1. Consequently Γ_1' is a lock deduction too. There are two possibilities:

a) Γ_1' is a lock refutation; as Γ_1' is also a lock deduction from $\mathcal{D}$ we are finished.

b) Γ_1' is not a refutation.

Because Γ_1' differs from Γ_1 merely by the surplus literal (L, r) in some clauses, and because the clauses are in N_{vl}-normal form (merging low) Γ_1' must be a lock deduction of (L, r). In this case Γ_1', Γ_2' is a lock refutation of $\mathcal{D}$. $\lozenge$

In order to lift ground lock deductions to general lock deductions we need a stronger version of $\le_s$ (like $\le_{sc}$ in the case of A-ordering).

Definition 3.4.6. *Let C, D be clauses in N_{vl}-normal form. We define $C \le_{sl} D$ if there exists a substitution ϑ such that $N_{vl}(C\vartheta) = D$.*

Definition 3.4.7. Let $\Gamma : C_1,\ldots,C_n$ and $\Delta : D_1,\ldots,D_n$ be two R_l-deductions. Then $\Gamma \leq_{sl} \Delta$ iff for all $i = 1,\ldots,n : C_i \leq_{sl} D_i$.

Like for the relations $\leq_s, \leq_{sc}$, the property $\Gamma \leq_{sl} \Delta$ for an R_l-refutation Δ implies that Γ is an R_l-refutation too.

In the completeness proof for lock deduction we use the technique of unlocking a clause. For this purpose we formally define a mapping "unlock" translating indexed clauses into ordinary ones:

$$
\begin{aligned}
unlock((L,r)) &= L \text{ for literals } L,\\
unlock(A \vee B) &= unlock(A) \vee unlock(B) \text{ for indexed clauses } A, B,\\
unlock(\square) &= \square.
\end{aligned}
$$

Theorem 3.4.1 (completeness of lock deduction). *Let C be an unsatisfiable set of indexed N_{vl}-clauses for some locking l. Then there exists an R_l-refutation of C.*

Proof. Because $unlock(C)$ is unsatisfiable there exists a finite, unsatisfiable set of ground instances $\mathcal{D}'$ from clauses in $unlock(C)$ (by Herbrand's theorem). Let C' be the finite, unsatisfiable set of indexed ground clauses corresponding to $\mathcal{D}'$. Let $\mathcal{D} = N_l(C')$ ($N_l(C')$ coincides with $N_{vl}(C')$ on the ground level); then $\mathcal{D}$ is a finite, unsatisfiable set of ground N_{vl}-clauses. By Lemma 3.4.1 there exists an R_l-refutation Δ of $\mathcal{D}$. We have to show the existence of an R_l-refutation Γ of C such that $\Gamma \leq_{sl} \Delta$. The outer structure of the lifting proof is exactly the same as in Theorem 3.3.1. We thus can reduce our argument to the "kernel" which is specific to lock deductions:

(*) Let C_1, C_2 be two variable-disjoint indexed clauses and D_1, D_2 be two indexed ground clauses in lock normal form such that $C_1 \leq_{sl} D_1$ and $C_2 \leq_{sl} D_2$. Let $D \in \rho_l(\{D_1,D_2\})$; then there exists a clause $C \in \rho_l(\{C_1,C_2\})$ fulfilling $C \leq_{sl} D$.

It remains to prove (*).

By $C_1 \leq_{sl} D_1$ and $C_2 \leq_{sl} D_2$ there are substitutions ϑ_1, ϑ_2 such that $N_l(C_1\vartheta_1) = D_1$ and $N_l(C_2\vartheta_2) = D_2$.

Because D is a lock resolvent of D_1 and D_2, D_1 and D_2 must be of the form

$$D_1 = A_1 \vee (L_1, r_1) \vee B_1, \quad D_2 = A_2 \vee (L_2, r_2) \vee B_2$$

where r_1 is a minimal index in D_1, r_2 in D_2 and

$$D = A_1' \vee B_1' \vee A_2' \vee B_2'.$$

The clauses A_i', B_i' are defined from A_i, B_i via lock ordering and merging low (note that D must be in N_{vl}-normal form).

By definition of instantiation in indexed clauses there must exist a literal (M_1, r_1) in C_i and similarly (M_2, r_2) in C_2 with $(M_1, r_1) \leq_s (L_1, r_1)$ and $(M_2, r_2) \leq_s (L_2, r_2)$. Note that indices minimal in the instantiated clause must appear as minimal indices in the uninstantiated clause too (here we need the principle of merging low). Let

$$\begin{aligned} \mathcal{L}_1 &= \{L \mid L \text{ in } C_1, unlock(L)\vartheta_1 = L_1\} \quad \text{and} \\ \mathcal{L}_2 &= \{L \mid L \text{ in } C_2, unlock(L)\vartheta_2 = L_2\}. \end{aligned}$$

As ϑ_1 and ϑ_2 are unifiers of $unlock(\mathcal{L}_1)$ and of $unlock(\mathcal{L}_2)$ respectively, they can be replaced by m.g.u.'s σ_1 of $unlock(\mathcal{L}_1)$ and σ_2 of $unlock(\mathcal{L}_2)$.

Let $unlock(\mathcal{L}_1)\sigma_1 = \{K_1\}$ and $unlock(\mathcal{L}_2)\sigma_2 = \{K_2\}$.

By definition of lock instantiation on indexed clauses (K_1, r_1) is in $C_1\sigma_1$ and (K_2, r_2) is in $C_2\sigma_2$. Because r_1 is minimal in C_1 and r_2 in C_2, $N_l(C_1\sigma_1)$ and $N_l(C_2\sigma_2)$ are lock factors of C_1 and C_2 (according to Definition 3.4.2). By the normalization operator N_l, all literals in $\mathcal{L}_1$ are merged to (K_1, r_1) and all literals in $\mathcal{L}_2$ to (K_2, r_2). Moreover only the leftmost occurrences of multiple literals are kept in $N_l(C_1\sigma_1)$ and $N_l(C_2\sigma_2)$.

Suppose now that

$$N_l(C_1\sigma_1) = E_1 \vee (K_1, r_1) \vee F_1, \quad N_l(C_2\sigma_2) = E_2 \vee (K_2, r_2) \vee F_2.$$

Because $K_1 \leq_s L_1$, $K_2 \leq_s L_2$ and $L_1^d = L_2$ there exists an m.g.u. τ of $\{K_1, K_2^d\}$. By Definition 3.4.3 the clause $C : (E_1' \vee F_1' \vee E_2' \vee F_2')\tau$ is a lock resolvent of C_1, C_2. Like for the resolvent D, E_i', F_i' are defined from the E_i, F_i by lock ordering and merging low.

We may assume $dom(\vartheta_1) \cap dom(\vartheta_2) = \emptyset$ as $V(C_1) \cup V(C_2) = \emptyset$. So we obtain

$$(\sigma_1 \cup \sigma_2)\tau \leq_s \vartheta_1 \cup \vartheta_2$$

by the principle of most general unification. The relations

$$E_1'\tau \leq_s A_1', \quad F_1'\tau \leq_s B_1'$$

etc., do not necessarily hold as $\vartheta_1 \cup \vartheta_2$ may unify more literals than those contained in $\mathcal{L}_1$ and $\mathcal{L}_2$. Nevertheless we obtain

$$(E_1' \vee F_1' \vee E_2' \vee F_2')\tau \leq_{sl} (A_1' \vee B_1' \vee A_2' \vee B_2')$$

i.e., $C \leq_{sl} D$ and $C \in \rho_l(\{C_1, C_2\})$. $\Diamond$

It is essential for the completeness of lock resolution that clauses with the same literals but different indices are not identified (resulting in the deletion of one of these clauses).

Example 3.4.4.
$$C = \{\overset{1}{P}(a) \vee \overset{2}{R}(a), \ \neg\overset{3}{R}(a) \vee \overset{4}{P}(a), \ \overset{5}{R}(a) \vee \neg\overset{6}{P}(a), \ \neg\overset{7}{P}(a) \vee \neg\overset{8}{R}(a)\}.$$

$$S_l^1(C) = C \cup \{\overset{2}{R}(a) \vee \neg\overset{8}{R}(a), \ \overset{4}{P}(a) \vee \neg\overset{6}{P}(a)\},$$

$$S_l^2(\mathcal{C}) = S_l^1(\mathcal{C}) \cup \{\overset{4}{P}(a) \vee \neg\overset{8}{R}(a), \ \neg\overset{6}{P}(a) \vee \neg\overset{8}{R}(a)\}.$$

If we forget the indices in $\overset{4}{P}(a) \vee \neg\overset{8}{R}(a)$, $\neg\overset{6}{P}(a) \vee \neg\overset{8}{R}(a)$ and represent the clauses in ordered form then we obtain

$$P(a) \vee \neg R(a), \ \neg P(a) \vee \neg R(a).$$

But these clauses are the same as the ordered, unlocked forms of the second and fourth clause in $\mathcal{C}$. By deleting the clauses in $S_l^2(\mathcal{C}) - S_l^1(\mathcal{C})$ according to the property above we obtain $S_l^2(\mathcal{C}) = S_l^1(\mathcal{C})$ and therefore $R_l(\mathcal{C}) = S_l^1(\mathcal{C})$. As $\mathcal{C}$ is unsatisfiable and $\square \notin R_l(\mathcal{C})$ (according to the deletion rule above) completeness is destroyed. We see that we may not identify different indexed clauses even if they are semantically identical. Note that in the clauses

$$\overset{4}{P}(a) \vee \neg\overset{8}{R}(a) \text{ and } \neg\overset{3}{R}(a) \vee \overset{4}{P}(a)$$

the index "status" of the literals $P(a)$ and $\neg R(a)$ is different ($P(a)$ is minimal in the first, but not in the second indexed clause).
By continuing correctly we obtain:

$$S_l^3(\mathcal{C}) = S_l^2(\mathcal{C}) \cup \{\neg\overset{8}{R}(a)\}, \ \neg\overset{6}{P}(a) \in S_l^4(\mathcal{C}), \ \overset{2}{R}(a) \in S_l^5(\mathcal{C}) \text{ and } \square \in S_l^6(\mathcal{C}).$$

So we obtain $\square \in R_l(\mathcal{C})$.

Lock resolution is very restrictive and may prune the search space in ATP-programs considerably. A disadvantage is its incompatibility with standard deletion methods; in particular we will show in Chapter 4 that tautology deletion and subsumption destroy the completeness of lock resolution (in fact Example 3.4.4 may serve for this purpose too).

3.5 Linear Refinements

The most natural way to refute a sentence is to start with the negated conclusion of a theorem and to continue the derivation in a top-down manner.

Let us consider, for illustration, a theorem of the form $A_1 \wedge \ldots \wedge A_n \to C$, where the A_i are generalized disjunctions of literals and C is of the form

$$(\exists \bar{x})(L_1 \wedge \ldots \wedge L_m)$$

($\bar{x}$ being a vector of variables in $L_1, \ldots, L_m$). For a proof by contradiction we consider the formula

$$A_1 \wedge \ldots \wedge A_n \wedge \neg C$$

which corresponds to a set of clauses $\mathcal{C} : \{C_1, \ldots, C_n, D\}$ for $D = L_1^d \vee \ldots \vee L_m^d$. In order to refute $\mathcal{C}$ it is quite reasonable to start with the clause D; note that deductions based on $A_1, \ldots, A_n$ only may produce theorems which are of no relevance to a proof of C (i.e., to a refutation of D). The form of deduction shown in Figure 3.3 guarantees that the negated conclusion D is "present" in every newly derived clause. In general the single axioms and the negated conclusion of a theorem are transformed into several different clauses. Even in this more general case we may prevent resolution among clauses obtained from the axioms; the resulting refinement is called *set of support* resolution [WRC65].

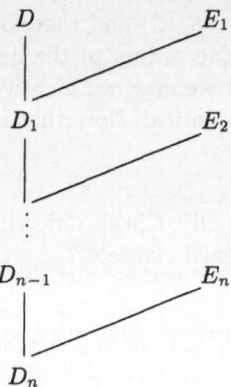

Fig. 3.3. A linear deduction

The clauses E_i in Figure 3.3 are either variants of clauses in $\mathcal{C}$ or are some D_j for $j < i$. We see that every derived clause, that is not a variant of a clause in $\mathcal{C}$, has D as an ancestor clause. This very natural form of deduction is also important in other computational calculi, particularly in those based on the *enumeration of proofs* rather than on the *enumeration of clauses*. The linear structure of derivations plays a major role in the connection calculus [Bib82] and in the tableau method [Fit90].

There are several possibilities to define linear resolution refinements. Originally this type of refinement was invented by Loveland [Lov70] and Luckham [Luc70]. Besides the restriction on the linear form of the deduction tree, ordering restrictions can be imposed additionally [Lov78] [CL73]. In using different normal form principles for clauses we can obtain even more types of linear refinements. In Section 3.3 on A-ordering we have used the strong principle of N_c-normalization. In this section we show that, in case of linear refinements, we may treat clauses as sequences of literals (i.e., we do not normalize); in one of the two clauses subjected to resolution, namely in the "center clause", only the rightmost literal may be cut out. The deduction principle introduced in this chapter is asymmetric, i.e., it might be the case that C, D is resolvable but D, C is not. Formally we define resolvents for pairs of clauses (C, D) instead for sets $\{C, D\}$. The factoring rule to be defined here is "asymmetric" to the resolution cut rule. While we allow resolution on rightmost literals only, we merge to the left in factoring.

Definition 3.5.1 (*l*-factor). *Let C be a clause of the form $C_0 \vee L_1 \vee C_1 \vee \ldots \vee L_n \vee C_n$ such that the L_i are literals, the C_i are clauses and $\{L_1, \ldots, L_n\}$ is unifiable by an m.g.u. σ. Then $(C_0 \vee L_1 \vee C_1 \vee \ldots \vee C_n)\sigma$ is called an l-factor of C. If $n = 1$ then the l-factor coincides with C and is called a trivial l-factor. l-factors of l-factors are also l-factors.*

In the *l*-factoring rule the unified literal always takes the leftmost position. This is a real restriction as we do not normalize the clauses (i.e., different sequences of literals define different clauses). In contrast to lock factoring (where the indices are global), the form of *l*-factors depends on the position of literals in a clause. These positions, in turn, may change the set of deductions due to the definition of resolution below.

Definition 3.5.2 (LRM-resolvent). *Let (C, D) be a pair of clauses of the form $C : C_1 \vee L$, $D : D_1 \vee M \vee D_2$ such that L, M are literals and C_1, D_1, D_2 are clauses; furthermore let η_1, η_2 be variable renamings such that $V(C\eta_1) \cap V(D\eta_2) = \emptyset$. Suppose that $\{L\eta_1, M^d\eta_2\}$ is unifiable by an m.g.u. σ. Then the clause $(C_1\eta_1 \vee D_1\eta_2 \vee D_2\eta_2)\sigma$ is called a binary LRM-resolvent of the pair (C, D) (LRM stands for left-rightmost). A (general) LRM-resolvent of (C, D) is a binary LRM-resolvent of (C', D) where C' is an l-factor of C.*

Example 3.5.1. $C = P(x) \vee Q(x)$, $D = \neg Q(y) \vee R(y) \vee R(a)$.
The clause $P(x) \vee R(x) \vee R(a)$ is an LRM-resolvent of (C, D). $P(a) \vee R(a)$ is a resolvent of C and D, but it is not an LRM-resolvent (factoring in D is not allowed). There exists no binary LRM-resolvent of (D, C) because the literal $R(a)$ cannot be resolved away. $D' : \neg Q(a) \vee R(a)$ is an l-factor of D (and the only one). Again the literal $R(a)$ stands at the rightmost position and thus (D', C) does not define a binary LRM-resolvent. Therefore there exists no LRM-resolvent of (C, D) at all. We see that the principle of LRM-resolution is not "symmetric". The principle of linear deduction is based on LRM-resolutions of pairs of clauses (C, D), where C is a clause having the top clause of the deduction as ancestor clause. It should be emphasized that the restriction to LRM-resolvents and the linear form of deductions are two different things. Many forms of linear deduction are not based on the LRM-restriction. (see [CL73], [Lov78]).

Definition 3.5.3 (LR-deduction). *Let C be a set of clauses and D be a clause in C. An R-deduction $\Gamma : D_0, E_1, D_1, \ldots, E_n, D_n$ (for $n \geq 0$) from C is called a linear R-deduction (LR-deduction) of D_n with top clause D from C if Γ is an R-deduction of D_n from C such that*

a) $D_0 = D$ and
b) for $i \in \{1, \ldots, n\} : D_i$ is LRM-resolvent of (D_{i-1}, E_i).

D is called the top clause, the E_i are called side clauses and the D_i are called center clauses of Γ.

Example 3.5.2.

$C = \{P(x) \vee Q(x),\ \neg P(x) \vee Q(f(y)),\ P(x) \vee \neg Q(f(x)), \neg P(x) \vee \neg Q(x)\}.$
We select $\neg P(x) \vee Q(f(y))$ as the top clause. Then the sequence:

$\Gamma :\ \neg P(x) \vee Q(f(y)),\ \ \neg P(x) \vee \neg Q(x),\ \ \neg P(u) \vee \neg P(f(y)),$

$\quad P(x) \vee Q(x),\ \ Q(f(y)),\ \ P(x) \vee \neg Q(f(x)),$

$\quad P(x),\ \ \neg P(x) \vee \neg Q(x),\ \ \neg Q(u),\ \ Q(f(y)),\ \ \square$

is an LR-refutation of C. The more illustrative tree representation of Γ is
shown in Figure 3.4; literals cut out by resolution are underlined (if more
than one literal is underlined we have a nontrivial l-factor). Note that the
clause $Q(f(y))$ occurs twice, first as center clause and then as side clause of
an LRM-resolution.

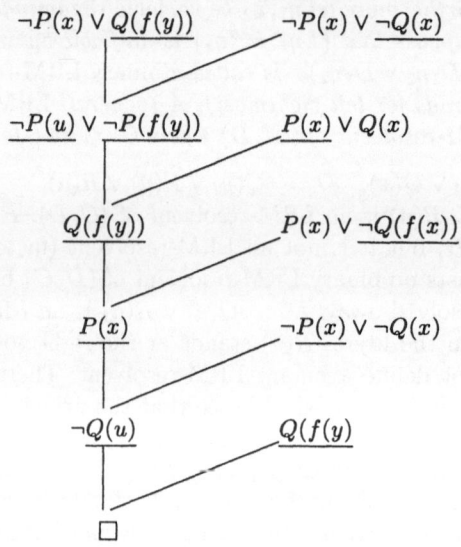

Fig. 3.4. A linear refutation

Using Example 3.5.2 it is easy to realize that a further restriction of the
resolution rule – namely resolving only the rightmost literals in both clauses
– must lead to incompleteness (Exercise 3.5.2). But even without further
restrictions, the wrong choice of a top clause can cause incompleteness.

Example 3.5.3. Let $C = \{C_1, C_2, C_3, C_4\}$ with $C_1 = P(x)$, $C_2 = \neg P(y) \vee R(y)$, $C_3 = \neg R(u)$ and $C_4 = Q(a)$.
C is unsatisfiable. But by selecting C_4 as top clause, the only linear deduction

from $\mathcal{C}$ is C_4 itself; as C_4 is not a refutation there exists no LR-refutation of $\mathcal{C}$ with top clause C_4. However we obtain a refutation by choosing

$$\neg R(u), \ \neg P(y) \vee R(y), \ \neg P(y), \ P(x), \ \square.$$

In Example 3.5.3 the subset $\{C_1, C_2, C_3\}$ is already unsatisfiable, but $\{C_1, C_2\}$ is satisfiable. So we may consider $\{C_1, C_2\}$ as a consistent set of axioms and C_3 as negated conclusion to be refuted on the basis of $\{C_1, C_2\}$. Considered as a set of axioms, $\{C_1, C_2, C_3\}$ represents an inconsistent theory. In the standard deduction systems for classical logic (Hilbert-type systems, natural deduction etc.) the principle *ex falso quodlibet* holds; i.e., once we have derived a contradiction we may also derive anything. Thus in such inference systems we can derive $\neg Q(a)$ from $\{C_1, C_2, C_3\}$ while by using resolution $\neg Q(a)$ cannot be derived. As the literal $\neg Q(\ldots)$ does not occur in $\{C_1, C_2, C_3\}$ there is no inference with top clause $Q(a)$; that $\{C_1, C_2, C_3\}$ is unsatisfiable does not change anything. Thus linear deduction shows the behavior of a weak, paraconsistent calculus.

A similar phenomenon appears in the set of support refinement [WRC65]; its "positive" use in paraconsistent logics was investigated by M.Baaz [Ba87].

Thus, contrary to A-ordering and lock refinement, we need some semantic information about the set of clauses in order to establish a complete deduction principle. Although the satisfiability problem is not even semidecidable [BJ74], the existence of models for common mathematical structures usually is apparent. Therefore it is realistic to postulate satisfiability of clause sets in many relevant problems.

Definition 3.5.4. *Let $\mathcal{C}$ be a set of clauses. A clause $D \in \mathcal{C}$ is called* relevant *(as top clause of an LR-deduction) if there exists a subset $\mathcal{D} \subseteq \mathcal{C}$ such that $\mathcal{D}$ is satisfiable, but $\mathcal{D} \cup \{D\}$ is unsatisfiable.*
In Example 3.5.3 the clauses C_1, C_2, C_3 are relevant, but C_4 is not. It is easy to verify that in every unsatisfiable set of clauses (not containing $\square$) there exists at least one relevant clause (Exercise 3.5.3). We will show now that LR-deduction with relevant top clauses is complete. The first step consists in the proof of ground completeness.

Lemma 3.5.1 (ground completeness of LR-deduction). *Let $\mathcal{C}$ be an unsatisfiable set of ground clauses (not containing $\square$) and C be a relevant clause in $\mathcal{C}$. Then there exists an LR-refutation of $\mathcal{C}$ with top clause C.*

Proof. We first describe the basic idea of the proof:

Take a clause $E : L \vee C$ in $\mathcal{C}$ and split $\mathcal{C}$ into two parts $\mathcal{C}_1$ and $\mathcal{C}_2$; $\mathcal{C}_1$ contains L (in place of E) and $\mathcal{C}_2$ contains C. As $\mathcal{C}_1$ and $\mathcal{C}_2$ are both unsatisfiable and smaller than $\mathcal{C}$ we may (inductively) construct LR-refutations Γ_1 of $\mathcal{C}_1$ with top clause L and Γ_2 of $\mathcal{C}_2$ with top clause C. Afterwards we replace every occurrence of C in Γ_2 by the original clause E, thus obtaining an LR-deduction Γ_2' of $L \vee \ldots \vee L$ from $\mathcal{C}$. Then we contract $L \vee \ldots \vee L$ to L and append the deduction Γ_1; Γ_2' and Γ_1 together define a refutation of $\mathcal{C}$.

Formally we proceed by induction on $occl(\mathcal{C})$, the number of occurrences of literals in a set of (ground) clauses $\mathcal{C}$. The cases $occl(\mathcal{C}) = 0$, $occl(\mathcal{C}) = 1$ are marginal as $occl(\mathcal{C}) = 0$ implies $\mathcal{C} = \{\Box\}$ and $occl(\mathcal{C}) = 1$ implies $\mathcal{C} = \{L\}$ for a unit clause L; in both cases there are no relevant clauses in $\mathcal{C}$.

Thus we start with $occl(\mathcal{C}) = 2$:

In this case $\mathcal{C}$ must be of the form $\{L, M\}$. The unsatisfiability of $\mathcal{C}$ enforces $M = L^d$. Both clauses L, M are relevant and $L, M, \Box$ $(M, L, \Box)$ are the corresponding LR-refutations. Note that for $\mathcal{C} = \{L \vee M\}$, $\mathcal{C}$ is satisfiable (thus this case can be excluded too).

(IH) Suppose that the lemma holds for all unsatisfiable sets of ground clauses $\mathcal{C}$ such that $occl(\mathcal{C}) \leq n$ (for some $n \geq 2$).

Now let $\mathcal{C}$ be an unsatisfiable set of ground clauses with $occl(\mathcal{C}) = n + 1$ (for $n \geq 2$) and C be an relevant clause in $\mathcal{C}$. We distinguish two cases:

a) $\mathcal{C}$ consists of unit clauses only.

   As R-deduction is complete, there must be two complementary clauses L, L^d in $\mathcal{C}$. Because C is relevant there exists a subset $\mathcal{D} \subseteq \mathcal{C}$ such that $\mathcal{D}$ contains no complementary unit clauses, but there exists an $M \in \mathcal{D}$ such that $M = L^d$ for $C = L$. Clearly $L, L^d, \Box$ is an LR-refutation of $\mathcal{C}$ with top clause C.

b) $\mathcal{C}$ contains nonunit clauses.

   b1) The admissible clause C is a unit clause.

   Then $C = L$ for some literal L. If $\mathcal{C} - \{L\}$ is unsatisfiable then there must be a proper subset $\mathcal{D} \subseteq \mathcal{C} - L$ such that $\mathcal{D} \cup \{L\}$ is unsatisfiable, but $\mathcal{D}$ is satisfiable. In this case $occl(\mathcal{D} \cup \{L\}) \leq n$ (note that $\Box \notin \mathcal{C}$) and, by induction hypothesis, there exists an LR-refutation Γ of $\mathcal{D} \cup \{L\}$ with top clause C.

   So we may asssume that $\mathcal{C} = \{L, C_1, \ldots, C_n\}$ such that $\{C_1, \ldots C_n\}$ is satisfiable. Let $\mathcal{D} : \{D_1, \ldots, D_k\}$ be the subset of all clauses in $\mathcal{C}$ that do not contain the literal L. $\mathcal{D}$ cannot be empty (otherwise $\mathcal{C}$ would be satisfiable). By the one-literal rule of Davis and Putnam (see Section 2.4) $\mathcal{C}$ is sat-equivalent to $\mathcal{D} \cup \{L\}$. If there exists a proper subset $\mathcal{D}' \subset \mathcal{D}$ such that $\mathcal{D}' \cup \{L\}$ is unsatisfiable then we may apply (IH) and obtain an LR-refutation Γ of $\mathcal{D}' \cup \{L\}$ with top clause L, which is also an LR-refutation of $\mathcal{C}$.

   We are left with the case that $\mathcal{D}' \cup \{L\}$ is satisfiable for all proper subsets $\mathcal{D}' \subset \mathcal{D}$:

   $\mathcal{D}$ must contain a clause D such that L^d occurs in D (otherwise the pure literal rule gives $\mathcal{C} \sim_{sat} \mathcal{C} - \{L\}$). Thus let D be a clause in $\mathcal{D}$ having the form $D_1 \vee L^d \vee D_2$. For such a D the sequence

$\Gamma : L, D, D_1 \vee D_2$ is an LR-deduction from C with top clause L. If $D_1 \vee D_2 = \square$ then Γ is the required refutation.
If $D_1 \vee D_2 \neq \square$ then consider the set of clauses

$$\mathcal{D}' : (\mathcal{D} - \{D\}) \cup \{D_1 \vee D_2\}.$$

Because $D_1 \vee D_2 \to D$ is valid, $\mathcal{D}' \cup \{L\}$ is unsatisfiable. Moreover $D_1 \vee D_2$ is relevant in $\mathcal{D}' \cup \{L\}$; this follows immediately from the fact that $(\mathcal{D} - \{D\}) \cup \{L\}$ must be satisfiable.
But $occl(\mathcal{D}' \cup \{L\}) \leq occl(\mathcal{C}) - 1 = n$ and, by the induction hypothesis, there exists an LR-refutation Δ of $\mathcal{D}' \cup \{L\}$ with top clause $D_1 \vee D_2$. We see that the deduction L, D, Δ is an LR-refutation of $\mathcal{C}$ with top clause L.

b2) The relevant clause C is a nonunit clause.
Then C is of the form $L \vee C_1$ for a clause C_1 with $C_1 \neq \square$. Again we may suppose that $\mathcal{C} - \{C\}$ is satisfiable. Otherwise we can find a proper subset $\mathcal{D} \subseteq \mathcal{C} - \{C\}$ such that $\mathcal{D}$ is satisfiable and $\mathcal{D} \subseteq \{C\}$ is unsatisfiable; in this case an LR-refutation can be found by applying the induction hypothesis.
As $\mathcal{C}$ is unsatisfiable the following sets of clauses are unsatisfiable by the splitting rule of Davis and Putnam:

$$\mathcal{C}_1 : \{L\} \cup (\mathcal{C} - \{C\}) \quad \text{and} \quad \mathcal{C}_2 : \{C_1\} \cup (\mathcal{C} - \{C\}).$$

L and C_1 are both strictly smaller than C and we obtain
$occl(\mathcal{C}_1) \leq n, \quad occl(\mathcal{C}_2) \leq n$.
Therefore, by induction hypothesis, there exists an LR-refutation of $\mathcal{C}_1$ with top clause L and an LR-refutation Γ_2 of $\mathcal{C}_2$ with top clause C_1 (L and C_1 are both relevant).
We will show the existence of an LR-deduction Γ_2, of a clause $L \vee \ldots \vee L$ (k times) for some $k \geq 1$ from $\mathcal{C}$ with top clause C. For this purpose we consider the refutation $\Gamma_2 : C_1, D_1, \ldots, C_n, D_n, \square$ and transform Γ_2 into an LR-deduction

$$\Gamma_2' : L \vee C_1, \ D_1', \ldots, C_n', \ D_n', \ L \vee \ldots \vee L \ (k \text{ times})$$

with top clause $L \vee C_1$. In Γ_2' the C_i' are of the form

$$L \vee C_{i_1} \vee \ldots L \vee C_{i_k} \text{ for } C_{i_1} \vee \ldots \vee C_{i_k} = C_i;$$

the D_i' are D_i or are of the form

$$L \vee D_{i_1} \vee \ldots \vee L \vee D_{i_k} \text{ for } D_{i_1} \vee \ldots \vee D_{i_k} = D_i.$$

To verify this form of the C_i', D_i' and that Γ_2' is an LR-deduction a simple induction argument does the job.

$L \vee C_1$ clearly is an LR-deduction and the clause (C_0') is of the required form.

For the induction step consider the pair (C_i', D_i') for some $i \le n$, such that

$$C_i' = L \lor C_{i_1} \lor \ldots \lor L \lor C_{i_k} \text{ for } C_{i_1} \lor \ldots \lor C_{i_k} = C_i$$

and

$$D_i' = D_i \text{ or } D_i' = L \lor D_{i_1} \lor \ldots \lor L \lor D_{i_k}$$

for $D_{i_1} \lor \ldots \lor D_{i_k} = D_i$.

By definition of Γ_2 there exists a clause C_{i+1} such that C_{i+1} is LRM-resolvent of (C_i, D_i). Let M be the rightmost literal in C; and $C_i = C' \lor M$.

First we consider the case $D_i' = D_i$:

By definition of l-factoring, the clause $L \lor C' \lor M$ is an l-factor of C_i'. Because C_{i+1} is an LRM-resolvent of (C_i, D_i), D_i must be of the form

$$D_{i_1} \lor M^d \lor D_{i_2} \text{ and } C_{i+1} = C' \lor D_{i_1} \lor D_{i_2}.$$

But $L \lor C' \lor D_{i_1} \lor D_{i_2}$ is a (binary) LRM-resolvent of $(L \lor C' \lor M, D_i)$. So we obtain $C_{i+1}' = L \lor C_{i+1}$ and C_{i+1}' is of the required form. Now we have to investigate the case

$$D_i' = L \lor D_{i_1} \lor \ldots \lor L \lor D_{i_k} \text{ such that } D_{i_1} \lor \ldots \lor D_{i_k} = D_i.$$

Again we consider the l-factor $L \lor C' \lor M$ of C_i'. Then there exists a binary LRM-resolvent of $(L \lor C' \lor M, D_i')$ of the form

$$C_{i+1}' : L \lor C' \lor L \lor D_{i_1} \lor \ldots \lor D_{i_r}' \lor \ldots \lor L \lor D_{i_k}$$

for $D_{i_r}' = D_{i_r} \setminus M^d$ (D_{i_r} is the segment containing the literal M^d). But C_{i+1} (the LRM-resolvent defined by (C_i, D_i) by cutting out M^d from D_i) must be of the form

$$C' \lor D_{i_r} \lor \ldots \lor D_{i_r}' \lor \ldots \lor D_{i_k}.$$

Again C_{i+1}' is of the required form.

By definition of Γ_2 we have $C_{n+1} \equiv \square$ and therefore $C_{n+1}' = L \lor \ldots \lor L$ (k times) for some $k \ge 1$. Thus Γ_2' is an LR-deduction of $L \lor \ldots \lor L$ (k times) from C with top clause C.

Note that, by the principle of l-factoring, a center clause of the form

$$L \lor C_{i_1} \lor \ldots \lor L \lor C_{i_k}$$

never blocks the required resolution (even in the case $C_{i_k} = \square$); by factoring L is shifted to the left and makes the rightmost literal in C_i free for resolution. All l-factors used for LRM-resolvents in Γ_2 can be simulated in Γ_2' (note that an l-factor of an l-factor is again an l-factor). Moreover all center clauses C_i' contain L as leftmost occurrence (due to the LRM-resolution principle).

Let

$\Gamma_1 = L, \Gamma'_1$ and $\Gamma'_2 = \Gamma''_2, L \vee \ldots \vee L$ (k times).
The deduction

$$\Gamma''_1 = L \vee \ldots \vee L \text{ (k times)}, \Gamma'_1$$

is an LR-refutation of $(\mathcal{C} - \{C\}) \cup \{L \vee \ldots \vee L \ (k \text{ times})\}$ with top clause $L \vee \ldots \vee L$ (k times). Γ_1 differs from Γ''_1 only in the first resolution (l-factoring is used in the top clause of Γ''_1). Then, obviously, $\Gamma''_2 \Gamma''_1$ is an LR-refutation of $\mathcal{C}$ with top clause C.

Therefore we have shown that for all $\mathcal{C}$ with $\mathcal{C}$ unsatisfiable, $occl(\mathcal{C}) = n+1$ and arbitrary admissible top clauses C in $\mathcal{C}$ there exists an LR-refutation of $\mathcal{C}$ with top clause C.   ◇

The case $\square \in \mathcal{C}$ has been omitted for purely technical reasons. Note that in case $\square \in \mathcal{C}$, $\square$ is an LR-refutation of $\mathcal{C}$ with top clause $\square$ (only $\square$ is relevant in $\mathcal{C}$). It remains to show that LRM-ground deductions can be lifted to (general) LRM-deductions.

Theorem 3.5.1 (completeness of LR-deduction). *Let $\mathcal{C}$ be an unsatisfiable set of clauses and let C be a relevant clause in $\mathcal{C}$. Then there exists an LR-refutation of $\mathcal{C}$ with top clause C.*

Proof. If $\square \in \mathcal{C}$ then only $\square$ is relevant in $\mathcal{C}$ and $\square$ is the required LR-refutation. Thus, from now on, we may assume $\square \notin \mathcal{C}$. So let $\mathcal{C}$ be unsatisfiable and C be relevant in $\mathcal{C}$. By definition of relevance there exists a (proper) subset $\mathcal{D} \subseteq \mathcal{C}$ such that $\mathcal{D}$ is satisfiable and $\mathcal{D} \cup \{C\}$ is unsatisfiable.

By Herbrand's theorem there exists a finite, unsatisfiable set of ground instances $\mathcal{D}' \cup \mathcal{F}'$ such that $\mathcal{D}'$ is a set of ground instances from clauses in $\mathcal{D}$ and $\mathcal{F}'$ is a set of ground instances from $\{C\}$ (note that, in general, more than one ground instance may be necessary to achieve unsatisfiability). Because $\mathcal{D}$ is satisfiable, $\mathcal{D}'$ must be satisfiable too. Thus we may select an relevant clause C' from the set $\mathcal{F}'$. By Lemma 3.5.1 there exists an LR-refutation Γ' of $\mathcal{D}' \cup \mathcal{F}'$ with top clause C'.

Let $\Gamma' = C', D'_1, C'_1, D'_2, \ldots, C'_n, D'_n, \square$. We have to find an LR-refutation Γ of $\mathcal{C}$ with top clause C such that $\Gamma \leq_s \Gamma'$.

We first define the segments Γ'_k of Γ' by:

$$\Gamma'_0 = C'$$
$$\Gamma'_{k+1} = \Gamma'_k, D'_{k+1}, C_{k+1} \text{ for } k < n \ (C'_{n+1} = \square).$$

Now we show by induction on k that there is an LR-deduction Γ_k of a clause C_k from $\mathcal{C}$ with top clause C such that $\Gamma_k \leq_s \Gamma'_k$. As direct consequence we obtain an LR-refutation $\Gamma(= \Gamma_{n+1})$ of $\mathcal{C}$ with top clause C.

$k = 0$:

$\Gamma'_0 = C'$. By definition of C', C' is a ground instance of C. We define $\Gamma_0 = C$; then Γ_0 is an LR-deduction of $C_0(= C)$ from $\mathcal{C}$ with top clause C and $\Gamma_0 \leq_s \Gamma'_0$.

(IH) Suppose that Γ_k has been constructed successfully.

If $k = n + 1$ then all is shown. So we may suppose that $k < n + 1$.

By definition of Γ'_{k+1} we obtain $\Gamma'_{k+1} = \Gamma'_k, D'_{k+1}, C'_{k+1} = \Gamma''_k, C'_k, D'_{k+1}, C'_{k+1}$ (for some deduction Γ''_k). Because Γ'_{k+1} is an LR-deduction, C'_{k+1} is an LRM-resolvent of (C'_k, D'_{k+1}).

By (IH) we have $C_k \leq_s C'_k$, i.e., C'_k is a ground instance of C_k. We will show that there exists a clause D_{k+1} such that $D_{k+1} \leq_s D'_{k+1}$ and either D_{k+1} is a variant of a clause in $\mathcal{C}$ or there exists a $j \leq k$ such that $D_{k+1} = C_j$ (for a center clause C_j in Γ_k).

If D'_{k+1} is in $\mathcal{D}' \cup \mathcal{F}'$ we define D_{k+1} as a clause in $\mathcal{C}$ having D'_{k+1} as a ground instance. If $D'_{k+1} = C'_j$ for some $j \leq k$ we define $D_{k+1} = C_j$ (for the clause C_j in Γ_k with $C_j \leq_s C'_j$). Thus we obtain clauses C_k, D_{k+1} such that $C_k \leq_s C'_k$ and $D_{k+1} \leq_s D'_{k+1}$. What remains is the construction of an LRM-resolvent C_{k+1} of (C_k, D_{k+1}) such that $C_{k+1} \leq_s C'_{k+1}$. We distinguish two cases:

a) C'_{k+1} is a binary LRM-resolvent of (C'_k, D'_{k+1}). Then

$$C'_k = E'_1 \vee L', \quad D'_{k+1} = F'_1 \vee L'^d \vee F'_2$$

(for a ground literal L' and appropriate ground clauses E'_1, F'_1, F'_2) and $C'_{k+1} = E'_1 \vee F'_1 \vee F'_2$.

By $C_k \leq_s C'_k$, $D_{k+1} \leq_s D'_{k+1}$ there exist ground substitutions ϑ and η such that $C'_k = (E_1 \vee L)\vartheta, D'_{k+1} = (F_1 \vee M \vee F_2)\eta$ for appropriate clauses E_1, F_1, F_2 and literals L, M such that

$$E_1\vartheta = E'_1, \quad L\vartheta = L', \quad F_1\eta = F'_1, \quad M\eta = L'^d, \quad F_2\eta = F'_2.$$

In particular we have $L\vartheta = M^d\eta$. By renaming

$$C_k, D_{k+1} \text{ to } C_k\tau_1, D_{k+1}\tau_2 \text{ such that } V(C_k\tau_1) \cap V(D_{k+1}\tau_2) = \emptyset$$

we can construct a single substitution λ such that

$$C'_k = C_k\tau_1\lambda \text{ and } D'_{k_1} = D_{k+1}\tau_2\lambda.$$

For this λ we obtain $L\tau_1\lambda = M^d\tau_2\lambda$ and λ is a unifier of $\{L\tau_1, M^d\tau_2\}$. Let σ be the m.g.u. of $\{L\tau_1, M^d\tau_2\}$. Then $\sigma \leq_s \lambda$ and

$$E_1\tau_1\sigma \vee F_1\tau_2\sigma \vee F_2\tau_2\sigma \leq_s E'_1 \vee F'_1 \vee F'_2.$$

But $(E_1\tau_1 \vee F_1\tau_2 \vee F_2\tau_2)\sigma$ is an binary LRM-resolvent of (C_k, D_{k+1}).

b) C'_{k+1} is obtained as an LRM-resolvent of (C'_k, D'_{k+1}) via nontrivial l-factoring in C'_k (note that factoring on the ground level is necessary, as we don't work with normal forms here). Because l-factors may be iterated (l-factors of l-factors are also l-factors) there may be several literals $L_1, \ldots, L_m$ in C'_k each of them appearing more than once which are contracted to their leftmost position. We only consider the case of one-step factoring in this proof; the general case follows by an easy induction on the number of factoring steps. So C'_k must be of the form

$$E'_0 \vee L \vee E'_1 \vee \cdots \vee L \vee E'_r$$

for some clauses $E_0,', \ldots, E'_r$ (possibly containing L) and the factor C''_k is of the form $E'_0 \vee L \vee E'_1 \vee \cdots \vee E'_r$. C_k must be of the form

$$E_0 \vee L_1 \vee E_1 \vee \cdots \vee L_r \vee E_r$$

such that $L_i\eta = L$ and $E_i\eta = E'_i$ for some ground substitution η. Then η is a unifier of $\{L_1, \ldots, L_r\}$. Let σ be an m.g.u. of $\{L_1, \ldots L_r\}$. Then, by Definition 3.5.1, the clause

$$E : (E_0 \vee L_1 \vee E_1 \vee \cdots \vee E_r)\sigma$$

is an l-factor of C_k. Thus $E \leq_s C''_k$, $D_{k+1} \leq_s D'_{k+1}$ and C'_{k+1} is a binary LRM-resolvent of (C''_k, D'_{k+1}).

By (a) we obtain a binary LRM-resolvent C_{k+1} of (E, D_{k+1}) such that $C_{k+1} \leq_s C'_{k+1}$. By Definition 3.5.2 C_{k+1} is an LRM-resolvent of (C_k, D_{k+1}).

In both cases a) and b) we define

$$\Gamma_{k+1} = \Gamma_k, D_{k+1}, C_{k+1}$$

and obtain $\Gamma_{k+1} \leq_s \Gamma'_{k+1}$. This concludes the proof of the induction case $k + 1$.   ◇

Note that the lifting in the proof of Theorem 3.5.1 is much easier to perform than that in the completeness proofs for the A-ordering and lock refinement. The reason is that there are no side effects (caused by normalization) in instantiations. Let us consider an LR-refutation $\Gamma : C_o, D_1, \ldots, C_{n-1}, D_n, \square$. As Γ is an R-deduction and C_{i+1} is a resolvent of C_i and D_{i+1}, either the side clauses D_i are variants of input clauses or they are center clauses C_j (for $j \leq i$) derived before. It is a natural question, whether LR-deductions can be refined further by stipulating that (all of the) D_i must be variants of input clauses. We will see that this further restriction leads to incompleteness, but completeness can be preserved by restricting the clause syntax to Horn form.

Definition 3.5.5 (linear input deduction). *An LR-deduction*

$$\Gamma : C_0, E_1, \ldots, C_{n-1}, E_n, C_n$$

from a set of clauses C *is called* linear input deduction *(LI-deduction) from* C *if all clauses* E_i *(for* $i = 1, \ldots, n$*) are variants of clauses in* C.
The following example shows that LI-deduction is incomplete.

Example 3.5.4. Let C be the set of clauses

$$\{P(x) \vee Q(x), \; \neg P(x) \vee Q(f(y)), \; P(x) \vee \neg Q(f(x)), \neg P(x) \vee \neg Q(x)\}$$

as in Example 3.5.2.

Let $\Gamma = C_0, E_1, \ldots, C_{n-1}, E_n, C_n$ be an arbitrary LI-deduction from C, where C_0 is a variant of some clause in C. Then E_n must be a variant of an input clause (i.e., of a clause in C) and C_n is LRM-resolvent of (C_{n-1}, E_n). Because E_n is not a unit clause and does not contain a unit factor, the resolvent C_n cannot be $\square$. We conclude that Γ cannot be a refutation of C. Even if we allow unrestricted factoring for all clauses C_i, E_i it is impossible to obtain a refutation (none of the input clauses possesses a unit factor).

The set of clauses C in Example 3.5.4 contains a clause with two positive literals; even under arbitrary renaming of signs (e.g., replace $P(\ldots)$ by $\neg P(\ldots)$ and vice versa) this property of C remains invariant. By restricting the syntax to clauses containing at most one positive literal we obtain Horn clauses (see Definition 2.2.4); on sets of clauses of this type LI-deduction is complete.

Definition 3.5.6 (Horn logic). *Let C be a clause of the form*

a) P *or*
b) $P \vee \neg Q_1 \vee \ldots \vee \neg Q_n$ *or*
c) $\neg Q_1 \vee \ldots \vee \neg Q_n$

such that $P, Q_1, \ldots, Q_n$ *are atom formulas. Then C is called a* Horn clause. *Clauses of the form a) (positive unit clauses) are called* facts, *of the form b)* rules *and of the form c)* goals *($\square$ is considered as goal). Horn logic is the class of all finite sets of Horn clauses. The set of all facts (rules, goals) in a set of Horn clauses C is denoted by* facts(C), *(*rules(C), *goals(C)).*
The terms 'facts', 'rules', and 'goals' come from the area of logic programming [Llo87]. There facts P are commonly written as $P \leftarrow$,
rules $P \vee \neg Q_1 \vee \cdots \vee \neg Q_n$ as $P \leftarrow Q_1, \ldots, Q_n$,
and goals $\neg Q_1 \vee \ldots \vee \neg Q_n$ as $\leftarrow Q_1, \ldots, Q_n$.

In logic programming a set of positive Horn clauses is interpreted as a program (logic program); goals serve as input to the program. The operational semantics of this programming language essentially coincides with LI-deductions from which total substitutions are extracted; these substitutions serve as program output.

Theorem 3.5.2 (completeness of LI-deduction on Horn logic). *Let C be a finite, unsatisfiable set of Horn clauses. Then for every relevant goal*

G in C there exists an LI-refutation of C with top clause G. Moreover there exists at least one relevant goal in C.

Proof. First we show that there must be an relevant goal in C. Let $\mathcal{D}$ be the set of all rules and facts of C. Then every clause in $\mathcal{D}$ contains (exactly) one positive literal. $\mathcal{D}$ possesses an H-model in which all ground atoms are set to TRUE. As a consequence every unsatisfiable subset $\mathcal{F}$ of C must contain at least one goal. Therefore there must be a relevant goal.

So let G be a relevant goal in C. By Theorem 3.5.1 there exists an LR-refutation Γ of C with top clause G. Let Γ be $C_0, E_1, C_1, \ldots, C_{n-1}, E_n, \square$ for $C_0 = G$. A goal (being purely negative) cannot be resolved with another goal, thus E_1 must be a fact or a rule. Because E_1 is a Horn clause the resolvent C_1 must again be a goal. By an easy induction argument we conclude that all center clauses in Γ must be goals. For all $i < n$, the pair (C_i, E_{i+1}) must possess an LRM-resolvent C_{i+1}. Because C_i is a goal, E_{i+1} must be a fact or a rule and therefore $E_{i+1} \neq C_j$ for all $j \leq i$. Because Γ is an R-deduction, E_{i+1} must be a variant of an input clause. Thus Γ is an LI-refutation of C with top clause G. ◊

It is easy to see (Exercise 3.5.6) that only one goal clause is needed to refute a set of Horn clauses.

Example 3.5.5.

$$C = \{P(f(x)) \vee \neg P(x), \ P(a), \ \neg P(f(f(a)))\}.$$

C is a set of Horn clauses with facts$(C) = \{P(a)\}$, rules$(C) = \{P(f(x)) \vee \neg P(x)\}$ and goals$(C) = \{\neg P(f(f(a)))\}$. $\neg P(f(f(a)))$ can serve as top clause of an LI-deduction (it is admissible). An LI-refutation with top goal $\neg P(f(f(a)))$ is shown in Figure 3.5.

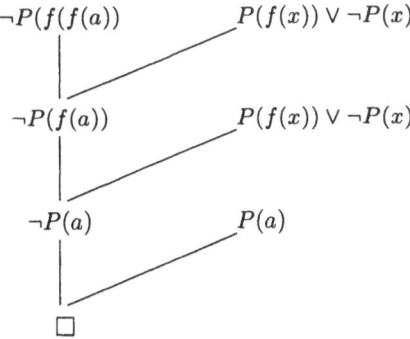

Fig. 3.5. LI-refutation (I)

In $\mathcal{C}$ also the rule $P(f(x)) \vee \neg P(x)$ is relevant. The LR-deduction shown in Figure 3.6 is also an LI-deduction (having a rule as top clause).

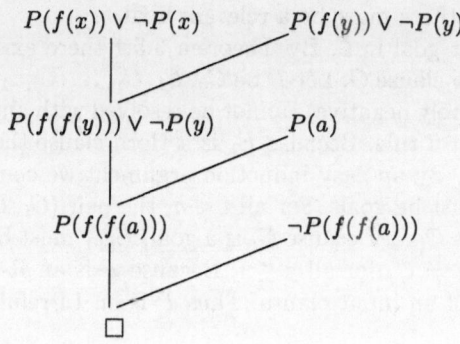

Fig. 3.6. LI-refutation (II)

The concept of LR-deduction is based on specific (linear) restrictions imposed on deduction trees. A-ordering and lock refinements are characterized by restrictions on the set of resolvents of *two* clauses; but they are local in the sense that the overall structure of deduction trees is not subjected to further restrictions. On the other hand, the property of being a center clause in a linear deduction is global, i.e., it depends on the whole derivation producing that clause. For this reason it is impossible to define linear refinements via resolution operators (according to Definition 3.1.3) unless we encode derivations into the clause syntax. Note that, for a refinement operator R_x, the $(i + 1)$-th generation of clauses is defined by

$$S_x^{i+1}(\mathcal{C}) = S_x^i(\mathcal{C}) \cup \rho_x(S_x^i(\mathcal{C})),$$

where ρ_x is an operator on sets of clauses without the "ability to recognize" the derivations leading to the clauses in the set. We see that linear deductions define a refinement in the sense of Definition 3.1.1, but not in the sense of Definition 3.1.3. Still (to be precise) we have to define an algorithmic method to select the top clause; as the satisfiability problem is undecidable there is no "direct" way to do this. However, any unsatisfiable set of clauses must contain at least one negative and one positive clause; therefore – in any case – there are positive and negative clauses which are relevant. Still the problem remains to select the right negative (positive) clauses (some of them might be irrelevant). Frequently (e.g., in Horn logic and in logic programming) only one negative clause is present and the choice becomes obvious. In general it is possible to turn linear resolution into an (algorithmic) theorem prover which is always complete: Select every clause in the set of clauses C as top clause and run the different linear deductions in parallel. If C is unsatisfiable there exists always a clause which is relevant and therefore a refutation is eventually found.

In this section we defined just one specific form of a linear refinement; it should be mentioned that a large variety of efficient and highly sophisticated forms of linear refinements exist (and are implemented). In some of them the history of a resolvent is stored by keeping the resolved literal and marking it as a so-called A-literal, which enables new types of global reductions. This method is called *model elimination* and is presented in depth in D. Loveland's book [Lov78]. Another important variant is SLD-resolution which is the proof-theoretic basis to logic programming; here we refer to J.W. Lloyd's book [Llo87].

Exercises

Exercise 3.5.1. Construct an LR-refutation of the set of clauses C in Example 3.5.2 with top clause $P(x) \vee \neg Q(f(x))$.

Exercise 3.5.2. Let BRM (both-rightmost) be a resolution principle in which only the rightmost literals (of both clauses) may be cut out by resolution. Show that LR-deduction based on BRM-resolution is incomplete.

Exercise 3.5.3. Show that in every unsatisfiable set of clauses there exists at least one relevant clause (remark: the empty set of clauses $\emptyset$ is satisfiable).

Exercise 3.5.4. A maximal l-factor C' of a clause C is an l-factor having only trivial l-factors itself. Show that LR-deduction based on maximal l-factoring is complete.

Exercise 3.5.5. Let $f^0(a) = a, f^{(n+1)}(a) = f(f^{(n)}(a))$ for an $f \in FS_1$ and an $a \in CS$. We define a sequence of Horn sets C_n by

$$C_n = \{P(a), P(f(x)) \vee \neg P(x), \neg P(f^{(n)}(a))\}.$$

Show that for all $n \geq 4$ there exists an LR-refutation of C_n which is not an LI-refutation.

Exercise 3.5.6. Show that a set of Horn clauses C is unsatisfiable iff there exists a $G \in goals(C)$ such that $facts(C) \cup rules(C) \cup \{G\}$ is unsatisfiable.

3.6 Hyperresolution

Hyperresolution is based on the idea of macro-inference, i.e., the principle of contracting a sequence of resolution steps into a single inference step. The advantage of a macro-inference over ordinary (binary) resolution lies in the fact that many (intermediary) resolvents do not appear in the search space; thus neither interactions among them nor interactions between them and other clauses have to be considered. In hyperresolution the internal resolutions within the macro-inferences define a linear deduction. Hyperresolution can be described as a special case of semantic clash resolution defined by J. Slagle [Sla67] and discussed at the end of this section. Our definition is close to Robinson's [Rob65a], but we lay special emphasis on normalization and factoring. The specific form of our definition of hyperresolution will be relevant to the resolution decision procedures in Chapter 5. The basic idea is to partition clauses into positive (i.e., all literals are positive) and nonpositive ones, to derive only positive clauses (or $\Box$) and to concentrate factoring and normalization on positive clauses. To make the analysis more transparent we introduce a weak form of normalization, where positive literals occur left of the negative ones.

Definition 3.6.1 (PN-form). *Let C be a clause of the form $C_1 \vee C_2$ such that C_1 contains only positive and C_2 contains only negative literals (C_1 or C_2, or both of them, may be $\Box$). Then C is in PN-form. C_1 is called the positive part of C and is denoted by C_P; similarly we call C_2 the negative part of C and denote it by C_N.*

Note that every clause in N_0-normal form (see Section 3.2) is in PN-form, but not vice versa. All clauses which are purely positive or purely negative are in PN-form; Horn clauses (according to Definition 3.5.6 are in PN-form too). First we define a restricted resolution principle, where one of the clauses to be resolved must be positive (we call clauses positive (negative) if they contain positive (negative) literals only).

Definition 3.6.2 (PRF-resolvent). *Let C, D be two PN-clauses such that D is positive and C is of the form $C' \vee \neg Q$ for some atom formula Q. Let D' be a variant of a factor of D such that $V(D') \cap V(C) = \emptyset$ and $D' = D_1 \vee P \vee D_2$ for an atom formula P and clauses D_1, D_2. Suppose that $\{P, Q\}$ is unifiable by an m.g.u. σ. Then $(D_1 \vee D_2 \vee C')\sigma$ is called a PRF-resolvent (positive-restricted factoring) of C and D.*

Note that every PRF-resolvent is also an ordinary resolvent (resolve D from the left) in PN-form. Factoring is limited to the positive clause and the resolution cut rule may only be applied to the rightmost literal of the nonpositive clause.

Example 3.6.1.
$\mathcal{C} = \{C_1, C_2, C_3, C_4\}$ for the clauses $C_1 : P(x) \vee P(y)$, $C_2 : P(x) \vee \neg R(x)$, $C_3 : \neg R(x) \vee \neg R(y) \vee \neg P(x)$, $C_4 : R(b)$. The resolvents

$\neg R(x) \vee \neg R(y)$ and $P(v) \vee \neg R(x) \vee \neg R(y)$

are PRF-resolvents of C_1 and C_3.

$\neg R(x) \vee \neg R(u) \vee \neg R(v)$ is a resolvent of C_2 and C_3 but (neither C_2 nor C_3 is positive) it is not a PRF-resolvent.

$\neg R(x)$ is a resolvent of C_1 and C_3, but it is not a PRF-resolvent (factoring has been applied to a non-positive clause).

$\neg P(b)$ is a resolvent of C_3 and C_4, but it is not a PRF-resolvent (the same holds for $\neg R(y) \vee \neg P(b)$ and $\neg R(x) \vee \neg P(x)$) – the cut literal in the nonpositive clause was not the rightmost one. In fact there is no PRF-resolvent of C_3 and C_4.

The idea of hyperresolution is to define a sequence of PRF-resolutions leading to a positive clause. The macro-inference step is based on a tuple of clauses containing one nonpositive and (several) positive clauses.

Definition 3.6.3 (clash sequence). *Let C be a nonpositive PN-clause and $D_1, \ldots, D_n$ be positive clauses. The tuple $\varrho : (C; D_1, \ldots, D_n)$ is called a* clash sequence. *If $\{C, D_1, \ldots, D_n\} \subseteq C$ then we say that ϱ is a clash sequence from C.*

Definition 3.6.4 (hyperresolvent). *Let $\varrho : (C; D_1, \ldots D_n)$ be a clash sequence. Let R_0 be C and R_{i+1} a PRF-resolvent of R_i and D_{i+1} (if such a resolvent exists) for $i < n$. If R_n exists and if it is positive then $N_c(R_n)$ (the condensed normal of form of R_n) is called* hyperresolvent *of ϱ.*

By definition of PRF-resolution a clash sequence $\varrho : (C; D_1, \ldots, D_n)$ can only define a hyperresolvent if C contains exactly n negative literals. Clash sequences may define one, several, or no hyperresolvent. We choose the condensed form of the hyperresolvents in order to guarantee that there are only finitely many ones (defined by one clash sequence); this finiteness condition will be relevant to the use of hyperresolution as a decision procedure in Chapter 5.

Example 3.6.2. $C = \{C_1, C_2, C_3\}$ for $C_1 : Q(x,y) \vee \neg P(f(x)) \vee \neg P(g(y))$, $C_2 : P(x) \vee R(x)$ and $C_3 : P(y) \vee P(f(y))$.

Let $\varrho = (C_1; C_2, C_3)$. Then, clearly, ϱ is a clash sequence. We define $R_0 = C_1$ and $R_1 = R(g(y)) \vee Q(x,y) \vee \neg P(f(x))$. There are two possibilities to define R_2, namely

$R_2^1 = P(x) \vee R(g(y)) \vee Q(x,y)$ and $R_2^2 = P(f(f(x))) \vee R(g(y)) \vee Q(x,y)$.

Both clauses R_2^1, R_2^2 define hyperresolvents; these are

$P(x_1) \vee Q(x_1, x_2) \vee R(g(x_2))$ and $P(f(f(x_1))) \vee Q(x_1, x_2) \vee R(g(x_2))$.

Note that the ordering of the positive clauses in a clash influences (the existence of) the hyperresolvent, e.g., the clash sequence $\varrho : (C_1, C_3, C_2)$ defines only one hyperresolvent, namely

$$P(f(g(x_1))) \lor Q(x_2, x_1) \lor R(f(x_2)).$$

The refinement of hyperresolution simply consists in the production of all hyperresolvents definable by clash sequences from the set of clauses derived so far. Although a hyperresolvent of a clash sequence $(C; D_1, \ldots, D_n)$ is a result of a linear input deduction from $\{C, D_1, \ldots, D_n\}$ with top clause C, the "outer" structure of the deductions is not subjected to further restrictions. Thus in contrast to the case of linear refinements, we may define the refinement of hyperdeduction by a resolution operator R_H.

Definition 3.6.5 (the operator of hyperresolution).
Let C be a set of PN-clauses. We define $\rho_H(C) = $ the set of all hyperresolvents definable by clash sequences from C. The operator of hyperresolution (R_H) is defined in the usual manner of level saturation. Let C_+ be the set of all positive clauses in C, then:

$$S_H^0(C) \;=\; N_c(C_+) \cup (C - C_+),$$

$$S_H^{i+1}(C) \;=\; S_H^i(C) \cup \rho_H(S_H^i(C)),$$

$$R_H(C) \;=\; \bigcup_{i=0}^{\infty} S_H^i(C).$$

The normalization of C_+ in the definition guarantees that all positive clauses in $R_H(C)$ are in N_c-normal form. A characteristic feature of R_H is the invariance of all nonpositive clauses, i.e., $R_H(C) - R_H(C)_+ = C - C_+$. Thus the clauses in $C - C_+$ may be interpreted as a fixed rule base that produces the set $R_H(C)_+$ out of C_+. The set of all newly derived clauses is $R_H(C)_+ - C_+$. By $R_H^+(C)$ we denote all clauses in $R_H(C)$ which do not contain negative literals (note that $R_H^+(C) \subseteq R_H(C)_+ \cup \{\Box\}$).

It is a natural question, why we have chosen the criterion of positivity in the definition of hyperresolution. Clearly, by simply changing all signs of the literals, we can define a similar principle where only negative clauses are produced; this principle is well known and called negative hyperresolution (the principle in Definition 3.6.5 is then called positive hyperresolution). More generally we can define hyperresolution with respect to every sign-renaming operator on C. We give a formal definition of sign-renaming now:

Definition 3.6.6 (sign-renaming). *Let $\mathcal{P} : \{P_1, \ldots, P_n\}$ be a set of predicate symbols and γ be a function with domain $\mathcal{P}$ such that for all $i = 1, \ldots, n$ $\gamma(P_i) = P_i$ or $\gamma(P_i) = \neg P_i$. Then γ is called a sign-renaming. We denote γ by the set $\{\gamma(P_1), \ldots, \gamma(P_n)\}$. γ can be extended to atoms, literals, clauses, and sets of clauses in the following way:*

$$\gamma(P(t_1, \ldots, t_n)) = \gamma(P)(t_1, \ldots, t_n)$$

for atom formulas,

$$\gamma(\neg P(t_1, \ldots, t_n)) = \gamma(P(t_1, \ldots, t_n))^d$$

for negative literals,

$$\gamma(C_1 \vee \ldots \vee C_n) = \gamma(C_1) \vee \ldots \vee \gamma(C_n)$$

for clauses and

$$\gamma(\mathcal{C}) = \bigcup_{C \in \mathcal{C}} \{\gamma(C)\}$$

for sets of clauses.

It is easy to verify that, for every sign-renaming γ, $\mathcal{C} \sim_{sat} \gamma(\mathcal{C})$. If $\mathcal{C}$ is a set of clauses containing n different predicate symbols then there are 2^n sign-renamings on $\mathcal{C}$. We define $LS(\mathcal{C}) = PS(\mathcal{C}) \cup \{\neg P | P \in PS(\mathcal{C})\}$ as the set of all literal symbols in $\mathcal{C}$; for clauses C we write $LS(C)$ instead of $LS(\{C\})$. Hyperresolution as defined in Definition 3.6.5 is a principle of generating only clauses C such that $LS(C) \subseteq PS(\mathcal{C})$. Instead of the set $PS(\mathcal{C}) : \{P_1, \ldots, P_n\}$ we may take $\{\gamma(P_1), \ldots, \gamma(P_n)\}$ for any sign-renaming γ. Instead of positive clauses we may take "γ-positive" clauses.

Definition 3.6.7. *Let γ be a sign-renaming on a set of clauses $\mathcal{C}$ and $\mathcal{M}(\gamma) = \{\gamma(P_1), \ldots \gamma(P_n)\}$ (for $PS(\mathcal{C}) = \{P_1, \ldots P_n\}$). A clause C is called γ-positive if $LS(C) \subseteq \mathcal{M}(\gamma)$.*

Example 3.6.3. $\mathcal{C} = \{P(x) \vee R(x), \neg P(x) \vee R(x), P(x) \vee \neg R(x), \neg P(x) \vee \neg R(x)\}, \gamma = \{P, \neg R\}$.

Then $P(x) \vee \neg R(x)$ is γ-positive; $P(x) \vee R(x)$ is positive, but not γ-positive. It is easy to see that for all four sign-renamings on $\mathcal{C}$ there exists exactly one γ-positive clause in $\mathcal{C}$. If $\gamma = \{\neg P_1, \ldots, \neg P_n\}$ then the γ-positive clauses are exactly the negative ones. In all Definitions 3.6.1–3.6.5 we may replace the term "positive" by "γ-positive" for a sign-renaming γ. If $\gamma = \{P_1, \ldots, P_n\} = PS(\mathcal{C})$ we obtain (positive) hyperresolution, if $\gamma = \{\neg P_1, \ldots, \neg P_n\}$ negative hyperresolution; in place of the operator R_H we have to consider operators $R_{H\gamma}$. In propositional logic sign-renamings can be interpreted as a notation for interpretations. So we may define $I(P) = $ FALSE if $\gamma(P) = P$ and $I(P) = $ TRUE if $\gamma(P) = \neg P$. Then the set of γ-positive clauses (in a set of propositional clauses) is exactly the set of clauses false in I. In predicate logic sign renamings represent specific Herbrand interpretations, so-called settings:

Definition 3.6.8 (setting). *Let $\mathcal{C}$ be a finite set of clauses, $PS(\mathcal{C}) = \{P_1, \ldots, P_n\}$ and γ be a sign renaming on $\mathcal{C}$. Let $\mathcal{P} = \{P | P \in PS(\mathcal{C}), \gamma(P) \neq P\}$ and $\text{atoms}(\mathcal{P}) = \{P(x_1, \ldots, x_n) | P \text{ is n-place}, P \in \mathcal{P}, n \in \mathbb{N}\}$.*
Then the Herbrand interpretation represented by $\{A' | A' \text{ is ground instance from } \text{atoms}(\mathcal{P})\}$ is called a setting (induced by γ).

Let $\mathcal{C}$ be a set of clauses, $PS(\mathcal{C}) = \{P_1, \ldots P_n\}$ and γ be the identity. Then the corresponding setting is $\mathcal{M} : \emptyset$, i.e., all atoms are interpreted as FALSE.

The clauses false in $\mathcal{M}$ are exactly the positive clauses. If $\gamma = \{\neg P_1, \ldots, \neg P_n\}$ then the setting $\mathcal{M} = AT(\mathcal{C})$, i.e., all atoms are set to TRUE. The clauses false in $\mathcal{M}$ are exactly the negative ones. Thus γ-hyperresolution (as a principle for producing γ-positive clauses) can be interpreted as a method for producing only false clauses with respect to a setting $\mathcal{M}$ (induced by γ). This semantic interpretation of hyperresolution leads to the term "semantic clash resolution". But it should be noted that Slagle's concept of semantic clash resolution is substantially more general than γ-hyperresolution, as it can be defined over arbitrary interpretations and is not limited to settings [Sla67]. We will focus on the more general concepts of semantic resolution and semantic clash resolution at the end of this section. The reasons for presenting the less general concept of γ-resolution in advance are the following ones:

1. Settings are a class of interpretations definable by simple syntactic means,
2. we use sign-renamings in the proof of ground completeness of semantic clash resolution, and
3. some syntax forms like PN-clauses and corresponding restrictions of inference are only possible for such syntactic interpretations.

Example 3.6.4. $\mathcal{C} = \{P(x) \vee P(f(x)), \neg P(x) \vee \neg P(f(x))\}$.

There are two sign-renamings over $\mathcal{C}$, $\gamma_1 : \{P\}$ and $\gamma_2 : \{\neg P\}$. γ_1 induces the setting $\emptyset$, γ_2 induces $AT(\mathcal{C}) = \{P(t) \mid t \in H\}$. Both γ_1 and γ_2 falsify $\mathcal{C}$ (γ_1 falsifies $P(x) \vee P(f(x))$ and γ_2 falsifies $\neg P(x) \vee \neg P(f(x))$. But $\mathcal{C}$ is satisfiable via the Herbrand model represented by $\mathcal{M} = \{P(a), P(f(f(a))), \ldots, P(f^{2n}(a)), \ldots\}$. However, $\mathcal{M}$ is not a setting. γ-hyperresolution is not substantially more general than hyperresolution defined in Definition 3.6.5; instead of producing γ-positive clauses in γ-hyperresolution for a sign-renaming γ we may apply the inverse mapping γ^{-1} to the set of clauses and then apply positive hyperresolution. We are showing now the completeness of (positive) hyperresolution:

Lemma 3.6.1 (ground completeness of hyperresolution). *Let $\mathcal{C}$ be an unsatisfiable set of ground PN-clauses, then $\square \in R_H(\mathcal{C})$.*

Proof. As in the case of linear deduction we proceed by induction on the number of literal occurrences $occl$ in a set of ground clauses $\mathcal{C}$.

Induction base $occl(\mathcal{C}) = 0$.

Here $\mathcal{C} = \{\square\}$ and $\square \in R_H(\mathcal{C})$.

(IH) Suppose that $\square \in R_H(\mathcal{C})$ for all unsatisfiable sets of PN-clauses $\mathcal{C}$ with $occl(\mathcal{C}) \leq n$.

Case $n+1$:
Let $\mathcal{C}$ be an unsatisfiable set of PN-clauses with $occl(\mathcal{C}) = n + 1$. We distinguish two cases:

a) All clauses in $\mathcal{C}$ are either (purely) positive or (purely) negative and all positive clauses are unit clauses.

Because $\mathcal{C}$ is unsatisfiable there must be a negative clause D in $\mathcal{C}$ such that $D = \neg P_1 \vee \ldots \vee \neg P_n$ and the unit clauses $P_1, \ldots, P_n$ are all contained in $\mathcal{C}$. Without the existence of such a clause D all negative clauses could be removed by the pure literal rule; the remaining clauses would all be positive and $\mathcal{C}$ would be satisfiable. Therefore such a clause D must exist and

$$(\neg P_1 \vee \ldots \vee \neg P_n; P_n, \ldots, P_1)$$

is a clash sequence possessing the hyperresolvent $\square$. By definition of R_H we get $\square \in R_H(\mathcal{C})$.

b) There are mixed clauses (clauses which are neither positive nor negative) in $\mathcal{C}$ or there is a nonunit positive clause in $\mathcal{C}$.

b1) There is a mixed clause C in $\mathcal{C}$.

Because C is a PN-clause it must be of the form $P \vee E$. By the splitting rule of Davis and Putnam the sets of clauses $\mathcal{C}_1 : (\mathcal{C} - \{C\}) \cup \{P\}$ and $\mathcal{C}_2 : (\mathcal{C} - \{C\}) \cup \{E\}$ are both unsatisfiable and $occl(\mathcal{C}_1), occl(\mathcal{C}_2) \leq n$.

b2) There exists a positive, nonunit clause D in $\mathcal{C}$. Then D is of the form $P \vee E$ for a positive clause E. As in case b1) we split $\mathcal{C}$ into

$\mathcal{C}_1 : (\mathcal{C} - \{C\}) \cup \{P\}$ and $\mathcal{C}_2 : (\mathcal{C} - \{C\}) \cup \{E\}$.

$\mathcal{C}_1, \mathcal{C}_2$ are both unsatisfiable and $occl(\mathcal{C}_1), occl(\mathcal{C}_2) \leq n$. In both cases b1) and b2) the induction hypothesis can be applied and we obtain $\square \in R_H(\mathcal{C}_1)$ and $\square \in R_H(\mathcal{C}_2)$.

We will show now that $R_H(\mathcal{C})$ differs from $R_H(\mathcal{C}_2)$ at most with respect to occurrences of the atom P in some clauses. For a formal treatment we define the relation $\leq_L$ on clauses (L being an arbitrary literal):

$$C \leq_L D \text{ if } N_c(C) = N_c(D) \text{ or } N_c(D) = N_c(C \vee L).$$

Clearly $\leq_L$ is reflexive and transitive.

If $\mathcal{C}, \mathcal{D}$ are sets of clauses we define $\mathcal{C} \leq_L \mathcal{D}$ if for every clause $C \in \mathcal{C}$ there exists a clause $D \in \mathcal{D}$ such that $C \leq_L D$. Using this quasiordering $\leq_L$ we will show that (in both cases b1) and b2)) $R_H(\mathcal{C}_2) \leq_P R_H(\mathcal{C})$ holds.

We use induction on the levels in the definition of R_H, i.e., we show:

$$S_H^i(\mathcal{C}_2) \leq_P S_H^i(\mathcal{C}) \text{ for all } i \in N.$$

The induction base $i = 0$:

$$N_c((\mathcal{C}_2)_+) \leq_P N_c(\mathcal{C}_+) \text{ and } (\mathcal{C}_2 - (\mathcal{C}_2)_+) \leq_P (\mathcal{C} - \mathcal{C}_+)$$

is trivial because $E \leq_P P \vee E$.

It is easy to see that the remaining part of the induction proof reduces to:

Let $\gamma_1 = (E; D_1, \ldots, D_n)$ and $\gamma_2 = (E'; D'_1, \ldots, D'_n)$ be clash sequences such that $E \leq_P E'$ and $D_i \leq_P D'_i$ for $i = 1, \ldots, n$ and the D_i, D'_i are in N_c-normal form. Let F be a hyperresolvent of γ_1 then there exists a hyperresolvent F' of γ_2 such that $F \leq_P F'$.

The last assertion can be further reduced to the preservation of the $\leq_P$-relation in the internal steps of the clash resolution. An internal resolvent is a resolvent of $E_1 \vee \neg Q$ and a positive clause D_i of the form $F_1 \vee Q \vee F_2$.

Thus suppose that $E_1 \vee \neg Q \leq_P G$ and $F_1 \vee Q \vee F_2 \leq_P H$. because P is positive and G is in PN-form we have $G = E'_1 \vee \neg Q$ for some clause E'_1 such that $E_1 \leq_P E'_1$. The clause H must contain Q and is of the form $F'_1 \vee Q \vee F'_2$ such that $F_1 \vee F_2 \leq_P F'_1 \vee F'_2$ (all these relations hold, no matter whether $P = Q$ or $P \neq Q$).

The PRF-resolvent of $E_1 \vee \neg Q$ and $F_1 \vee Q \vee F_2$ is $F_1 \vee F_2 \vee E_1$ (because the positive clauses are in N_c-form we don't need reduction). Similarly $F'_1 \vee F'_2 \vee E'_1$ is PRF-resolvent of $E'_1 \vee \neg Q$ and $F'_1 \vee Q \vee F'_2$. Because, in general, $C_1 \leq_P C_2$ and $D_1 \leq_P D_2$ implies $C_1 \vee D_1 \leq_P C_2 \vee D_2$, we obtain $F_1 \vee F_2 \vee E_1 \leq_P F'_1 \vee F'_2 \vee E'_1$.

It follows that, for the hyperresolvents of the clashes, $F \leq_P F'$ holds. Completing the induction argument we eventually obtain

$$R_H(\mathcal{C}_2) \leq_P R_H(\mathcal{C}).$$

For the above argument it is important that adding further positive atoms to a clause can never block possibilities of resolution steps existing before (there are no ordering constraints in the positive clauses). By (IH) we know that $\square \in R_H(\mathcal{C}_2)$. By definition of $\leq_P$ there must be a clause C in $R_H(\mathcal{C})$ such that $\square \leq_P C$. Thus either $C = \square$ or (by the N_c-normal form of positive clauses in $R_H(\mathcal{C})$) $C = P$.

In the first case we obtain $\square \in R_H(\mathcal{C})$ and all is shown. In the second case, $C = P$, we observe the following:

$$\mathcal{C}_1 = (\mathcal{C} - \{C\}) \cup \{P\} \text{ and } \mathcal{C} - \{C\} \subseteq \mathcal{C} \subseteq R_H(\mathcal{C});$$

by $P \in R_H(\mathcal{C})$ we obtain $\mathcal{C}_1 \subseteq R_H(\mathcal{C})$. By monotonicity and idempotence of R_H we get $R_H(\mathcal{C}_1) \subseteq R_H(\mathcal{C})$ and therefore $\square \in R_H(\mathcal{C})$ ($\square \in R_H(\mathcal{C}_1)$ by (IH)).

This concludes the proof of the case $occl(\mathcal{C}) = n + 1$. $\Diamond$

Theorem 3.6.1 (completeness of hyperresolution). *Let C be an unsatisfiable set of PN-clauses, then $\square \in R_H(\mathcal{C})$.*

Proof. By Herbrand's theorem there exists a finite, unsatisfiable set C' of ground instances from C. By Lemma 3.6.1 we know that $R_H(\mathcal{C}')$ contains $\square$. Thus it is enough to show that for every $C' \in R_H(\mathcal{C}')$ there exists a $C \in R_H(\mathcal{C})$ such that $C \leq_{sc} C'$ (we write $R_H(\mathcal{C}) \leq_{sc} R_H(\mathcal{C}')$). Because $R_H(\mathcal{C}) = \bigcup_{i=0}^{\infty} S_H^i(\mathcal{C})$ we may proceed by level induction.

$i = 0$:

C' consists of instances of clauses in $\mathcal{C}$ and for every $C \in \mathcal{C}$ and every substitution σ we have $N_c(C) \leq_{sc} N_c(C\sigma)$; thus also $S_H^0(\mathcal{C}') \leq_{sc} S_H^0(\mathcal{C}')$ holds.

(IH) Suppose that $S_H^i(\mathcal{C}) \leq_{sc} S_H^i(\mathcal{C}')$.

Case $i + 1$: The proof of this case can be reduced to the proof of:

(I) Let $\gamma_1 = (C; D_1, \ldots, D_n)$ and $\gamma_2 = (C'; D_1', \ldots, D_n')$ be two clashes such that $C \leq_s C'$ and $D_i \leq_{sc} D_i'$ for $i = 1, \ldots, n$ and E' is a clash resolvent of γ_2. Then there exists a clash resolvent E of γ_1 such that $E \leq_{sc} E'$.

Note that we face the delicate situation that the positive clauses are in N_c-normal form, but the other clauses are not. In resolving through a clash we append positive clauses to the left (these are parts of clauses in N_c-normal form) and the negative part is gradually reduced (this part is not normalized). To describe the corresponding substitution ordering of related internal clash resolvents we introduce some auxiliary notion:

Let C, C' be PN-clauses. We write $C \leq_{scs} C'$ if there exists a substitution ϑ such that $N_c(C_P\vartheta) = N_c(C_P')$ and $C_N\vartheta = C_N'$.

It is easy to see that (I) can be reduced to

(II) Let $C' = C_1' \lor \neg Q'$, $D' = D_1' \lor Q' \lor D_2'$ be two PN-ground clauses (D' being in N_c-normal form) and E' be the PRF-resolvent $D_1' \lor D_2' \lor C_1'$. Let C, D be two variable-disjoint PN-clauses such that $C \leq_{scs} C'$ and $D \leq_{sc} D'$. Then there exists a PRF-resolvent E of C and D with $E \leq_{scs} E'$.

Note that D' is in normal form, thus we do not need to consider factoring (factoring coincides with clause reduction on the ground level). If the resolvent E is already the clash resolvent then $E_N = \square$ and we obtain $E \leq_{sc} E'$.

By $C \leq_{scs} C'$ we conclude that C is of the form $C_1 \lor \neg Q$ such that $C_1 \leq_{scs} C_1'$ and $Q \leq_s Q'$.

By $D \leq_{sc} D'$ there exists a substitution η with $N_c(D\eta) = D'$ (note that $N_c(D') = D'$). By $V(C) \cap V(D) = \emptyset$ we can construct a common substitution ϑ such that

$$N_c(C_P\vartheta) = N_c(C_P'), \ C_N\vartheta = C_N' \text{ and } N_c(D\vartheta) = D'.$$

As $N_c(D\vartheta) = D'$ and $D' = D_1' \lor Q' \lor D_2'$ there exist literals $L_1, \ldots, L_m$ in D such that $D = D_1 \lor L_1 \lor \ldots \lor D_n \lor L_n \lor D_{n+1}$ such that $\{L_1, \ldots, L_n\}\vartheta = \{Q'\}$ and

$$N_c((D_1 \lor \ldots \lor D_{n+1})\vartheta) = D_1' \lor D_2'.$$

Then ϑ is a unifier of $\{L_1, \ldots, L_n\}$ and there exists an m.g.u. σ of $\{L_1, \ldots, L_n\}$. σ defines the factor $D_1\sigma \lor L_1\sigma \lor \ldots \lor D_n\sigma \lor D_{n+1}\sigma$ of D.

Because $Q\vartheta = Q'$, ϑ is also unifier of $\{L_1, \ldots, L_n, Q\}$. By definition of PRF-resolution, the clause

$$E : (D_1 \vee \ldots \vee D_{n+1})\sigma\mu \vee C_1\mu$$

is PRF-resolvent of C and D (μ being an m.g.u. of $\{L_1\sigma, Q\}$). As $\{L_1\sigma, Q\}$ is unifiable, the substitution μ must exist and $\sigma\mu$ is m.g.u. of $\{L_1, \ldots, L_n, Q\}$. We thus obtain $\sigma\mu \leq_s \vartheta$. Because σ is a factoring substitution of D (defining a G-instance) we have $dom(\sigma) \cap V(C) = \emptyset$ and therefore

$$E = (D_1 \vee \ldots \vee D_{n+1} \vee C_1)\sigma\mu.$$

But $(D_1 \vee \ldots \vee C_1)_N = (C_1)_N$ and $E'_N = (C'_1)_N$. By $C_N\vartheta = C'_N$ and by definition of C_1 we also obtain

$$((C_1)_N)\vartheta = (C'_1)_N.$$

Moreover $N_c(C_P\vartheta) = N_c(C'_P)$ and $N_c((D_1 \vee \ldots D_{n+1})\vartheta) = D'_1 \vee D'_2$ implies

$$N_c((D_1 \vee \ldots \vee D_{n+1} \vee (C_1)_P)\vartheta) = N_c((C'_1)_P \vee D'_1 \vee D'_2).$$

By definition of $\leq_{scs}$ we eventually obtain

$$E_P \leq_{sc} E'_P \quad \text{and} \quad E_N \leq_s E'_N, \text{ i.e., } E \leq_{scs} E'.$$

$\Diamond$

Restricting the R_H-refinement to Horn logic we obtain a resolution refinement without any factoring. Note that in PRF-resolution only positive clauses are subjected to factoring. Because in Horn logic all positive clauses are unit clauses only trivial factors can be produced. The condensing normalization of positive clauses reduces to renaming normalization. Thus R_H for Horn logic can also be defined as

$$S_H^0(\mathcal{C}) \quad = \quad N_v(facts(\mathcal{C})) \cup rules(\mathcal{C}) \cup goals(\mathcal{C}),$$

$$S_H^{i+1}(\mathcal{C}) \quad = \quad S_H^i(\mathcal{C}) \cup \rho'_H(S_H^i(\mathcal{C})),$$

$$R_H(\mathcal{C}) \quad = \quad \bigcup_{i=0}^{\infty} S_H^i(\mathcal{C}),$$

where ρ'_H differs from ρ_H in the substitution of N_v for N_c.

Let $\varrho = (P \vee \neg Q_1 \vee \ldots \neg Q_n; P_1, \ldots, P_n)$ be a clash sequence in a set of Horn clauses. Then, clearly, ϱ defines at most one hyperresolvent R; if R exists it must be of the form $P\vartheta$ for some substitution ϑ. Thus for every derived fact $Q \in R_H(\mathcal{C})$ there exists a fact $P \in facts(\mathcal{C})$ or a head P of a rule in $rules(\mathcal{C})$ such that $P \leq_s Q$. That means every clause in $R_H(\mathcal{C})_+ - \{\Box\}$ is a substitution instance of an atom in a fixed, finite set of atoms $\mathcal{P}$ such that

$$\mathcal{P} = facts(\mathcal{C}) \cup ruleh(\mathcal{C}) \text{ where } ruleh(\mathcal{C}) = \{P | (\exists C \in \mathcal{C}) P \vee C \in rules(\mathcal{C})\}.$$

Moreover R_H possesses an important semantic interpretation: If C is a satisfiable set of Horn clauses then facts($R_H(C)$) denotes a Herbrand model of C. Herbrand models, that can be specified in this manner possess some property of minimality (see Exercise 3.6.4) which is important to the semantics of logic programs [Llo87].

In Section 2.3 (remark after Example 2.3.3) we have introduced a representation of Herbrand models Γ of the form $\mathcal{M} = \{P_1, \ldots, P_n, \ldots\}$, where the set $\mathcal{M}$ consists of exactly those ground atoms of the atom set that are true in Γ. So we obtain a characterization of Γ by a set of ground atoms. The following definition introduces a more general notion of H-model representations.

Definition 3.6.9 (atom representations of H-interpretations). *Let Γ be an H-interpretation of a set of clauses C and $\mathcal{M} = \{A | A \in AT(C), v_\Gamma(A) = \mathbf{T}\}$. Then $\mathcal{M}$ is called a* ground representation *(GR) of Γ.*
Let $\mathcal{P}$ be a set of atoms over the signature of C and let $\mathcal{N}$ be the set of all ground instances of elements in $\mathcal{P}$ (over C). Then $\mathcal{P}$ is called an atom representation *(AR) of the model represented by (the ground representation) $\mathcal{N}$.*

The following theorem states that R_H can be interpreted as a model building procedure in Horn logic.

Theorem 3.6.2. *Let C be a satisfiable set of Horn clauses. Then $R_H^+(C)$ is an atom representation of a Herbrand model of C.*

Proof. Let $\mathcal{P} = R_H^+(C)$. Then $\mathcal{P}$ is a (possibly infinite) set of atoms ($R_H^+(C)$ does not contain $\square$ by the satisfiability of C). Let Γ be an arbitrary Herbrand model of C. Then $v_\Gamma(\{P\}) = \mathbf{T}$ for all $P \in facts(C)$ (all unit clauses must be true in Γ). As resolution is strongly correct (i.e., it is model preserving) every clash produces hyperresolvents true in Γ. Thus we obtain $v_\Gamma(\{P\}) = \mathbf{T}$ for all $P \in \mathcal{P}$.
Let $\mathcal{M}_0 = \{P' | (\exists P \in \mathcal{P})P \leq_s P', P' \text{ ground }\}$. Because $v_\Gamma(\{P\}) = \mathbf{T}$ for all $P \in \mathcal{P}$ we obtain $v_\Gamma(P\vartheta) = \mathbf{T}$ for $P \in \mathcal{P}$ and arbitrary ground substitutions ϑ and therefore

$$\mathcal{M}_0 \subseteq \{A | A \in AT(C), v_\Gamma(A) = \mathbf{T}\}.$$

That means $\mathcal{P}$ is an atom representation of an H-interpretation Γ_0 of C, which verifies at most the ground atoms verified by Γ. It remains to show that Γ_0 is a Herbrand model (!) of C. We assume that Γ_0 falsifies C and derive a contradiction.

Γ_0 falsifies C iff the (possibly infinite) set $C \cup \mathcal{M}_0 \cup \{\neg P / P \in AT(C) - \mathcal{M}_0\}$ is unsatisfiable. By the compactness theorem [BJ74] in Chapter 12 there exists a finite subset $\mathcal{D} \subseteq \mathcal{M}_0 \cup \{\neg P / P \in AT(C) - \mathcal{M}_0\}$ such that $C \cup \mathcal{D}$ is unsatisfiable. $\mathcal{D}$ must be a set of ground unit clauses $\{P_1', \ldots, P_n', \neg Q_1', \ldots, \neg Q_m'\}$.

It follows by the definition of $\mathcal{M}_0$ that the Q_i' are not ground instances of atoms in $\mathcal{P}$, i.e., for all $Q' \in \{Q_1', \ldots, Q_m'\}$, for all $P \in \mathcal{P} : P \not\leq_s Q'$. Thus no positive clause P – derivable from $\mathcal{C}$ via hyperresolution – can be unified with such a Q'. Because $\mathcal{C} \cup \{P'_1, \ldots, P'_n\}$ is satisfiable (because $\mathcal{C} \cup \mathcal{P}$ is satisfiable) we have $\square \notin R_H(\mathcal{C} \cup \{P'_1, \ldots, P'_n\})$.

Because negative clauses can only serve as central clauses in clashes we obtain:

$$R_H^+(\mathcal{C} \cup \mathcal{D}) \subseteq R_H^+(\mathcal{C} \cup \{P'_1, \ldots, P'_n\}) \cup \{\square\}.$$

We show now that $R_H^+(\mathcal{C} \cup \mathcal{D})$ cannot contain $\square$. Because $\square \notin R_H(\mathcal{C} \cup \{P'_1, \ldots, P'_n\})$, the assumption $\square \in R_H(\mathcal{C} \cup \mathcal{D})$ implies the existence of a $P \in R_H^+(\mathcal{C} \cup \{P'_1, \ldots, P'_n\})$ such that $\varrho = (\neg Q', P)$ is a clash resolving to $\square$ for some $\neg Q' \in \{\neg Q'_1, \ldots, \neg Q'_m\}$. Because Q' is ground there must exist a $P \in \mathcal{P}$ such that $P \leq_s Q'$ ($\{P, Q'\}$ must be unifiable). By definition of of $\mathcal{M}_0$ we get $Q' \in \mathcal{M}_0$.

On the other hand we have assumed $Q' \in AT(\mathcal{C}) - \mathcal{M}_0$. So we obtain a contradiction and $\mathcal{C} \cup \mathcal{M}_0 \cup \{\neg P | P \in AT(\mathcal{C}) - \mathcal{M}_0\}$ must be satisfiable. But this implies that Γ_0 is an H-model of $\mathcal{C}$.   ◊

The model construction via R_H is of particular interest if $R_H(\mathcal{C})$ is finite. Because Horn logic is undecidable (the halting problem for Turing machines and unsolvable word problems in equational theories are recursively equivalent to the satisfiability problem of Horn logic), we cannot expect to obtain always finite sets $R_H(\mathcal{C})$ for finite sets of Horn clauses $\mathcal{C}$. As in Proposition 3.1.1 we infer the existence of a finite set of Horn clauses $\mathcal{C}$ such that $R_H(\mathcal{C})$ is infinite. In Section 5.4 we will investigate subclasses Ω of Horn logic such that for all $\mathcal{C} \in \Omega$ $R_H(\mathcal{C})$ is finite; for such classes we can obtain finite model representations, i.e., representations of Herbrand models in the form of predicate logic formulas.

Suppose that $\mathcal{C}$ is a finite, satisfiable set of Horn clauses and $R_H(\mathcal{C})$ is finite too. Then

$$R_H(\mathcal{C}) = rules(\mathcal{C}) \cup goals(\mathcal{C}) \cup \{P_1, \ldots, P_n\}$$

for a finite set of atom formulas $\mathcal{P} : \{P_1, \ldots, P_n\}$. The set $\mathcal{P}$ is an atom representation of a Herbrand model Γ and corresponds to a PL-formula of the form

$$(\forall \bar{x}_1)P_1 \wedge \ldots \wedge (\forall \bar{x}_n)P_n.$$

Now there exists an easy algorithm for deciding whether an atom $A \in AT(\mathcal{C})$ is true in Γ:

$$v_\Gamma(A) = \mathbf{T} \text{ iff there exists a } P \in \mathcal{P} \text{ such that } P \leq_s A.$$

In order to test $v_\Gamma(A) = \mathbf{T}$ we simply have to try all matchings $P_i \leq_s A$ for $i = 1, \ldots, n$.

We have seen that hyperresolution under sign-renamings can be interpreted as a kind of semantic inference with respect to specific Herbrand interpretations, called settings (see Definition 3.6.8). Example 3.6.4, however, demonstrated that there are very simple Herbrand interpretations which are not settings. Moreover it makes sense to define model-based refinements over interpretations which are not of Herbrand type (such as finite models). The idea to define semantic refinements over arbitrary interpretations was first expressed by Slagle [Sla67]. These refinements represent inference principles restricting deduction by use of countermodels.

Example 3.6.5.
Let $\mathcal{C} = \{C_1, C_2, C_3, C_4\}$ for

$$C_1 = P(f(x)), C_2 = P(g(x)), C_3 = \neg P(x) \vee R(x), C_4 = \neg R(f(x)) \vee \neg P(g(x))$$

and $\mathcal{M} : (D, \Phi, I)$ be an interpretation with

$$D = \{0,1\}, \Phi(f) = \phi, \Phi(g) = \gamma, \Phi(P) = \pi \text{ and } \Phi(R) = \rho$$

such that

$$\phi(0) = \phi(1) = 0; \gamma(0) = \gamma(1) = 1;$$
$$\pi(0) = \mathbf{F}, \pi(1) = \mathbf{T}; \rho(0) = \rho(1) = \mathbf{F}.$$

We now evaluate the clauses of $\mathcal{C}$ under $\mathcal{M}$:

C_1 is false in $\mathcal{M}$, C_2 is true in $\mathcal{M}$. C_3 is false in $\mathcal{M}$ because of the "instance" $or(not(\pi(1)), \rho(1)) = \mathbf{F}$; but note that $or(not(\pi(0)), \rho(0)) = \mathbf{T}$. C_4 is true in $\mathcal{M}$ by $\rho(\alpha) = \mathbf{F}$ for all $\alpha \in D$.

Therefore $\mathcal{C}$ can be decomposed into the set of true clauses $\mathcal{C}_1 : \{C_2, C_4\}$ and the set of false clauses $\mathcal{C}_0 : \{C_1, C_3\}$. As in the case of setting-based hyperresolution our aim is to derive only clauses which are false in $\mathcal{M}$. As resolution is (strongly) correct we may not resolve clauses which are both true in $\mathcal{M}$; this immediately excludes resolutions within $\mathcal{C}_1$.

C_3 and C_4 are not both true and we construct the resolvent $C_5 : \neg P(f(x)) \vee \neg P(g(x))$. C_5, however, is true in $\mathcal{M}$ and thus is not "admissible"; but by further resolving C_5 with C_1 we obtain the clause $C_6 : \neg P(g(x))$ which is false in $\mathcal{M}$. The resolvent of C_6 and C_2 is $\square$ which is false in $\mathcal{M}$.

Note that C_1 and C_3 are both false in $\mathcal{M}$ but, nevertheless, are resolvable (in the case of settings this is impossible because (here) two false clauses cannot define complementary pairs of literals!). Resolution of a true and a false clause can result in a true clause (like C_5) or in a false clause (the resolvent of C_2 and C_3 is $C_7 : R(g(x))$ which is false in $\mathcal{M}$).

Example 3.6.5 shows that the truth status of resolvents with respect to arbitrary interpretations is not so easily predictable as in the case of settings. What we need in general is a computational procedure to evaluate clauses over given interpretations $\mathcal{M}$. Such an evaluation is straightforward for interpretations with finite domains; a method to evaluate clauses over arbitrary atom representations of Herbrand models is given in [FL96]. In Example 3.6.5 we resolved the true clause C_2 with the false clause C_3 and obtained the true clause C_5. According to to the principle that true clauses cannot be resolved with other true clauses we need a "false" partner for C_5. Such a clause is C_1 which, together with C_5, produces the false resolvent C_6. If we arrange the clauses in a tuple $S : (C_2; C_3, C_1)$ and proceed as in the case of hyperresolution then we obtain C_5 as intermediary resolvent and C_6 as *clash resolvent* of S. The PN-normal form for hyperresolution was specific to the rigid type of settings (in a setting the truth value of a clause is equal to the truth value of all its instances). In case of arbitrary interpretations we work without normalization (although condensing normalization is admissible) in order to make the problem more transparent).

Definition 3.6.10 (semantic resolution). *Let C be a set of clauses and M be an interpretation of C. Let C and D be clauses over the signature of C such that at least one of C, D is false in M. Then every resolvent of C and D is called* semantic M-resolvent *or simply M-resolvent.*

For every set of clauses D with the same signature as C we define

$$\rho_M(D) = \varphi\{E|E \text{ is semantic resolvent from } D\}$$

where φ is the variable normalizer defined in Section 3.1. The resolution operator R_M can be defined in the usual way:

$$
\begin{aligned}
S_M^0(C) &= C, \\
S_M^{i+1}(C) &= S_M^i(C) \cup \rho_M(S_M^i(C)), \\
R_M(C) &= \bigcup_{i \in \mathbb{N}} S_M^i(C).
\end{aligned}
$$

Definition 3.6.11 (semantic clash resolution).
  Let M be an interpretation of a set of clauses C and $C, D_1, \ldots, D_k$ be clauses over the signature of C such that the D_i are false in M for $i = 1, \ldots, k$. Let

  $R_0 = C$ *and*
  R_{i+1} *be a resolvent of R_i and D_{i+1} if $i < k$ and such a resolvent exists.*

If R_k exists and is false in M then we call $S : (C; D_1, \ldots, D_k)$ a semantic clash and R_k a semantic clash resolvent of S (with respect to M). C is called the nucleus of S.
As in Definition 3.6.10 we set

$$\rho_{cM}(D) = \varphi\{E|E \text{ is semantic clash resolvent from } D\}$$

The operators S_{cM}^i and R_{cM} are defined like S_M^i and R_M.

Example 3.6.6. Let M be the interpretation defined in Example 3.6.5 and

$$C = \neg R(f(x)) \vee \neg P(g(x)), \ D_1 = \neg P(x) \vee R(x), \ D_2 = P(f(x)).$$

Then $\neg P(g(x)) \vee \neg P(f(x))$ is a semantic resolvent of C and D_1, but $(C; D_1)$ is not a semantic clash.

$(C; D_1, D_2)$ is a semantic clash and gives the clash resolvent $\neg P(g(x))$. Note that (D_1, D_2) is also a semantic clash, because the resolvent is $R(f(x))$ which is false in $\mathcal{M}$. We see that the nucleus of a clash may itself be false in $\mathcal{M}$, but it must contain true "instances", i.e., it sometimes evaluates to true.

It remains to show that semantic clash resolution is complete over arbitrary interpretations.

Lemma 3.6.2 (ground completeness of semantic clash resolution).
Let C be a set of clauses, $\mathcal{M}$ be an interpretation of C and $\mathcal{D}$ be an unsatisfiable set of ground instances from C. Then $\square \in R_{c\mathcal{M}}(\mathcal{D})$.

Proof. We reduce the problem to the ground completeness of hyperresolution which was shown in Lemma 3.6.1.

Let $\mathcal{A}: \{A_1, \ldots, A_n\}$ be the set of atoms occurring in $\mathcal{D}$. Let

$$\mathcal{B}_0 = \{A | A \in \mathcal{A}, v_{\mathcal{M}}(A) = \mathbf{F}\} \text{ and } \mathcal{B}_1 = \{A | A \in \mathcal{A}, v_{\mathcal{M}}(A) = \mathbf{T}\}.$$

Define a sign renaming γ on $\mathcal{A}$ such that $\gamma(A) = A$ for $A \in \mathcal{B}_0$ and $\gamma(A) = \neg A$ for $A \in \mathcal{B}_1$ (note that we may consider $\mathcal{D}$ as a set of propositional clauses and thus may assign different symbols to different atoms). Then the set $\gamma(\mathcal{D})$ is unsatisfiable. By definition of γ a clause $D \in \mathcal{D}$ is false in $\mathcal{M}$ iff $\gamma(D)$ is positive. By Lemma 3.6.1 $\square \in R_H(\gamma(D))$. To be precise we have to mention that Lemma 3.6.1 holds for PN-forms and here we do not normalize clauses. We only have to extend R_H to sets of clauses which are not in PN-form (we give up the fixed order for the resolution of positive literals in order to maintain completeness).

It suffices to prove that $\gamma(S^i_{c\mathcal{M}}(\mathcal{D})) = S^i_H(\gamma(\mathcal{D}))$ for all $i \in \mathbb{N}$. Then $\square \in S^i_H(\gamma(\mathcal{D})$ and $\gamma^{-1}(\square) = \square$ imply $\square \in S^i_{c\mathcal{M}}(\mathcal{D})$ and therefore $\square \in R_{c\mathcal{M}}(\mathcal{D})$. We proceed by induction on i.

$i = 0$: trivial.

(IH) Assume that $\gamma(S^i_{c\mathcal{M}}(\mathcal{D})) = S^i_H(\gamma(\mathcal{D}))$

Let $\varrho : (C; D_1, \ldots, D_m)$ be a clash sequence over $S^i_H(\gamma(\mathcal{D}))$ and R be a resolvent of ϱ. Then, by (IH), there are clauses $E_1, \ldots, E_m \in S^i_{c\mathcal{M}}(\mathcal{D})$ such that $\gamma(E_i) = D_i$ for $i = 1, \ldots, m$. By definition of γ $v_{\mathcal{M}}(E_i) = \mathbf{F}$ iff $\gamma(E_i)$ is positive. But the D_i must be positive (by definition of a clash sequence) and we infer that $v_{\mathcal{M}}(E_i) = \mathbf{F}$ for all $i = 1, \ldots, m$. An easy induction argument shows that for all intermediary resolvents R_l of ϱ there are intermediary resolvents R'_l of $S : (\gamma^{-1}(C), E_1, \ldots, E_m)$ such that $\gamma(R'_l) = R_l$. Eventually we obtain a resolvent R' such that $\gamma(R') = R$. Because R is positive $\gamma(R')$ must be false in $\mathcal{M}$. Therefore S is a semantic clash and R' is a semantic clash resolvent of S with $\gamma(R') = R$.

The other direction (i.e., the proof of $\gamma(S^i_{c\mathcal{M}}(\mathcal{D})) \subseteq S^i_H(\gamma(\mathcal{D})))$ is completely symmetric.   ◊

Theorem 3.6.3 (completeness of semantic clash resolution).
Let C be an unsatisfiable set of clauses and $\mathcal{M}$ be an interpretation of C. Then $\square \in R_{c\mathcal{M}}(C)$.

Proof. By Herbrand's theorem there exists an unsatisfiable set of ground instances $\mathcal{D}$ from C. By Lemma 3.6.2 we know that $\square \in R_{c\mathcal{M}}(\mathcal{D})$. Therefore it suffices to prove the following lifting property:

(*) If $\mathcal{D}$ is a set of ground clauses from C then $R_{c\mathcal{M}}(C) \leq_s R_{c\mathcal{M}}(\mathcal{D})$.

The proof of (*) can be reduced to that of the lifting property of single steps, i.e.,

(+) If D_1, D_2 are ground instances of clauses C_1, C_2 and F is a semantic $\mathcal{M}$-resolvent of D_1, D_2 than there exists a semantic $\mathcal{M}$-resolvent E of C_1, C_2 such that $E \leq_s F$.

Because F is a semantic resolvent, at least one of D_1, D_2 must be false in $\mathcal{M}$. So let us assume without loss of generality that $v_{\mathcal{M}}(D_1) = \mathbf{F}$. D_1 is an instance of C_1 and so $F(\{C_1\}) \to F(\{D_1\})$ is valid; consequently C_1 is also false in $\mathcal{M}$. By Definition every resolvent of C_1, C_2 is a semantic resolvent. By the lifting theorem (Theorem 2.7.1) there exists a resolvent E of C_1 and C_2 such that $E \leq_s F$; but E is also semantic $\mathcal{M}$-resolvent of C_1 and C_2.   ◊

Exercises

Exercise 3.6.1.

a) Define a refinement Ψ_H (according to Definition 3.1.1) such that $Der(\Psi_H(C)) = R_H(C)$ for all $C \in CL$.
b) Find a refutation Γ of

$$C : \{P(x) \vee Q(x),\ \neg P(x) \vee Q(f(y)),\ P(x) \vee \neg Q(f(x)), \neg P(x) \vee \neg Q(x)\}$$

such that $\Gamma \in \Psi_H$.

Exercise 3.6.2. Let R_U be the operator of unit resolution, i.e., R_U is defined via ρ_U such that $\rho_U(C) = $ set of all resolvents from clauses $C, D \in C$ such that at least one of C, D is a unit clause.

a) Show that R_U is complete on Horn logic.
b) What is the relation among R_U and R_H on Horn logic?

Exercise 3.6.3. Let Ω be the class of all finite sets of clauses C such that all positive clauses in C are unit clauses (Ω contains Horn clause logic). Show that unit resolution is incomplete on Ω.

Exercise 3.6.4. Let $\mathcal{M}$ be a ground representation of a Herbrand model Γ of a set of clauses C. Γ is called the *least Herbrand model* of C if the following property holds: If $\mathcal{M}'$ is a ground representation of an arbitrary H-model Γ' of C then $\mathcal{M} \subseteq \mathcal{M}'$.

Let C be a finite, satisfiable set of Horn clauses. Show that $R_H^+(C)$ is an atom representation of a least H-model of C.

(Hint: Trace the proof of Theorem 3.6.2).

Exercise 3.6.5. Give an example of a (non-Horn) set of clauses that does not possess a least Herbrand model.

Exercise 3.6.6. Let $\mathcal{P} = \{P_1, \ldots, P_n\}$ be a finite atom representation of a H-interpretation Γ over a signature Σ. Let C be an arbitrary, finite set of clauses over Σ such that $R_H(C)$ is finite. Construct an algorithm that decides whether Γ is a model of C.

(Hint: Use the fact that $R_H^+(C)$ represents a least H-model).

Exercise 3.6.7. In the proof of Theorem 3.6.2 we used the property

$$R_H^+(C) \leq_s R_H^+(C \cup \{P_1', \ldots, P_n'\})$$

for ground atoms P_i' which are ground instances from elements in $R_H^+(C)$, where C is a set of Horn clauses.

Prove the following (somewhat more general) property: Let C be a set of Horn clauses and $\{C_1, \ldots, C_n\}$ be a set of clauses such that $R_H(C) \leq_s \{C_1, \ldots, C_n\}$ (i.e., for every C_i there exists a $D \in R_H(C)$ such that $D \leq_s C_i$). Then $R_H(C) \leq_s R_H(C \cup \{C_1, \ldots, C_n\})$.

Exercise 3.6.8. Define an interpretation $\mathcal{M}$ and two clauses C and D such that C and D are both false in $\mathcal{M}$, but there exists a resolvent of C and D that is true in $\mathcal{M}$.

3.7 Refinements: A Short Overview

In the preceeding sections we have presented refinements based on orderings, linear deduction, semantic inference, hyperdeduction etc. There are much more refinements (even types of refinements) which are of theoretical and of practical interest. Our specific selection was motivated mainly by the applications to decision theory (Chapter 5) and by the complexity analysis in Chapter 6. Moreover even the refinements described so far can be combined in many different ways (e.g. ordering with semantic resolution and ordering with linear deduction). Many quite sophisticated combinations of different refinement techniques can be found in D. Loveland's book [Lov78].

Roughly we may distinguish refinements generating sets of clauses from refinements generating deductions. Remember that linear deduction cannot simply be defined as an operator on sets of clauses; instead the history of a derived clause strongly influences its potential resolution partners. In model elimination, which is a variant of linear deduction, the history of a clause is even stored within the clause; for this purpose resolved literals are kept in the clause, but they are typed (A-type) to distinghish them from the ordinary literals (B-type) [Lov78]. Strictly speaking model eliminiation is not a refinement in the sense of Definition 3.1.1 but rather a variant of a refinement. Linear input deductions on Horn logic even yield a new quality of information, namely so-called answer substitutions; these substitutions on variables of the top clause (computed by a simultaneous m.g.u. of a deduction) define the output of logic programs [Llo87].

On the other hand, ordering refinements can conveniently be described by set operators: the ordering restriction only applies to pairs of clauses and their resolvents, otherwise the form of deductions does not matter. From an abstract point of view we may classify this kind of theorem proving (we are interested in sets of derivable clauses rather than in their derivations) as saturation-based. The advantage of this type of inference lies in a higher potential to apply deletion methods. Moreover saturation-based methods are superior in decision procedures and automated model building (see Chapter 5). An abstract and elegant analysis of saturation-based deduction can be found in [BG94]. More recent developments in the area of ordering refinements were strongly stimulated by investigations of resolution decision procedures. Besides further variants of so-called π-orderings [FLTZ93] (dating back to S.Y. Maslov) completely new types of orderings emerged. We just mention orderings defined by resolution games introduced in [Niv96]; these orderings need not fulfil the lifting property (A3) in Definition 3.3.1 and can be used as decision procedures for clause classes.

There are variants of resolution imposing global structures on the set of derived clauses by links connecting complementary literals. Such a combination of link construction and resolution is the characteristic feature of connection graph resolution [Kow75]. This method defines a natural "link" to nonresolution theorem proving, e.g., to the connection method [Bib82]. Several refinements, in particular linear ones, can be formulated and applied within the connection method. In fact the concept of refinement plays an important rôle also in nonclausal theorem proving and is crucial to any kind of calculus designed for proof search. We just mention the use of A-ordering restrictions in the tableau calculus for the construction of counter examples (see [KH94]).

Finally we would like to emphasize that refinements, a specific invention within the field of automated deduction, are not only important in the practice of theorem proving but also shed some more light on the very nature of deduction itself. Answer substitutions extracted from linear input deduction and (counter-)models generated by hyperresolution are examples of a new quality of information produced by inference systems.

4. Redundancy and Deletion

4.1 The Problem of Proof Search

The problems of showing the existence of proofs and finding proofs of given theorems mark the borderline between mathematical logic and computer science. So far we merely proved results about the existence of refutations for unsatisfiable sets of clauses under various types of refinements; we have not spoken about how to (really) obtain refutations. In our formalism the situation can be described in the following way: Let Ψ be an arbitrary complete refinement and C be an unsatisfiable set of clauses. By the completeness of Ψ there exists a refutation $\Gamma \in \Psi(C)$; finding such a Γ (within reasonable computing time) is the main problem of automated deduction. At this point we face the problem of *search* which is of central importance to all fields of Artificial Intelligence. With regard to a resolution refinement, search is an algorithmic method for producing the elements of $\Psi(C)$ until a refutation is found (in principle we can try to find all Ψ-refutations of C, but such a procedure usually is nonterminating). The computational cost of proof search is, in practice, the main obstruction to automated theorem proving. For this reason, several techniques have been invented to reduce search. We list three of them:

1. Refinements of resolution:
   Let Ψ_1, Ψ_2 be two refinements such that $\Psi_1(C) \subseteq \Psi_2(C)$ for all $C \in CL$. Then Ψ_1 is "more refined" than Ψ_2 in the sense that it contains fewer derivations. Thus, for Ψ_1, we have fewer derivations to search through, which may be an advantage in reducing search. It has to be noted, however, that by the (possible) increase of the minimal length of a refutation in Ψ_1 versus Ψ_2, an opposite effect may take place as well. We will explain this phenomenon in more detail using the concept of search complexity.

2. Redundancy test:
   Refinements, such as defined in Chapter 3, reduce the set of all derivations, but still contain redundancies. Redundancy can appear in the form of circular derivations or in that of tautological clauses. Removing redundant derivations results in a further reduction of the set of derivations defined by a refinement. Furthermore (in all relevant cases) the minimal length of a refutation is not increased by techniques eliminating redun-

dancy. Nevertheless there are some problems in applying redundancy tests in practice created by the costs for testing redundancy itself.

3. Heuristics:

In producing the set of derivations $\Psi(\mathcal{C})$ during the search for a refutation, the order according to which derivations are produced has a strong influence on the cost of the search. A rough, but generally useful, heuristic consists in priorizing deductions containing "small" clauses only. In selecting clauses according to their size, priority may be given to clauses containing fewer literals or to clauses of smaller term depth. Different types of clause complexity may be combined in a weight function, resulting in a preference for clauses having smaller weight. Although heuristics are of high practical value, it is hard to give mathematical criteria for selecting the "right" ones.

Efficient theorem proving programs use 1, 2 and 3 simultaneously. But 1 and 2 may "clash" in the sense that the completeness of specific refinements (such as lock resolution) may be destroyed by redundancy methods (such as subsumption). As 3 merely is concerned with ordering the derivations in $\Psi(\mathcal{C})$, it does not affect the set $\Psi(\mathcal{C})$ itself; consequently heuristics are of a "nonlogical" nature and thus cannot destroy completeness. Note that the restriction of producing *only* clauses of low complexity (e.g., clauses containing ≤ 3 literals) defines an (incomplete) refinement rather than a heuristic.

Example 4.1.1. Let $\mathcal{C}$ be the following set of N_c-clauses:

$$\{P(x_1), P(f(x_1)) \vee \neg P(x_1), R(x_1) \vee \neg P(x_1), \neg R(f(x_1))\}.$$

$\mathcal{C}$ is unsatisfiable and thus $\square \in R_x(\mathcal{C})$ for every complete refinement R_x.

Let $R_\emptyset$ be the operator of unrestricted resolution on N_c-clauses. Then $R_\emptyset(\mathcal{C})$ is infinite ($P(f^{(n)}(x_1)) \in R_\emptyset(\mathcal{C})$ for all $n \geq 0$). For every $n \geq 0$ the deduction

$\Gamma_n : P(x_1), P(f(x_1)) \vee \neg P(x_1), P(f(x_1)), \ldots, P(f^{(n)}(x_1)), R(x_1) \vee \neg P(x_1),$
$R(f^{(n)}(x_1)), \neg R(f(x_1)), \square$

is an $N_c R$-refutation of $\mathcal{C}$. As a measure for the size of clauses we take *comp*, defined by

$$comp(C) = |C| + \tau(C).$$

By the normalization of variables there are only finitely many clauses C (over the signature of $\mathcal{C}$) such that $comp(C) \leq k$ for $k \in \mathbb{N}$.

We now use *comp* as a basis of our heuristics. Let us try first to find a refutation containing only clauses with $comp \leq 1$. It is easy to verify that such a refutation cannot exist (only the first clause in $\mathcal{C}$ could be used). By increasing *comp* to ≤ 2 we obtain the following derivations:

$\Gamma \quad : \quad P(x_1), R(x_1) \vee \neg P(x_1), R(x_1), \neg R(f(x_1)), \square$ and
$\Gamma' \quad : \quad P(x_1), \neg R(f(x_1)), R(x_1) \vee \neg P(x_1), \neg P(f(x_1)), \square.$

(there are other N_cR-refutations of clause complexity comp ≤ 2).

According to a preference ordering in the set of clauses $\mathcal{C}$ we first find Γ or Γ' (or another refutation of complexity ≤ 2).

In any case we find a refutation not containing the clause $P(f(f(x_1)))$ (note that $comp(P(f(f(x_1)))) = 3$).

Now let us consider the A-ordering refinement $R_{<_d}$ defined in Section 3.3. Then

$$R_{<_d}(\mathcal{C}) = \mathcal{C} \cup \{R(x_1), \ \neg P(f(x_1)), \ R(f(x_1)) \vee \neg P(x_1), \ \neg P(x_1), \ \square\}.$$

Whatever kind of search will be applied to find a $R_{<_d}$-refutation, it will be efficient ($R_{<_d}(\mathcal{C})$ is finite, while $R_\emptyset(\mathcal{C})$ is infinite).

Finally we consider the aspect of redundancy.

The clauses $P(x_1)$ and $P(f(x_1)) \vee \neg P(x_1)$ together produce the (infinite) sequence of resolvents $C_n : P(f^{(n)}(x_1))$. All clauses C_n are instances of $P(x_1)$ ($P(x_1)\lambda_n = C_n$ for $\lambda_n = \{x_1 \leftarrow f^{(n)}(x_1)\}$) and thus are less general than $P(x_1)$ itself. By the validity of the formula

$$F(\{P(x_1)\}) \rightarrow F(\{P(f(x_1)) \vee \neg P(x_1)\})$$

the clause $P(f(x_1)) \vee \neg P(x_1)$ can be deleted (in the presence of $P(x_1)$). By removing the second clause we thus obtain

$$\mathcal{C}' = \{P(x_1), \ R(x_1) \vee \neg P(x_1), \ \neg R(f(x_1))\}.$$

After this deletion even $R_\emptyset(\mathcal{C}')$ is finite and search may be reduced considerably.

In Example 4.1.1 we have illustrated how heuristics, refinement and deletion can influence the search for a refutation. So far, however, we have not spoken about standard search methods. We will present two basic types, namely breadth-first search and depth-first search. There are much more (and more refined) search methods used in AI and defined in contexts independent of automated theorem proving; for more information we refer to [GH93], [BF81], and [Pea84]. For the sake of simplicity we concentrate our investigations on operator-based refinements R_x.

(I) Breadth-first search (BFS):

Let $R_x(\mathcal{C}) = \bigcup_{i=0}^\infty S_x^i(\mathcal{C})$ such that all clauses are normalized under N_v (N_s or N_c respectively). Then $S_x^i(\mathcal{C})$ is finite for every $i \in \mathbb{N}$. In order to find $\square$ we successively generate all sets $S_x^i(\mathcal{C})$ until we find a level j such that $\square \in S_x^j(\mathcal{C})$. Obviously breadth-first search is complete, i.e., if $\square \in R_x(\mathcal{C})$ then $\square$ is eventually found by the search method.

(II) Depth-first search (DFS):

This is characterized by preferring depth, which means going to deeper levels before exhausting the former ones. In depth-first search the set of all resolvents between a single clause C and a set of clauses $\mathcal{C}$ is important. Thus we have to modify the operator ρ_x (see Section 3.1) to $\hat{\rho}_x$:

$\hat{\rho}_x(C, \mathcal{C}) =$ the subset of $\rho_x(\{C\} \cup \mathcal{C})$ obtained by "using" C.

This rather informal definition can easily be made precise for refinements of binary type, i.e., if

$$\begin{aligned} \rho_x(\mathcal{C}) &= \{E | E \in \rho_x(\{C, D\}), \ C, D \in \mathcal{C}\} \text{ then} \\ \hat{\rho}_x(C, \mathcal{C}) &= \{D | D \in \rho_x(\{C, E\}), \ E \in \mathcal{C}\}. \end{aligned}$$

In case of hyperresolution "using" C can be defined as C being an element in a clash sequence from $\mathcal{C}$ (for positive C).

The algorithm for depth-first search is illustrated in Figure 4.1.

Depth-first search (DFS):
    {$\mathcal{C}$ is a nonempty set of clauses }
    <u>begin</u>
        Order $\mathcal{C}$ and construct a list L out of $\mathcal{C}$;
        For all C in L <u>do</u> used$(C) \leftarrow$ <u>false</u>;
        exhaust$(L) \leftarrow$ <u>false</u>;
        <u>while</u> $\square$ is not in L <u>and</u> exhaust(L)=false
            <u>do</u> <u>begin</u>
                search for the first clause C in L such that used(C)=<u>false</u>;
                <u>if</u> C does not exist <u>then</u> exhaust(L) $\leftarrow$ <u>true</u>
                <u>else</u> <u>begin</u>
                    $\mathcal{D} \leftarrow \hat{\rho}_x(C, \text{clauses}(L))$;
                    used$(C) \leftarrow$ <u>true</u>;
                    order $\mathcal{D}$ into a list L';
                    $L \leftarrow L'L$
                <u>end</u>
        <u>end</u> {while}
        <u>if</u> $\square$ is in L <u>then</u> refutation $\leftarrow$ <u>true</u>
            <u>else</u> refutation $\leftarrow$ <u>false</u>
    <u>end</u> {DFS}

Fig. 4.1. Depth-first search

If in DFS (in Figure 4.1) $\mathcal{D} = \emptyset$, i.e., C does not define resolvents with the clauses in the list L, then L' is empty and the list L is not extended. As C is already marked as used we have to go "back" in order to select another (unused) clause from L; this phenomenon is called *backtracking*. Even in case $\square \in R_x(\mathcal{C})$ it might happen that depth-first search does not terminate and does not find $\square$; in this case the search generates a proper subset of $R_x(\mathcal{C})$ only. Indeed it may be the case that the statement producing L' is executed "too often" and we never return to the clauses in some former level ($S_x^k(C)$) that are required for a contradiction. There is an alternative definition of DFS: Instead of stopping under the condition $\Box$ in L we require $\Box$ to appear at the beginning of the list L; this definition leads to a different (in fact worse) termination behavior.

Example 4.1.2. Let C: $\{P(a),\ P(f(x_1)) \vee \neg P(x_1),\ \neg P(f(f(a)))\}$ be a set of N_s- clauses and R_0 be the operator of unrefined resolution (on N_s-clauses). Breadth-first search successively produces the sets $S_\emptyset^i(C)$:

$$\begin{aligned} S_\emptyset^0(C) &= C, \\ S_\emptyset^1(C) &= C \cup \{P(f(a)), \neg P(f(a))\}. \end{aligned}$$

So $\Box \in S_\emptyset^2(C)$ and BFS stops at level 2. For DFS we turn C into the list $L : P(a), P(f(x_1)) \vee \neg P(x_1), \neg P(f(f(a)))$. Then DFS proceeds as follows

$\mathcal{D}_1 = \hat{\rho}_\emptyset(P(a), C) = \{P(f(a))\},\ \text{used}(P(a)) = \text{true},$
$L_1 = P(f(a)), L_1.$
$\mathcal{D}_2 = \hat{\rho}_\emptyset(P(f(a)),\ \text{clauses}(L_1)) = \{P(f(f(a)))\},\ \text{used}(P(f(a))) = \text{true},$
$L_2 = P(f(f(a))), L_1.$
$\mathcal{D}_3 = \hat{\rho}_\emptyset(P(f(f(a))),\ \text{clauses}(L_2)) = \{P(f^{(3)}(a)), \Box\},\ \text{used}(P(f(f(a))))$
$= \text{true},$
$L_3 = P(f^{(3)}(a)), \Box, L_2.$

As $\Box$ is contained in L_3 DFS stops, the while loop terminates and we obtain refutation = true.

The clauses produced by BFS are different from those of DFS. While BFS finds $\Box$ on level 2, the refutation found by DFS corresponds to level 3. It is obvious that BFS always finds $\Box$ on the minimal level, while DFS in general does not. Moreover we were lucky to find a refutation by DFS at all. To illustrate the problem of nontermination we modify C to

$$C_1 : \{P(a),\ P(f(x_1)) \vee \neg P(x_1),\ R(a) \vee \neg P(f(a)),\ \neg R(a)\}.$$

Clearly C_1 is unsatisfiable and (again) $\Box \in S_\emptyset^2(C_1)$. For DFS we transform C_1 into the list

$$L : P(a),\ P(f(x_1)) \vee \neg P(x_1),\ R(a) \vee \neg P(f(a)),\ \neg R(a).$$

It is easy to verify that DFS produces the following sequence of lists L_i (in resolving from left to right):

$$\begin{aligned} L_1 &= P(f(a)),\ L \\ L_2 &= P(f^{(2)}(a)),\ R(a),\ P(f(a)),\ L \\ &\vdots \\ L_n &= P(f^{(n)}(a)), \ldots, P(f(2)(a)),\ R(a),\ P(f(a)),\ L \\ &\vdots \end{aligned}$$

Thus the list is properly extended infinitely many times (the sets $\mathcal{D}$ in DFS are always nonempty) and we never come to the point of resolving $R(a)$ and $\neg R(a)$ (although both clauses are in L_i for $i \geq 2$). If, on the other hand, we turn $\mathcal{C}_1$ into the list

$$L' : \neg R(a),\ R(a) \vee \neg P(f(a)),\ P(f(x_1)) \vee \neg P(x_1),\ \neg R(a)$$

then DFS stops and yields $\square$.

So, unlike BFS, DFS is an incomplete search method and its success depends on the ordering of clauses in the lists.

Although the order-dependence of (termination of) depth-first search might make it appear pathological, it can be practically useful if subjected to additional restrictions. So we may restrict depth first search to levels $\leq k$ or apply iterative deepening [BF81]; using the first restriction we either find $\square$ or we produce $S_x^k(\mathcal{C})$ as a whole (but in a order different from that defined by BFS). We then can obtain a complete search procedure by successively increasing k.

A measure characterizing the efficiency of a theorem proving program is the number of clauses generated before $\square$ is found. This number can be very high (even if relatively short refutations exist) and its reduction is one of the central problems in automated theorem proving. Obviously such a measure must depend on the search method, the refinement, and the deletion methods. We give a formal definition of search complexity on the basis of breadth-first search:

Definition 4.1.1 (search complexity). *Let R_x be a complete resolution operator and $\mathcal{C}$ be an unsatisfiable set of clauses. We define $d_x(\mathcal{C}) = min\{i | \square \in S_x^i(\mathcal{C})\}$ (the minimal refutation level). Then the* search complexity *of R_x with respect to $\mathcal{C}$ is defined as $CS_x(\mathcal{C}) = |S_x^{d_x(\mathcal{C})}(\mathcal{C})|$.*

CS_x is the number of clauses produced by breadth-first search by exhausting the set of all clauses appearing on or below the minimal refutation level. CS_x is not the exact number of clauses generated during (a reasonably implemented) BFS because we might stop before the whole level $d_x(\mathcal{C})$ is exhausted. In our definition, however, we added all clauses appearing on the minimal refutation level to avoid order dependence. Thus CS_x is in fact an upper bound to the search expense created by BFS.

If R_x is more refined than R_y (i.e., $S_x^i(\mathcal{C}) \subseteq S_y^i(\mathcal{C})$ for all $\mathcal{C} \in CL$ and $i \in \mathbb{N}$) then, clearly, fewer clauses are generated by R_x than by R_y. However, as a kind of counterbalance, we may obtain $d_x(\mathcal{C}) > d_y(\mathcal{C})$ for some unsatisfiable set of clauses $\mathcal{C}$. In such a case the beneficial effect of the stronger refinement may be destroyed by the increase of the minimal refutation level. This means that strengthening refinements does not always reduce search complexity.

Example 4.1.3. Let $\mathcal{C}$ be the following set of N_s-clauses

$$\{P(x_1) \vee R(x_1), \ R(x_1) \vee \neg P(x_1), \ P(x_1) \vee \neg R(x_1), \ \neg P(x_1) \vee \neg R(x_1)\}.$$

We first compute $CS_{\emptyset}$, the search complexity under unrefined resolution. For this purpose we have to compute the sets $S_{\emptyset}^i$.

$$S_{\emptyset}^1(\mathcal{C}) = \mathcal{C} \cup \{R(x_1), \ P(x_1), \ \neg R(x_1), \ \neg P(x_1), \ P(x_1) \vee \neg P(x_1), \ R(x_1) \vee \neg R(x_1)\}.$$

Moreover

$$\begin{aligned}
S_{\emptyset}^2 &= S_{\emptyset}^1(\mathcal{C}) \cup \{\Box\} \text{ and thus} \\
CS_{\emptyset}(\mathcal{C}) &= |S_{\emptyset}^2(\mathcal{C})| = 11.
\end{aligned}$$

We now apply a locking l to $\mathcal{C}$ and compute the sets S_l^i. Clearly $\mathrm{unlock}(S_l^i(\mathcal{C})) \subseteq S_{\emptyset}^i(\mathcal{C})$ for all $i \in \mathbb{N}$ (note that by the different forms of clause representation we need the operator unlock in order to compare $R_{\emptyset}$ and R_l). The locked input set $\mathcal{C}$ is represented by $\mathcal{C}_1$.

$$\mathcal{C}_1 = \{\overset{1}{R}(x_1) \vee \overset{2}{P}(x_1), \ \overset{3}{\neg P}(x_1) \vee \overset{6}{R}(x_1), \ \overset{4}{P}(x_1) \vee \overset{7}{\neg R}(x_1), \ \overset{2}{\neg R}(x_1) \vee \overset{8}{\neg P}(x_1)\}.$$

$$S_l^1(\mathcal{C}_1) = \mathcal{C}_1 \cup \{\overset{5}{P}(x_1) \vee \overset{8}{\neg P}(x_1), \ \overset{6}{R}(x_1) \vee \overset{7}{\neg R}(x_1)\}.$$

$$S_l^2(\mathcal{C}_1) = S_l^1(\mathcal{C}_1) \cup \{\overset{6}{R}(x_1) \vee \overset{8}{\neg P}(x_1), \ \overset{7}{\neg R}(x_1) \vee \overset{8}{\neg P}(x_1)\}.$$

$$S_l^3(\mathcal{C}_1) = S_l^2(\mathcal{C}_1) \cup \{\overset{8}{\neg P}(x_1)\}.$$

$$S_l^4(\mathcal{C}_1) = S_l^3(\mathcal{C}_1) \cup \{\overset{7}{\neg R}(x_1)\}.$$

$$S_l^5(\mathcal{C}_1) = S_l^4(\mathcal{C}_1) \cup \{\overset{5}{P}(x_1)\} \text{ and, eventually,}$$

$$S_l^6(\mathcal{C}_1) = S_l^5(\mathcal{C}_1) \cup \{\overset{6}{R}(x_1), \Box\}.$$

Hence we obtain $CS_l(\mathcal{C}_1) = |S_l^6(\mathcal{C}_1)| = 13$.

While, by unrestricted resolution, $\Box$ is found on level 2, level 6 is required by lock resolution. Thus although $\mathrm{unlock}(S_l^i(\mathcal{C})) \subseteq S_{\emptyset}^i(\mathcal{C})$ for $i \le 5$, lock resolution (in this example!) causes higher search expenses than unrestricted resolution. That does not mean that refinements are altogether useless, but only that we face the problem of choosing the "right one". For $\mathcal{C}$, hyperresolution is a "good" refinement; we show this by computing $CS_H(\mathcal{C})$:

$$S_H^1(\mathcal{C}) = \mathcal{C} \cup \{R(x_1), \ P(x_1)\}.$$

As the clash $(\neg P(x_1) \vee \neg R(x_1), R(x_1), P(x_1))$ gives $\Box$ we obtain

$$S_H^2(\mathcal{C}) = S_H^1(\mathcal{C}) \cup \{\Box\} \text{ and therefore}$$

$$CS_H(\mathcal{C}) = |S_H^2(\mathcal{C})| = 7.$$

While refinements may increase the minimal refutation level, typical redundancy methods (like subsumption and elimination of tautologies) do not. We will see that some of these redundancy methods even define refinements (in the formal sense such as defined in Section 3.1). Indeed methods like subsumption always lead to a decrease of search complexity, but the problem of expensive redundancy tests may be a serious one. We illustrate the phenomenon of redundancy elimination by presenting tautology-elimination. The principle is very simple: If a tautological clause is produced during deduction then it is deleted immediately. We will see later that this deletion technique preserves completeness for most of the relevant refinements.

Again let us denote the operator of unrestricted resolution on N_c-clauses by $R_\emptyset$. By $\text{TAUT}(\mathcal{C})$ we denote the set of all tautological clauses in $\mathcal{C}$ (a clause is a tautology iff it contains a pair of complementary literals L, L^d). Elimination of tautologies is described by the operator

$$\text{TAUTEL}(\mathcal{C}) = \mathcal{C} - \text{TAUT}(\mathcal{C}).$$

Then $R_\emptyset$ may be refined to $R_{\emptyset_t}$ by including the elimination of tautologies:

$$S_{\emptyset_t}^0(\mathcal{C}) = \text{TAUTEL}(\mathcal{C}), \quad S_{\emptyset_t}^{i+1}(\mathcal{C}) = S_{\emptyset_t}^i(\mathcal{C}) \cup \text{TAUTEL}(N_s(\rho_\emptyset(S_t^i(\mathcal{C})))),$$

$$R_{\emptyset_t}(\mathcal{C}) = \bigcup_{i=0}^\infty S_{\emptyset_t}^i(\mathcal{C}).$$

If $\mathcal{C}$ is reduced under elimination of tautologies, i.e., $\text{TAUTEL}(\mathcal{C}) = \mathcal{C}$ then $R_{\emptyset_t}$ is a refinement operator such as defined in Section 3.1 and $S_{\emptyset_t}^i(\mathcal{C}) \subseteq S_\emptyset^i(\mathcal{C})$ for all $i \in \mathbb{N}$. Note that we have to ensure the completeness of $R_{\emptyset_t}$ (this property will be shown in Section 4.4). We will use the former Example 4.1.3 to illustrate the reduction of search space by elimination of tautologies.

Example 4.1.4.
$\mathcal{C} = \{P(x_1) \vee R(x_1),\ R(x_1) \vee \neg P(x_1),\ P(x_1) \vee \neg R(x_1),\ \neg P(x_1) \vee \neg R(x_1)\}.$

In Example 4.1.3 we computed $CS_\emptyset$ and found $CS_\emptyset(\mathcal{C}) = |S_\emptyset^2(\mathcal{C})| = 11$.

Among these 11 clauses are the tautologies $P(x_1) \vee \neg P(x_1)$ and $R(x_1) \vee \neg R(x_1)$. These clauses are not contained in $R_{\emptyset_t}(\mathcal{C})$. In particular we obtain

$$S_{\emptyset_t}^1(\mathcal{C}) = \{R(x_1),\ P(x_1),\ \neg R(x_1),\ \neg P(x_1)\} \cup \mathcal{C}$$

by $\text{TAUTEL}(S_\emptyset^1(\mathcal{C})) = S_\emptyset^1(\mathcal{C}) - \{P(x_1) \vee \neg P(x_1),\ R(x_1) \vee \neg R(x_1)\}$.

Moreover $S_{\emptyset_t}^2(\mathcal{C}) = S_{\emptyset_t}^1(\mathcal{C}) \cup \{\Box\}$. We thus obtain $CS_{\emptyset_t}(\mathcal{C}) = 9$.

In Example 4.1.4 the practical value of tautology-elimination is questionable

(even in determining $R_{\emptyset_t}$ we have to compute the tautologies before eliminating them). In general, however, tautologies may produce further redundant (not necessarily tautological) clauses, which may lead to a substantial
increase of search space. In automated theorem proving programs the elimination of tautologies is always profitable, provided it preserves completeness;
the reason is twofold:

Testing the TAUT-property is fast and
$CS_{xt}(\mathcal{C}) \subseteq CS_x(\mathcal{C})$ for most relevant refinements (including hyperresolution
and A-ordering).

We will see later that lock resolution does not admit the elimination of
tautologies (without loss of completeness). Thus we cannot reduce the search
complexity of lock resolution in Example 4.1.3 by eliminating the tautologies.
Thus an improvement, as in Example 4.1.4 for $R_{\emptyset}$, is impossible for lock
resolution.

4.2 The Subsumption Principle

Unlike tautologies that are "absolutely redundant", subsumed clauses are
redundant with respect to other clauses. During deduction it may happen
that clauses are derived which are instances of previously derived clauses (or
contain instances of such clauses). Because the philosophy of resolution is
to work on the most general level only, less general clauses can be regarded
as redundant; as a consequence such clauses can be deleted without loss of
completeness. Moreover we will see that removing subsumed clauses increases
neither proof- nor search complexity.

Example 4.2.1. Let $\mathcal{C}$ be the set of clause $\{C_1, \ldots, C_5\}$ with

$$C_1 = P(x, f(x)) \vee R(x),\ C_2 = Q(x) \vee P(x, y),\ C_3 = \neg R(f(x)) \vee Q(f(x)),$$
$$C_4 = \neg Q(y) \vee \neg R(y)\ \text{and}\ C_5 = \neg P(x, f(x)).$$

Consider the R-refutation Γ of $\mathcal{C}$ in Figure 4.2 and the clause C_6 appearing in
Γ. It is easy to verify that $F(\{C_2\}) \to F(\{C_6\})$ is valid. In fact C_6 can be obtained from C_2 by first applying the substitution $\vartheta : \{x \leftarrow f(y), y \leftarrow f(f(y))\}$
to C_2 and then commuting the literals. In this sense C_6 is an instance of C_2
and we obtain a shorter refutation Γ_1 in replacing C_6 by C_2 (see Figure 4.3).
The refutation Γ_1 is not only shorter than Γ, it is also more general; note
that every clause in Γ (from C_6 downwards) is an instance of a clause in Γ_1,
where C_7, C_8 are proper instances of C_7', C_8'.

By the subsumption principle (to be defined formally below) we recognize the
redundancy of C_6 (with respect to C_2) and we do not continue a derivation
containing C_6. There are different ways to modify derivations: Either we try

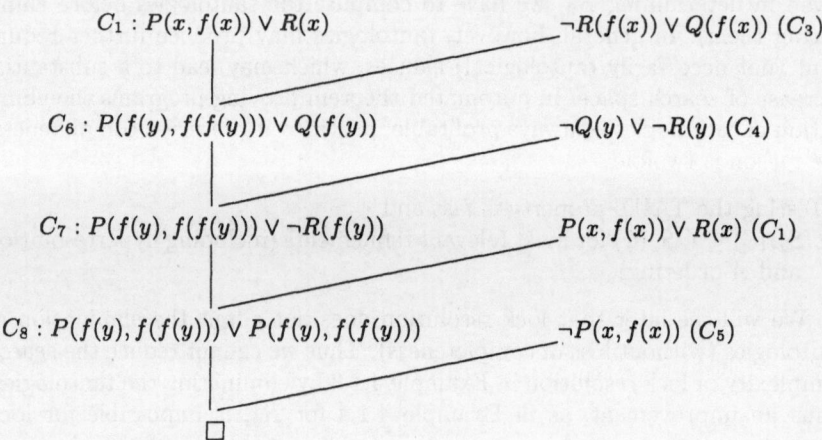

Fig. 4.2. The refutation Γ

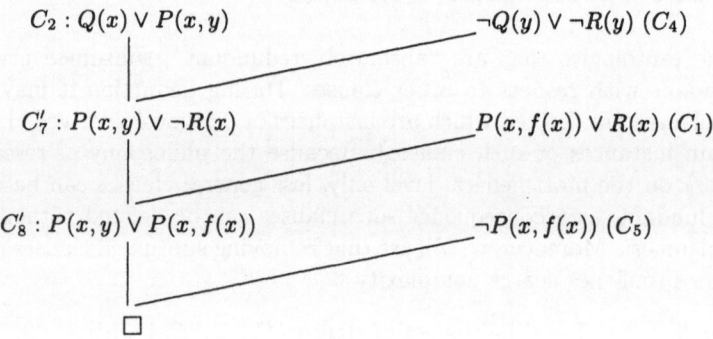

Fig. 4.3. The refutation Γ_1

to derive a new resolvent using the clause C_1 or we replace C_6 by C_2 and then start a new derivation.

In general we consider clauses as redundant not only if they *are* instances of other clauses, but also if they contain instances of other clauses.

Definition 4.2.1 (subsumption). *A clause C subsumes a clause D if there exists a substitution ϑ such that $\mathrm{LIT}(C)\vartheta \subseteq \mathrm{LIT}(D)$ (we write $C \leq_{ss} D$).*

Example 4.2.2. Let $C_1 = P(x) \vee P(f(x))$, $C_2 = P(f(y)) \vee P(f(f(y))) \vee R(y)$, $C_3 = P(a) \vee P(f(f(a)))$ and $C_4 = P(f(y)) \vee P(y) \vee P(f(z))$.
$C_1 \leq_{ss} C_2$ because

$$\{P(x), P(f(x))\}\vartheta \subseteq \{P(f(y)), P(f(f(y))), R(y)\}$$

by $\vartheta = \{x \leftarrow f(y)\}$.

$C_1 \not\leq_{ss} C_3$: Indeed there is no ϑ such that $\{P(x), P(f(x))\}\vartheta \subseteq \{P(a), P(f(f(a)))\}$.

For suppose $\{P(x), P(f(x))\}\vartheta \subseteq \mathrm{LIT}(C_3)$. Then $P(x)\vartheta \in \mathrm{LIT}(C_3)$. There are two possibilities:

a) $P(x)\vartheta = P(a)$: Then $P(f(x))\vartheta = P(f(a)) \notin \mathrm{LIT}(C_3)$,
b) $P(x)\vartheta = P(f(f(a)))$: Then $P(f(x))\vartheta = P(f^{(3)}(a)) \notin \mathrm{LIT}(C_3)$.

Thus we conclude that C_1 does not subsume C_3. This last example shows that the subsumption test cannot be "parallelized" (note that $P(x) \leq_{ss} C_3$ and $P(f(x)) \leq_{ss} C_3$, but $P(x) \vee P(f(x)) \not\leq_{ss} C_3$).

It is easy to see that neither C_2 nor C_3 subsumes C_1. For C_1, C_4 we even get $C_1 =_{ss} C_4$ (i.e., $C_1 \leq_{ss} C_4$ and $C_4 \leq_{ss} C_1$). $C_1 \leq_{ss} C_4$ is trivial via $\vartheta = \{x \leftarrow y\}$. Now let $\vartheta = \{y \leftarrow x, z \leftarrow x\}$. We then obtain

$$\mathrm{LIT}(C_4)\vartheta = \{P(f(y)), P(y), P(f(z))\}\vartheta = \{P(x), P(f(x))\} = \mathrm{LIT}(C_1)$$

and thus $C_4 \leq_{ss} C_1$.

Here we see that bigger clauses may subsume smaller ones by internal unification of literals. In fact ϑ is a unifier of $\{P(f(y)), P(f(z))\}$ giving $\{P(f(x))\}$.

We now discuss some mathematical properties of $\leq_{ss}$ (e.g. its decidability) and show some relations among $\leq_{ss}, \leq_s$, and $\leq_{sc}$.

Proposition 4.2.1. $\leq_{ss}$ *fulfills the following properties:*

a) *reflexivity*
b) *transitivity,*
c) *If $C \leq_{ss} D$ then $C \leq_{ss} D\vartheta$ for all substitutions ϑ,*
d) *If $C \leq_{ss} D$ then $F(\{C\}) \rightarrow F(\{D\})$ is valid.*

Proof. a) Choose the empty substitution.
b) If $\mathrm{LIT}(C)\vartheta \subseteq \mathrm{LIT}(D)$ and $\mathrm{LIT}(D)\eta \subseteq \mathrm{LIT}(E)$ then also $\mathrm{LIT}(C)\vartheta\eta \subseteq \mathrm{LIT}(E)$.
c) If $\mathrm{LIT}(C)\eta \subseteq \mathrm{LIT}(D)$ then $\mathrm{LIT}(C)\eta\vartheta \subseteq \mathrm{LIT}(D\vartheta)$.
d) Clearly $F(\{C\}) \rightarrow F(\{C\vartheta\})$ is valid for every ϑ. Moreover $F(\{C\vartheta\}) \rightarrow F(\{D\})$ is valid for $C\vartheta \subseteq D$ (note that clauses represent universally quantified disjunctions).   ◇

Proposition 4.2.2. *Let C, D be clauses and $V(D) = \{x_1, \ldots, x_m\}$. Choose new constant symbols $c_1, \ldots, c_m$ which are not contained in $\{C, D\}$ and define $\vartheta = \{x_1 \leftarrow c_1, x_m \leftarrow c_m\}$. Then $C \leq_{ss} D$ iff $C \leq_{ss} D\vartheta$.*

Proof. a) Suppose that $C \leq_{ss} D$ holds. From Proposition 4.2.1(c) we obtain $C \leq_{ss} D\vartheta$.

b) Let us assume $C \leq_{ss} D\vartheta$.

Then there exists a substitution γ fulfilling $\mathrm{LIT}(C)\gamma \subseteq D\vartheta$. Let $V(C) = \{y_1, \ldots, y_n\}$ and $\gamma = \{y_1 \leftarrow t_1, \ldots, y_n \leftarrow t_n\}$ (there is no loss of generality in restricting $dom(\gamma)$ to $V(C)$). Note that γ must replace all variables in C because $D\vartheta$ is ground. Let $s_i = t_i[c_1/x_1, \ldots, c_n/x_n]$ (where s/t denotes the replacement of all occurrences of the term s by the term t) for $i = 1, \ldots, n$ and

$$\mu = \{y_1 \leftarrow s_1, \ldots, y_n \leftarrow s_n\}.$$

Clearly

$$\mathrm{LIT}(C)\mu \subseteq \mathrm{LIT}(D\vartheta[c_1/x_1, \ldots, c_n/x_n]).$$

But $(D\vartheta[c_1/x_1, \ldots, c_n/x_n]) = D$ (here it is important that the c_i are indeed new symbols). So we obtain $\mathrm{LIT}(C)\mu \subseteq \mathrm{LIT}(D)$, i.e., $C \leq_{ss} D$.

◊

Proposition 4.2.2 tells us that, in testing $C \leq_{ss} D$, we may assume D to be ground without loss of generality. This assumption makes formulation and analysis of subsumption algorithms easier. Another application can be found in the proof of the following proposition.

Proposition 4.2.3. *Subsumption is decidable.*

Proof. In order to test $C \leq_{ss} D$ we may replace D by some "ground version" D' according to Proposition 4.2.2. For technical reasons we assume that D is already ground.

Let $d = \tau(D)$ (= the maximal depth of a term occurring in D) and let H be the Herbrand universe of D. Let

$$\Theta = \{\vartheta | \vartheta \in \mathrm{SUBST}, dom(\vartheta) = V(C), rg(\vartheta) \subseteq H\}.$$

Trivially, $C \leq_{ss} D$ iff there exists a $\vartheta \in \Theta$ such that $\mathrm{LIT}(C)\vartheta \subseteq \mathrm{LIT}(D)$. If $\vartheta \in \Theta$ and $\tau(rg(\vartheta)) > d$ then clearly $\tau(C\vartheta) > d$ and $\mathrm{LIT}(C\vartheta) \not\subseteq \mathrm{LIT}(D)$.

Let us define $\Theta_d = \{\vartheta | \vartheta \in \Theta; \tau(rg(\vartheta)) \leq d\}$. Then $C \leq_{ss} D$ iff there exists a $\vartheta \in \Theta_d$ such that $\mathrm{LIT}(C)\vartheta \subseteq \mathrm{LIT}(D)$. But the set Θ_d is finite. We thus obtain a decision procedure for $\leq_{ss}$ by testing $\mathrm{LIT}(C)\vartheta \subseteq \mathrm{LIT}(D)$ for $\vartheta \in \Theta_d$.   ◊

The decision procedure for $\leq_{ss}$ which can be extracted from the proof of Proposition 4.2.3 is very inefficient and not suited for practical use. In Section 4.3 we will present more efficient subsumption algorithms which can actually be applied within theorem proving programs.

We now compare the subsumption relation $\leq_{ss}$ with the relations $\leq_s$ (Definition 2.7.7) and $\leq_{sc}$ (Definition 3.3.7).

Definition 4.2.2. *Let $\leq_1$ and $\leq_2$ be reflexive and transitive relations on the set of all clauses. We call $\leq_1$ stronger than $\leq_2$ if for all clauses C, D $C \leq_2 D$ implies $C \leq_1 D$. $\leq_1$ is called strictly stronger than $\leq_2$ if $\leq_1$ is stronger than $\leq_2$, but not vice versa.*

Let $\leq_r$ be an arbitrary binary relation which is reflexive and transitive. We write $C =_r D$ iff $C \leq_r D$ and $D \leq_r C$.

Proposition 4.2.4. *$\leq_{ss}$ is strictly stronger than $\leq_s$ and $\leq_{sc}$.*

Proof. As $\leq_{sc}$ is strictly stronger than $\leq_s$ it suffices to prove that $\leq_{ss}$ is strictly stronger than $\leq_{sc}$.

Recall that $C \leq_{sc} D$ iff there exists a substitution ϑ such that $N_c(C\vartheta) = N_c(D)$, where N_c is the normalization operator defined as $N_s \circ \gamma$ where γ is the condensation function from Definition 3.2.2.

We first prove that $\leq_{ss}$ is stronger than $\leq_{sc}$, i.e., that $C \leq_{sc} D$ implies $C \leq_{ss} D$.

So let us assume $C \leq_{ss} D$: By definition of the condensation function γ,

$$\text{LIT}(\gamma(C\vartheta)) \subseteq \text{LIT}(C\vartheta) \text{ and } \gamma(C\vartheta) \text{ is a factor of } C\vartheta.$$

Thus there exists a factoring substitution σ such that $\text{LIT}(C\vartheta\sigma) = \text{LIT}(\gamma(C\vartheta))$. By definition of the operator N_s (recall $N_s = N_r \circ N_v \circ N_0$),

$$N_s(C) = N_s(D) \quad \text{iff} \quad \text{LIT}(C) = \text{LIT}(D).$$

Thus
$$N_s(\gamma(C\vartheta)) = N_s(\gamma(D))$$

implies
$$\text{LIT}(\gamma(C\vartheta)) = \text{LIT}(\gamma(D)).$$

But
$$\text{LIT}(\gamma(C\vartheta)) = \text{LIT}(C\vartheta\sigma)$$

and therefore
$$\text{LIT}(C\vartheta\sigma) = \text{LIT}(\gamma(D)).$$

By definition of γ we have $\text{LIT}(\gamma(D)) \subseteq \text{LIT}(D)$. Combining all the arguments above we get $\text{LIT}(C\vartheta\sigma) \subseteq \text{LIT}(D)$, i.e., $C \leq_{ss} D$.

It remains to show that $\leq_{ss}$ is strictly stronger:

$$P(x) \leq_{ss} P(a) \vee Q(a) \text{ but not } P(x) \leq_{sc} P(a) \vee Q(a).$$

$\Diamond$

The simplest use of subsumption is that of preprocessing, that means to reduce a set of clauses under subsumption *before* resolution deduction actually takes place. The following proposition shows that such a reduction is semantically justified. For technical reasons we extend $\leq_{ss}$ from clauses to sets of clauses.

Definition 4.2.3. *Let C, D be sets of clauses. We say that C subsumes D (and write $C \leq_{ss} D$) iff for all $D \in D$ there exists a $C \in C$ such that $C \leq_{ss} D$.*

Proposition 4.2.5. *Let C be a set of clauses and $C \in C$ such that $C - \{C\} \leq_{ss} \{C\}$. Then $F(C - \{C\})$ is logically equivalent to $F(C)$.*

Proof. By definition of the operator F, $F(C) \to F(C - \{C\})$ is trivially valid.

By Definition 4.2.3 $C - \{C\} \leq_{ss} \{C\}$ implies the existence of a $D \in C - \{C\}$ with $D \leq_{ss} C$. By Proposition 4.2.1(d) we know that $F(\{D\}) \to F(\{C\})$ is valid.

The sentence $F(C - \{C\}) \to F(C - \{C\})$ is trivially valid and the validity of $F(C - \{C\}) \to F(\{C\})$ follows from the argument above. Therefore $F(C - \{C\}) \to F(C)$ is valid too.   ◊

By Proposition 4.2.5 we may delete subsumed clauses from the (input) set of clauses C until we obtain a set of clauses C' which is subsumption-reduced, i.e., for every $C_1, C_2 \in C'$, $C_1 \leq_{ss} C_2$ implies $C_1 = C_2$. Besides the advantage that C' is in general smaller than C (and therefore $R_x(C') \subseteq R_x(C)$ for every refinement operator R_x), removing subsumed clauses does not increase the minimal proof-length. Therefore, replacing C by C' always reduces proof search. Moreover, we never have problems concerning completeness (note that $\Box \in R_x(C')$ iff $\Box \in R_x(C)$).

More important (and more delicate) than subsumption as preprocessing is its use *during* deduction. In combining subsumption and resolution we (roughly) have the following alternatives:

1. *Forward subsumption,*
2. *backward subsumption,* and
3. (total) *replacement.*

Forward subsumption means removing newly derived clauses which are subsumed by clauses derived before.

In backward subsumption a clause C which was derived before, but is subsumed by a clause C' obtained afterwards, is rendered inactive. If the output of the search procedure should be a deduction (or a deduction tree) then we cannot simply delete C without destroying the structure of the (newly produced) deduction. Instead C is reactivated if C' is no longer an ancestor of the actual clause in the search process.

In case of replacement the set of derived clauses – as a whole – is (periodically) reduced under subsumption. If used in this way, subsumption essentially depends on dynamical features of deductions (the terms "before" and "afterwards" point to this phenomenon); thus the effect of subsumption essentially depends on the method of search. In this chapter we use breadth-first search as basis, a method which is naturally compatible with the operator formalism for resolution. Only forward subsumption and replacement will be treated in detail. It will turn out that forward subsumption always defines a refinement (in the sense of Section 3.1), while replacement does not (instead it is an iterated reduction method).

We first analyze forward subsumption and investigate its impact on the completeness of refinements and on search complexity.

Definition 4.2.4 (forward subsumption). *Let C, D be sets of clauses. We first define the subset of clauses in D which are not subsumed by C:*

$$sf(C, D) = \{D | D \in D, C \not\leq_{ss} \{D\}\}.$$

Let R_x be a resolution refinement operator on N-clauses for some normalization principle N. We define

$$S_{xs}^0(C) \;=\; C,$$

$$S_{xs}^{i+1}(C) \;=\; S_{xs}^i(C) \cup sf(S_{xs}^i(C), \rho_x(S_{xs}^i(C))),$$

$$R_{xs}(C) \;=\; \bigcup_{i=0}^{\infty} S_{xs}^i(C).$$

R_{xs} is called the refinement R_x "under forward subsumption".

By definition of R_{xs} and of sf we always obtain

$$R_{xs}(C) \subseteq R_x(C) \text{ for } C \in CL.$$

Therefore R_{xs} is always more refined than R_x. The question remains whether R_{xs} still is complete (provided R_x is) and whether forward subsumption improves search complexity. We will see that, in most relevant cases, we can guarantee these beneficial properties.

For the completeness results to be proved in this chapter it is of central importance that subsumption is "preserved" under resolution.

Lemma 4.2.1. *Let $C_1, D_1, C_2,$ and D_2 be clauses such that $C_1 \leq_{ss} D_1$ and $C_2 \leq_{ss} D_2$. Let D be a resolvent of D_1 and D_2. Then one of the following properties holds:*

(a) $C_1 \leq_{ss} D$, or
(b) $C_2 \leq_{ss} D$, or
(c) there exists a resolvent C of C_1 and C_2 such that $C \leq_{ss} D$.

Proof. Because D is resolvent of D_1, D_2 there exist (variable-disjoint variants of) factors $E_1 \vee M \vee F_1$ of D_1 and $E_2 \vee N \vee F_2$ of D_2 such that $\{M, N^d\}$ is unifiable by an m.g.u. σ. So D is of the form:

$$D : (E_1 \vee F_1 \vee E_2 \vee F_2)\sigma.$$

Now we have $C_1 \leq_{ss} D_1$, $C_2 \leq_{ss} D_2$ and $E_1 \vee M \vee F_1$, $E_2 \vee N \vee F_2$ are factors of D_1 and D_2 respectively. By Exercise 4.2.6 we obtain

$$C_1 \leq_{ss} E_1 \vee M \vee F_1 \text{ and } C_2 \leq_{ss} E_2 \vee N \vee F_2.$$

By $C_1 \leq_{ss} E_1 \vee M \vee F_1$ and by definition of $\leq_{ss}$ there exists a substitution ϑ_1 such that

$$\mathrm{LIT}(C_1)\vartheta_1 \subseteq \mathrm{LIT}(E_1 \vee M \vee F_1).$$

Similarly there exists a substitution ϑ_2 with

$$\mathrm{LIT}(C_2)\vartheta_2 \subseteq \mathrm{LIT}(E_2 \vee N \vee F_2).$$

case a) : $M \notin \mathrm{LIT}(C_1)\vartheta_1$ or $N \notin \mathrm{LIT}(C_2)\vartheta_2$.
Let us suppose $M \notin \mathrm{LIT}(C_1)\vartheta_1$. Then $\mathrm{LIT}(C_1)\vartheta_1 \subseteq \mathrm{LIT}(E_1 \vee F_1)$ and thus $C_1 \leq_{ss} E_1 \vee F_1$. Moreover we have

$$E_1 \vee F_1 \leq_{ss} (E_1 \vee F_1)\sigma \leq_{ss} (E_1 \vee F_1 \vee E_2 \vee F_2)\sigma = D.$$

By transitivity of $\leq_{ss}$ we get $C_1 \leq_{ss} D$ and Lemma 4.2.1(a) holds.
If $N \notin \mathrm{LIT}(C_2)\vartheta_2$ a completely analogous argument yields $C_2 \leq_{ss} E_2 \vee F_2$ and $C_2 \leq_{ss} D$ (Lemma 4.2.1(b)).

case b) : $M \in \mathrm{LIT}(C_1)\vartheta_1$ and $N \in \mathrm{LIT}(C_2)\vartheta_2$. Let $\mathcal{L}_1 = \{L_1, \ldots, L_m\}$ be the set of all literals in $\mathrm{LIT}(C_1)$ with $L\vartheta_1 = M$.
In the same way we define $\mathcal{L}_2$ for C_2 and ϑ_2. Then ϑ_1 is a unifier of $\mathcal{L}_1$ and ϑ_2 is a unifier of $\mathcal{L}_2$. By the unification theorem there exist m.g.u.'s λ_1 of $\mathcal{L}_1$ and λ_2 of $\mathcal{L}_2$. Therefore the clauses $C_1\lambda_1$ and $C_2\lambda_2$ are G-instances of C_1 and of C_2, respectively. Moreover there are corresponding factors (i.e., p-reducts of G-instances)
$G_1 \vee R \vee H_1$ of C_1 and $G_2 \vee S \vee H_2$ of C_2
fulfilling the following properties:
$R \leq_s M,\ S \leq_s N$ and

$$G_1 \vee H_1 \leq_{ss} E_1 \vee F_1,\ G_2 \vee H_2 \leq_{ss} E_2 \vee F_2.$$

We even know that $R \leq_s M$ and $G_1 \vee H_1 \leq_{ss} E_1 \vee F_1$ via a common substitution (the same holds for $S \leq_s N, G_2 \vee H_2 \leq_{ss} E_2 \vee F_2$).
So let η_1, η_2 be defined by

$$\lambda_1\eta_1 = \vartheta_1 \quad \text{and} \quad \lambda_2\eta_2 = \vartheta_2.$$

By definition of the resolvent D, σ is m.g.u. of $\{M, N^d\}$. Let $\eta = \eta_1 \cup \eta_2$ (note that such a definition is possible as $E_1 \vee M \vee F_1$ and $E_2 \vee N \vee F_2$ are variable-disjoint). By $R\eta = M$ and $S\eta = N$ the set $\{R, S^d\}$ is unifiable by the substitution $\eta\sigma$.
By the unification theorem there exists an m.g.u. τ of $\{R, S^d\}$. Because τ is m.g.u. there must be a substitution ρ such that

$$\tau\rho = \eta\sigma.$$

We thus obtain the following relations:

$$R\tau\rho = M\sigma,\ S\tau\rho = N\sigma \text{ and}$$

$\mathrm{LIT}(G_1 \vee H_1)\tau\rho \subseteq \mathrm{LIT}(E_1 \vee F_1)\sigma, \ \mathrm{LIT}(G_2 \vee H_2)\tau\rho \subseteq \mathrm{LIT}(E_2 \vee F_2)\sigma.$

But the clause
$$C : (G_1 \vee H_1 \vee G_2 \vee H_2)\tau$$
is a resolvent of C_1 and C_2 and

$$\mathrm{LIT}(G_1 \vee H_1 \vee G_2 \vee H_2)\tau\rho \subseteq \mathrm{LIT}(E_1 \vee F_1 \vee E_2 \vee F_2)\sigma.$$

But that means $C \leq_{ss} D$ (Lemma 4.2.1(c)). $\Diamond$

A refinement type compatible with forward subsumption is $R_{<_A}$ for arbitrary A-orderings $<_A$.

Theorem 4.2.1. *Any A-ordering refinement is complete under forward subsumption, or more formally: Let $\mathcal{C}$ be an unsatisfiable set of N_c-clauses and $<_A$ be an arbitrary A-ordering. Then $\square \in R_{<_A^s}(\mathcal{C})$.*

Proof. It suffices to show that, for all i, the i-th level of clauses generated by $R_{<_A^s}$ subsumes that generated by $R_{<_A}$:

(*) For all $i \in \mathbb{N} : S^i_{<_A^s}(\mathcal{C}) \leq_{ss} S^i_{<_A}(\mathcal{C}).$

Suppose that (*) has already been proved. By Theorem 3.3.1 we know that $R_{<_A}$ is complete and thus $\square \in R_{<_A}(\mathcal{C})$. By definition of $R_{<_A}$ there exists a $k \in \mathbb{N}$ such that $\square \in S^k_{<_A}(\mathcal{C})$. From (*) we get $S^k_{<_A^s}(\mathcal{C}) \leq_{ss} S^k_{<_A}(\mathcal{C})$. But $\square$ can only be subsumed by $\square$ itself and therefore $\square \in S^k_{<_A^s}(\mathcal{C})$. So we obtain $\square \in S^k_{<_A^s}(\mathcal{C})$ and it remains to prove (*). We do so by induction on i.

$i = 0$:
$S^0_{<_A^s}(\mathcal{C}) = S^0_{<_A}(\mathcal{C}) = \mathcal{C}$ and $S^0_{<_A^s}(\mathcal{C}) \leq_{ss} S^0_{<_A}(\mathcal{C})$ trivially holds.

(IH): Suppose that $S^i_{<_A^s}(\mathcal{C}) \leq_{ss} S^i_{<_A}(\mathcal{C})$.

Let $D \in S^{i+1}_{<_A}(\mathcal{C})$.
We have to find a $C \in S^{i+1}_{<_A^s}(\mathcal{C})$ such that $C \leq_{ss} D$. By definition of the S_i we have:
$$S^{i+1}_{<_A}(\mathcal{C}) = S^i_{<_A}(\mathcal{C}) \cup \rho_{<_A}(S^i_{<_A}(\mathcal{C})).$$
If $D \in S^i_{<_A}(\mathcal{C})$ then the required C exists by (IH). So we may assume $D \in \rho_{<_A}(S^i_{<_A}(\mathcal{C}))$.

By definition of $\rho_{<_A}$, $D \in \rho_{<_A}(S^i_{<_A}(\mathcal{C}))$ iff there are clauses $D_1, D_2 \in S^i_{<_A}(\mathcal{C})$ such that $D \in \rho_{<_A}(\{D_1, D_2\})$. Then $D = N_c(D_0)$ such that D_0 is a resolvent of D_1 and D_2 with the following property:
There exists no literal L in D_0 such that $A >_A L$, where A is the resolved atom of the (corresponding) resolution of D_1 and D_2.

By (IH) we have $S^i_{<_A^s}(\mathcal{C}) \leq_{ss} S^i_{<_A}(\mathcal{C})$. Consequently there exist clauses $C_1, C_2 \in S^i_{<_A^s}(\mathcal{C})$ such that

$$C_1 \leq_{ss} D_1 \text{ and } C_2 \leq_{ss} D_2.$$

As D_0 is resolvent of D_1, D_2 we know by Lemma 4.2.1 that one of the following properties must hold:

a) $C_1 \leq_{ss} D_0$ or
b) $C_2 \leq_{ss} D_0$ or
c) There exists a resolvent C of C_1, C_2 such that $C \leq_{ss} D_0$

If a) or b) holds then clearly $S^i_{<_A}(C) \leq_{ss} \{D_0\}$ and we are done (as D was an arbitrary A-resolvent from $S^i_{<_A}(S)$ we conclude $S^i_{<_A}(C) \leq_{ss} S^{i+1}_{<_A}(C)$ under restriction a) or b)).

So it remains to analyze case c). In fact we need more than just c); we must ensure that (in case a and b both fail) there exists a resolvent C_0 of C_1, C_2 such that $C_0 \leq_{ss} D_0$ and $N_c(C_0) \in \rho_{<_A}(\{C_1, C_2\})$. Only if the last property is guaranteed we have that

$$S^{i+1}_{<_A}(C) \leq_{ss} \{D_0\} \text{ (and thus } S^{i+1}_{<_A}(C) \leq_{ss} \{D\}).$$

So it remains to show that (provided a and b do not hold) there exists a resolvent C_0 of C_1, C_2 such that

$$N_c(C_0) \in \rho_{<_A}(\{C_1, C_2\}).$$

Let C_0 be like C in the proof of Lemma 4.2.1, i.e.,

$$C_0 = (G_1 \vee H_1 \vee G_2 \vee H_2)\tau$$

$$D_0 = (E_1 \vee F_1 \vee E_2 \vee F_2)\sigma$$

such that R, S are the cut-literals of C_1, C_2 and M, N of D_1, D_2.

We have to show that the resolvent C_0 constructed in Lemma 4.2.1 must be an A-resolvent (assuming that D_0 is an A-resolvent). As $R\tau$ is the resolved literal in the resolution of C_1 and C_2 we have to show:

There is no literal L in $(G_1 \vee H_1 \vee G_2 \vee H_2)\tau$ such that $R\tau <_A L$.

So let us assume that $L \in \text{LIT}((G_1 \vee H_1 \vee G_2 \vee H_2)\tau)$ such that $R\tau <_A L$ and derive a contradiction. By property (A3) in Definition 3.3.1 we obtain $R\tau\mu <_A L\mu$ for all substitutions μ.

Let ρ be the substitution defined in the proof of Lemma 4.2.1, i.e., $\tau\rho = \eta\sigma$, $R\tau\rho = M\sigma$, $S\tau\rho = N\sigma$. Then we get $\mathcal{R}\tau\rho <_A L\rho$.

By $\text{LIT}(G_1 \vee H_1 \vee G_2 \vee H_2)\tau\rho \subseteq \text{LIT}(D_0)$ we also have $L\rho \in \text{LIT}(D_0)$.

$\text{at}(M\sigma)$ is the resolved atom in the resolution of D_1 and D_2. Moreover from $R\tau\rho = M\sigma$ we obtain $M\sigma <_A L\rho$ and $L\rho$ is a literal in D_0. But this contradicts the assumption that $D_0 \in \rho_{<_A}(\{D_1, D_2)\})$.

If C_0 is the R-resolvent of C_1, C_2 corresponding to D_0 (and constructed according to the proof of Lemma 4.2.1) then there exists no L in C_0 such that $R_T < L$. Therefore $N_c(C_0) \in \rho_{<_A}(\{C_1, C_2\}) \subseteq \rho_{<_A}(S^i_{<^s_A}(C))$. Thus, for $C = N_c(C_0)$, we obtain

$$C \in S^{i+1}_{<^s_A}(C) \text{ and } C \leq_{ss} D$$

Note that $C_0 \leq_{ss} D_0$ implies $C \leq_{ss} D$ (by Exercise 4.2.7). Putting a), b), c) together we have

$$S^{i+1}_{<^s_A}(C) \leq_{ss} S^{i+1}_{<_A}(C).$$

$\Diamond$

The argument in the proof of Theorem 4.2.1 can also be used to prove the completeness of unrestricted resolution under forward subsumption (Exercise 4.2.8). Note that this result cannot be derived directly from the fact that $R_{<_A}$ is a refinement of $R_\emptyset$ ($C_1 \subseteq D_1$ and $C_2 \subseteq D_2$ does not imply $sf(C_1, C_2) \subseteq sf(D_1, D_2)$).

In Section 4.1 we have seen that search complexity can be increased by choosing a stricter refinement. The following proposition shows that in refining A-ordering by forward subsumption such an effect is impossible.

Proposition 4.2.6. *Forward subsumption never increases the search complexity of A-ordering refinements, or more formally: Let $<_A$ be an A-ordering and C be a (finite) unsatisfiable set of N_c-clauses then $CS_{<^s_A}(C) \leq CS_{<_A}(C)$.*

Proof. By $sf(C, D) \subseteq D$ and by the definitions of $R_{<_A}$ and of $R_{<^s_A}$ we obtain $S^i_{<^s_A}(C) \subseteq S^i_{<_A}(C)$ for all $i \in \mathbb{N}$. In the proof of Theorem 4.2.1 we have shown that for all $i \in \mathbb{N}$:

$$S^i_{<^s_A}(C) \leq_{ss} S^i_{<_A}(C).$$

In particular $\square \in S^i_{<_A}(C)$ implies $\square \in S^i_{<^s_A}(C)$ and thus

$$d_{<^s_A}(C) = d_{<_A}(C).$$

Putting things together we obtain

$$S^{d_{<^s_A}(C)}_{<^s_A}(C) \subseteq S^{d_{<_A}(C)}_{<_A}.$$

But that means $CS_{<^s_A}(C) \leq CS_{<_A}(C)$. $\Diamond$

By an argument similar to that in the proof of Proposition 4.2.6 we obtain $CS_{\emptyset_s}(C) \leq CS_\emptyset(C)$, i.e., forward subsumption decreases the search complexity of unrestricted resolution (Exercise 4.2.9). Forward subsumption can lead to incompleteness if combined with certain refinements. A typical example is lock resolution.

Proposition 4.2.7. *Lock resolution with forward subsumption is incomplete.*

Proof. In order to make the above statement more precise we have to define subsumption for indexed clauses. We define $C \leq_{ss} D$ for indexed clauses C, D iff $unlock(C) \leq_{ss} unlock(D)$.

Let $C = \{(R(x_1), 1) \vee (P(x_1), 5), (\neg P(x_1), 3) \vee (R(x_1), 6), (P(x_1), 4) \vee (\neg R(x_1), 7),$
$(\neg R(x_1), 2) \vee (\neg P(x_1), 8)\}$.

(C is identical to the set of N_c-clauses defined in Example 4.1.3).

By definition of the lock operator R_l, we obtain

$$R_l(C) = \bigcup_{i=0}^{\infty} S_l^i(C) \text{ for } S_l^{i+1}(C) = S_l^i(C) \cup \rho_l(S_l^i(C)),$$

where $\rho_l(D)$ denotes the set of all lock resolvents from D (with respect to the locking l). As C is unsatisfiable and $\mathcal{R}_l$ is complete we have $\square \in R_l(C)$. We show now that $\square \notin R_{ls}(C)$. For this purpose we compute the i-th levels $S_l^i(C)$.

$$S_l^1(C) = C \cup \{(P(x_1), 5) \vee (\neg P(x_1), 8), \ (R(x_1), 6) \vee (\neg R(x_1), 7)\}.$$

The clauses in $S_l^1(C) - C$ are not subsumed by clauses in C and therefore

$$sf(S_l^0(C), \rho_l(C)) = \rho_l(C) \text{ and } S_l^1(C) = S_{ls}^1(C).$$

$$S_l^2(C) = S_l^1(C) \cup \{(R(x_1), 6) \vee (\neg P(x_1), 8), (\neg R(x_1), 7) \vee (\neg P(x_1), 8)\}.$$

Thus

$$\rho_l(S_l^1(C)) = \rho_l(S_{ls}^1(C)) = \{(R(x_1), 6) \vee (\neg P(x_1), 8), \ (\neg R(x_1), 7) \vee (\neg P(x_1), 8)\}.$$

But

$$(\neg P(x_1), 3) \vee (R(x_1), 6) \leq_{ss} (R(x_1), 6) \vee (\neg P(x_1), 8))$$

and

$$(\neg P(x_1), 2) \vee (\neg P(x_1), 8) \leq_{ss} (\neg R(x_1), 7) \vee (\neg P(x_1), 8)).$$

That means $C \leq_{ss} \rho_l(S_{ls}^1(C))$ and particularly $sf(S_{ls}^1(C), \rho_l(S_{ls}^1(C))) = \emptyset$. We obtain $S_{ls}^2(C) = S_{ls}^1(C)$ and, by the nature of operator definitions, $R_{ls}(C) = S_{ls}^1(C)$. As $\square \notin S_{ls}^1(C)$ we obtain $\square \notin R_{ls}(C)$. But C is unsatisfiable and thus R_{ls} is incomplete.   ◊

We can overcome the loss of completeness, however, by avoiding the technique of unlocking (Exercise 4.2.10). Forward subsumption can also be defined for refinements like linear input deduction, which do not admit an operator definition. For such refinements we have to restrict the deduction principle to a "nonredundant" type. Our next subject is to define forward subsumption for linear input deductions. Because subsumption does not respect the order of literals within a clause, the LI-deduction refinement of Section 3.5 is inappropriate. Therefore we will use a less restricted principle of linear input deduction, where every literal in a center clause may be resolved.

Definition 4.2.5 (unrestricted LI-deduction). *Let C be a set of (unnormalized) clauses, $C_0, E_1, E_2, \ldots, E_n$ in C and $C_{i+1} \in Res(C_i, E_{i+1})$ for $i < n$. Then the R-deduction*

$$\Gamma : C_0, \ E_1, \ C_1, \ \ldots, E_n, \ C_n$$

is called an unrestricted LI-deduction *(for short: ULI-deduction) of C_n from C with top clause C_0.*

Remember that LI-deductions (see Definition 3.5.5) were subjected to additional restrictions such as 1) resolving only the rightmost literal in a center clause and 2) factoring with the leftmost literal in a center clause only. In the definition of ULI-deductions we omit all these restrictions. As LI-deduction is more restricted than ULI-deduction, the latter is complete relative to the former; in particular ULI-deduction is complete on Horn logic (but is not complete on the whole of clause logic).

In ULI-deductions (and in LI-deductions) infinite cycling may occur, a quite unpleasant feature in automated deduction. Cycling means that infinitely many instances of a center clause may be produced in the deduction procedure (more exactly, there are arbitrarily many repetitions of instances of a center clause in ULI-deductions). Fortunately, cycling can be avoided under preservation of completeness. This result is an immediate consequence of the completeness result concerning nonredundant ULI-deductions.

Definition 4.2.6 (SULI-deduction). *Let $\Gamma : C_0, E_1, \ldots, C_{n-1}, E_n, C_n$ be an ULI-deduction from a set of clauses C. Γ is called* subsumption-reduced *(or a SULI-deduction) if*

1) $C \not\leq_{ss} C_i$ for $i = 1, \ldots, n$ and
2) If $i, j \leq n$ and $i < j$ then $C_i \not\leq_{ss} C_j$.

SULI-deductions model the principle of forward subsumption within ULI-deductions. Any clause derived "later" (i.e., a C_j for $j \neq 1$) cannot be subsumed by an input clause or by a clause derived "before" (these are the $C_i's$ for $i < j$). Our goal is to prove the completeness of SULI-deduction relative to ULI-deduction. As linear input deduction is incomplete, SULI-deduction is incomplete too. However we still can obtain *relative completeness*. Generally a refinement Ψ is complete relative to a refinement Ω if, for all unsatisfiable sets of clauses C, $\Psi(C)$ contains a refutation if $\Omega(C)$ contains one. To show relative completeness we require a technical lemma that replaces the lifting by the subsumption property. A natural complexity measure for ULI-deductions is depth:

Definition 4.2.7 (depth of ULI-deductions).
Let $\Gamma : C_0, E_1, C_1, \ldots, E_n, C_n$ be a ULI-deduction. Then the depth *of Γ is defined by $\delta(\Gamma) = n$.*

It is obvious that $\delta(\Gamma) = \frac{1}{2}(l(\Gamma) - 1)$ for all ULI-deductions (where $l(\Gamma)$ is the length of Γ).

Lemma 4.2.2. *Let C be a set of clauses and Γ' be an ULI-deduction of a clause D' from C with top clause C'. Let C be a clause with $C \leq_{ss} C'$. Then there exists an ULI-deduction Γ of a clause D from $(C - \{C'\}) \cup \{C\}$ with top clause C such that $D \leq_{ss} D'$ and $\delta(\Gamma) \leq \delta(\Gamma')$.*

Proof. We proceed by induction on $\delta(\Gamma')$.

$\delta(\Gamma') = 0$: In this case Γ' must be C'. We simply define $\Gamma = C$ and obtain $\delta(\Gamma) = \delta(\Gamma') = 0$ and $C \leq_{ss} C'$.

(IH) Suppose that for all Γ' fulfilling the conditions of Lemma 4.2.2 and $\delta(\Gamma') = n$ there exists an appropriate deduction Γ.

case $\delta(\Gamma') = n + 1$: Γ' must be of the form Δ', G', E', D' such that $\Lambda' : \Delta', G'$ is an ULI-deduction of G' from C with top clause C' and $\delta(\Delta', G') = n$.

By (IH) there exists an ULI-deduction Λ of a clause G from $(C - \{C'\}) \cup \{C\}$ with top clause C such that $G \leq_{ss} G'$ and $\delta(\Lambda) \leq n$. By Lemma 4.2.1 we conclude that a) $G \leq_{ss} D'$, or b) $E' \leq_{ss} D'$, or c) there exists a resolvent D of G and E' such that $D \leq_{ss} D'$. In fact we need slightly more than that, namely that always a) or c) holds. Indeed the conditions are more specific than in Lemma 4.2.1 and we can obtain a stronger result (see Exercise 4.2.17) namely:

Let C_1, C_2, D, E be clauses such that $C_1 \leq_{ss} C_2$ and $E \in Res(C_2, D)$. Then either $C_1 \leq_{ss} E$ or there exists a resolvent F of C_1 and D such that $F \leq_{ss} E$.

Setting $C_1 : G$, $C_2 : G'$ and $D : E'$ we obtain $G \leq_{ss} D'$ or the existence of an appropriate resolvent D of G and E'. Therefore a) or c) must hold. In case a) we define $\Gamma = \Lambda$. Γ fulfils the required properties and particularly $\delta(\Gamma) < \delta(\Gamma')$. In case c) we set $\Gamma = \Delta, G, E', D$. Then Γ is an ULI-deduction of D from $(C - \{C'\}) \cup \{C\}$ with topclause C such that $D \leq_{ss} D'$ and

$$\delta(\Gamma) = \delta(\Lambda) + 1 \leq \delta(\Lambda') + 1 = \delta(\Gamma').$$

$\Diamond$

Looking carefully at the proof of Lemma 4.2.2 we can derive the more general result: If $C \leq_{ss} C'$ and Γ' is an ULI-deduction of C' from C' then we can find an ULI-deduction Γ of a clause C from C such that $C \leq_{ss} C'$ and $\delta(\Gamma) \leq \delta(\Gamma')$ (Exercise 4.2.11). But we do not need this more general version for the proof of the next theorem.

Theorem 4.2.2. *SULI-deduction is complete relative to ULI-deduction; moreover the following property is fulfilled: Let Γ be an unrestricted linear input refutation of a set of clauses C. Then there exists a subsumption-reduced, unrestricted linear input refutation Δ of C such that $\delta(\Delta) \leq \delta(\Gamma)$.*

Proof. If Γ is already a SULI-deduction then the theorem trivially holds.
So let us assume that Γ is an ULI- but not a SULI-refutation of C and

$$\Gamma = C_0, E_1, \ldots, C_{n-1}, E_n, \square.$$

Then either

a) There exists a clause $D \in C$ and a C_i for $i \geq 1$ such that $D \leq_{ss} C_i$

or

b) There are i, j such that $i < j \leq n$ and $C_i \leq_{ss} C_j$.

case a) By definition of ULI-deductions, $\Pi : C_i, E_{i+1}, \ldots, C_{n-1}, E_n, \square$ is an
ULI-refutation of $C \cup \{C_i\}$ (Π may collapse to $\square$). Because $D \leq_{ss} C_i$
we may apply Lemma 4.2.2 and obtain an ULI-refutation Δ of C with
topclause D and $\delta(\Delta) \leq \delta(\Pi)$. From $\delta(\Pi) < \delta(\Gamma)$ we infer $\delta(\Delta) < \delta(\Gamma)$.
That means we obtain an ULI-refutation Δ of a depth strictly smaller
than that of Γ. Note that Δ still may contain "redundancies" and need
not be a SULI-deduction (yet).

case b) $\Gamma = C_0, E_1, \ldots, C_i, E_{i+1}, \ldots, C_j, \ldots, \square$ such that $C_i \leq_{ss} C_j$.
C_j cannot be $\square$ (otherwise C_i would already be $\square$). But, clearly,
$\Pi : C_j, E_{j+1}; \ldots, \square$ is an ULI-refutation of $C \cup \{C_j\}$. By $C_i \leq_{ss} C_j$ and
Lemma 4.2.2 we find an ULI-refutation Λ of $C \cup \{C_i\}$ with top clause C_i
such that $\delta(\Lambda) \leq \delta(\Pi)$.

Let $\Lambda = C_i, \Lambda'$. Then $\Delta : C_0, E_1, \ldots, C_i, \Lambda'$ is a ULI-refutation of C with
top clause C_0. But $\delta(\Delta) = i + \delta(\Lambda') < j + \delta(\Pi) = \delta(\Gamma)$ and therefore
$\delta(\Delta) < \delta(\Gamma)$.
Again we obtain a ULI-refutation of strictly smaller depth. So we can
define a transformation Ψ with the following property:
    If Γ is an ULI-refutation, but not a SULI-refutation, then $\Psi(\Gamma)$
    is an ULI-refutation such that $\delta(\Psi(\Gamma)) < \delta(\Gamma)$.

Because every deduction is of finite depth only, Ψ can be applied only
finitely many often. Therefore, for each ULI-refutation Γ there exists a
$k \in \mathbb{N}$ such that $\Psi^{(k)}(\Gamma)$ is a SULI-refutation and $\delta(\Psi^{(k)}(\Gamma)) \leq \delta(\Gamma)$
(note that equality holds only if Γ is already a SULI-refutation and $k = 0$). $\diamond$

Using forward subsumption in linear input deductions is not only profitable with respect to proof complexity (the length of a shortest proof cannot
be increased), but also with respect to search complexity. So far we have defined search complexity for operator-based refinements. For ULI-deduction it

can naturally be defined as the number of all deductions within the minimal depth of a refutation.

Definition 4.2.8 (search complexity of linear input deduction). *Let $\mathcal{C}$ be a set of clauses which is ULI-refutable and let $C \in \mathcal{C}$ such that C is admissible as top clause. We define $\delta_{ULI}(\mathcal{C}, C)$ as the minimal depth of a ULI-refutation of $\mathcal{C}$ with top clause C. $\delta_{SULI}(\mathcal{C}, C)$ is defined in the same way, using SULI-refutations instead of ULI-refutations.*

Let $\Delta_{ULI}(\mathcal{C}, C, k)$ $(\Delta_{SULI}(\mathcal{C}, C, k))$ be the set of all ULI-deductions (SULI-deductions) from $\mathcal{C}$ with top clause C and depth $\leq k$. We define the search complexity of ULI-deduction (SULI-deduction) as $CS_{ULI}(CS_{SULI})$:

$$CS_{ULI}(\mathcal{C}, C) = |\Delta_{ULI}(\mathcal{C}, C, \delta_{ULI}(\mathcal{C}, C))| \text{ and}$$
$$CS_{SULI}(\mathcal{C}, C) = |\Delta_{SULI}(\mathcal{C}, C, \delta_{SULI}(\mathcal{C}, C))|.$$

Proposition 4.2.8. *The search complexity of subsumption-reduced ULI-deduction is always less or equal than that of ULI-deduction.*

Proof. Let $\mathcal{C}$ be a set of clauses, $C \in \mathcal{C}$ and let us assume that $\mathcal{C}$ is ULI-refutable with top clause C.

Because the set of all SULI-deductions is a subset of the set of all ULI-deductions we get:

$$\Delta_{SULI}(\mathcal{C}, C, k) \subseteq \Delta_{ULI}(\mathcal{C}, C, k) \text{ for all } k \in \mathbb{N}.$$

By Theorem 4.2.2 we know that, given an ULI-refutation, there is always a SULI-refutation of equal or smaller depth. Particularly we obtain

$$\delta_{SULI}(\mathcal{C}, C) = \delta_{ULI}(\mathcal{C}, C).$$

Putting things together we obtain

$$\Delta_{SULI}(\mathcal{C}, C, \delta_{SULI}(\mathcal{C}, C)) \subseteq \Delta_{ULI}(\mathcal{C}, C\delta_{ULI}(\mathcal{C}, C)).$$

$\Diamond$

In forward subsumption we only allow resolvents which are not subsumed by clauses derived on a lower level. By such a deletion technique we obtain a refinement (in the sense of Section 3.1) which contains redundancy-free derivations only. In fact, if R_x is a refinement and $\square \in R_{xs}(\mathcal{C})$ we can construct an R_x-refutation of $\mathcal{C}$ by using the clauses in $R_{xs}(\mathcal{C})$. The situation changes if we apply the subsumption principle "backward", which means we delete clauses on lower levels if they are subsumed by clauses on higher levels. By such a deletion technique we may destroy the "history" of some clauses derived so far and, as a consequence, break the structure of the deductions. Deleting every subsumed clause, no matter how and when it was derived, is called the principle of *replacement*. While refinements are of "deduction" type, replacement is essentially a *reduction* method.

Example 4.2.3. Let

$$\mathcal{C} = \{P(x_1) \vee Q(x_1), \ P(x_1) \vee \neg Q(x_1), \ Q(x_1) \vee \neg P(x_1),$$
$$\neg P(x_1) \vee \neg Q(x_1)\}.$$

By applying N_c-resolution (without any additional restriction) we obtain the set of N_c-resolvents:

$$\mathcal{D}: \{P(x_1), \ \neg P(x_1), Q(x_1), \ \neg Q(x_1), \ P(x_1) \vee \neg P(x_1), \ Q(x_1) \vee \neg Q(x_1)\}.$$

In the refinement $R_\emptyset$ we simply add $\mathcal{D}$ to $\mathcal{C}$ and we obtain:

$$S_\emptyset^1(\mathcal{C}) \ = \ \mathcal{C} \cup \mathcal{D}.$$

But, once we have $\mathcal{D}$, all clauses in $\mathcal{C}$ are rendered redundant. In fact we obtain $\mathcal{D} \leq_{ss} \mathcal{C}$. But $\mathcal{D}$ is even internally redundant (the tautologies in $\mathcal{D}$ are subsumed by other clauses in $\mathcal{D}$) and thus reduces to $\mathcal{D}'$: $\{P(x_1), \neg P(x_1), Q(x_1), \neg Q(x_1)\}$.

So let us replace $\mathcal{C}$ by $\mathcal{D}'$. Then $\rho_\emptyset(\mathcal{D}') = \{\Box\}$. But, in the presence of $\Box$, all other clauses are redundant and we may delete $\mathcal{D}'$ too. So, finally, we are left with the set $\{\Box\}$ only. The reduction of $\mathcal{C}$ to $\mathcal{D}'$ and of $\mathcal{D}'$ to $\{\Box\}$ are called replacement steps. So we may write S_r^i for the "generations" of replacement and obtain:

$$S_r^0(\mathcal{C}) = sub(\mathcal{C}), \ S_r^1(\mathcal{C}) = \mathcal{D}', \ S_r^2(\mathcal{C}) = \{\Box\}.$$

We see that the aim of replacement is to reduce unsatisfiable sets of clauses to $\{\Box\}$.

We can proceed similarly in using replacement for hyperresolution. In such a case we obtain

$$S_H^1(\mathcal{C}) = \mathcal{C} \cup \{P(x_1), Q(x_1)\} \ \text{ and } \ S_H^2(\mathcal{C}) = S_H^1(\mathcal{C}) \cup \{\Box\}.$$

Using replacement we obtain

$$S_{Hr}^0(\mathcal{C}) = \mathcal{C}$$

($\mathcal{C}$ is reduced under subsumption) and

$$S_{Hr}^1(\mathcal{C}) = \{P(x_1), Q(x_1), \neg P(x_1) \vee \neg Q(x_1)\}$$

(note that $P(x_1) \leq_{ss} P(x_1) \vee Q(x_1)$ and $Q(x_1) \leq_{ss} Q(x_1) \vee \neg P(x_1)$).

But $\delta_H(S_{Hr}^1(\mathcal{C})) = \{\Box\}$ and, like before, we get $S_{Hr}^2(\mathcal{C}) = \{\Box\}$.

Definition 4.2.9 (reduction under subsumption). *Let $\mathcal{C}$ be a set of clauses. $\mathcal{C}$ is called subsumption-reduced if for all $C, D \in \mathcal{C} : C \leq_{ss} D$ implies $C = D$.*

There may be more than just one subsumption-reduced form of a set of clauses (i.e., clauses C, D such that $C =_{ss} D$). But by equipping the set of clauses with a linear ordering (e.g., the lexicographic one) and by keeping the smallest clause of every $=_{ss}$-equivalence class we can obtain a unique reduced form. Thus we can consider subsumption-reduction as a function (which we denote by $sub()$). The sub-operator (on a set of N-clauses) enjoys the following obvious properties:

a) $sub(\mathcal{C}) \subseteq \mathcal{C}$,
b) $sub(sub(\mathcal{C})) = sub(\mathcal{C})$ and
c) If $\mathcal{C} \leq_{ss} \mathcal{D}$ then $sub(\mathcal{C} \cup \mathcal{D}) =_{ss} sub(\mathcal{C})$.

Definition 4.2.10 (replacement sequence). *Let R_x be a resolution refinement operator defined on N-clauses for some normalization operator N. We define*

$$S_{xr}^0(\mathcal{C}) \;=\; sub(\mathcal{C}) \quad \text{and}$$

$$S_{xr}^{i+1}(\mathcal{C}) \;=\; sub(S_{xr}^i(\mathcal{C}) \cup \rho_x(S_{xr}^i(\mathcal{C})))$$

for $i \in \mathbb{N}$. Then the sequence $(S_{xr}^i(\mathcal{C}))_{i \in \mathbb{N}}$ is called the R_x-replacement sequence for $\mathcal{C}$ (in short: R_{xr}-sequence).

An R_x-replacement sequence $(S_{xr}^i(\mathcal{C}))_{i \in \mathbb{N}}$ is called convergent *if there exists a $k \in \mathbb{N}$ such that for all $l \geq k : S_{xr}^l(\mathcal{C}) = S_{xr}^k(\mathcal{C})$; otherwise it is called* divergent. *It is called a* refutation-sequence *if there exists a $k \in \mathbb{N}$ such that $S_{xr}^k(\mathcal{C}) = \{\Box\}$.*

Example 4.2.4. Let

$$\mathcal{C} = \{P(x_1) \vee Q(x_1), P(x_1) \vee \neg Q(x_1), Q(x_1) \vee \neg P(x_1), \neg P(x_1) \vee \neg Q(x_1)\}$$

as in example 4.2.3.
Then $\mathcal{C}, \{P(x_1), Q(x_1), \neg P(x_1), \neg Q(x_1)\}, \Box$ is a $R_{\emptyset r}$-refutation sequence (note that not only all original clauses are deleted in the second step but also the two tautological resolvents). $\mathcal{C}, \{P(x_1), Q(x_1)\}, \{\Box\}$ is an R_{Hr}-refutation sequence.
The set of clauses $\mathcal{D}$, defined by $\mathcal{D} = \mathcal{C} - \{\neg P(x_1) \vee \neg Q(x_1)\}$, obviously is satisfiable.
The R_{Hr}-sequence for $\mathcal{D}$ is

$$\Psi : \mathcal{D}, \{P(x_1), Q(x_1)\}, \{P(x_1), Q(x_1)\}, \ldots$$

Ψ clearly is convergent.
We will see that for unsatisfiable sets of N_c-clauses $\mathcal{C}$ the sequence $(S_{Hr}^i(\mathcal{C}))_{i \in \mathbb{N}}$ is a refutation sequence (i.e., R_H-replacement is complete). But in case of satisfiable sets $\mathcal{C}$ divergent sequences must exist (note that clause logic is undecidable).

Example 4.2.5. $\mathcal{C} = \{P(a), P(f(x_1)) \vee \neg P(x_1)\}$.

The R_H-replacement sequence for $\mathcal{C}$ is:

$$\Psi : \mathcal{C}, \mathcal{C} \cup \{P(f(a))\}, \ \mathcal{C} \cup \{P(f(a)), P(f^{(2)}(a))\}, \ldots.$$

Ψ is divergent as $P(f^{(i)}(a)) \in S^i_{Hr}(\mathcal{C}) - S^{i-1}_{Hr}(\mathcal{C})$ for $i > 0$. If $\mathcal{C}$ is changed to $\mathcal{D} : \{P(f^{(2)}(x_1)), P(f(x_1)) \vee \neg P(x_1)\}$ then R_{Hr}-converges. In fact $S^1_{Hr}(\mathcal{D}) = S^0_{Hr}(\mathcal{D}) = \mathcal{D}$ because $\{P(f^{(3)}(x_1))\} = \rho_H(\mathcal{D})$ and $P(f^{(2)}(x_1)) \leq_{ss} P(f^{(3)}(x_1))$.
Note that by omitting replacement we obtain $P(f^{(k)}(x_1)) \in R_H(\mathcal{D})$ for all $k \geq 2$.

Still the question remains whether the replacement operators R_{xr} are admissible, i.e., whether they are correct and complete. The correctness ($R_{xr}(\mathcal{C})$ is refutation sequence implies that $\mathcal{C}$ is unsatisfiable) easily follows from the fact that $\mathcal{C}$ is logically equivalent to $sub(\mathcal{C})$; thus all members of a reduction sequence are logically equivalent (Exercise 4.2.13). It remains to investigate completeness. As in the case of forward subsumption we prove that the i-th member in a reduction sequence $(S^i_{xr}(\mathcal{C}))_{i \in N}$ subsumes the i-th level $S^i_x(\mathcal{C})$ of the corresponding refinement R_x. As refinement we choose hyperresolution.

Lemma 4.2.3. *The replacement sequences for hyperresolution subsume the level-sets generated by hyperresolution (without subsumption). More formally, let $\mathcal{C}$ be a set of N_c-clauses; then for all $k \in \mathbb{N}$: $S^k_{Hr}(\mathcal{C}) \leq_{ss} S^k_H(\mathcal{C})$.*

Proof. We show first (to facilitate the argumentation below) that we can reduce the problem to subsumption-reduced sets of clauses. Indeed $S^k_{Hr}(\mathcal{C}) = S^k_{Hr}(sub(\mathcal{C}))$ for all $k \in \mathbb{N}$. From $sub(\mathcal{C}) \subseteq \mathcal{C}$ we derive $S^k_H(sub(\mathcal{C})) \subseteq S^k_H(\mathcal{C})$ and thus
$S^k_{Hr}(sub(\mathcal{C})) \leq_{ss} S^k_H(\mathcal{C})$ implies $S^k_{Hr}(\mathcal{C}) \leq_{ss} S^k_H(\mathcal{C})$ for all $k \in \mathbb{N}$.

So we may assume, without loss of generality, that $\mathcal{C}$ is subsumption reduced. We proceed further by induction on k:

For $k = 0$ we obtain $S^0_H(\mathcal{C}) = \mathcal{C}$ and $S^0_{Hr}(\mathcal{C}) = sub(\mathcal{C}) = \mathcal{C}$, and so $S^k_{Hr}(\mathcal{C}) \leq_{ss} S^k_H(\mathcal{C})$ trivially holds.

(IH) Suppose that we already have $S^k_{Hr}(\mathcal{C}) \leq_{ss} S^k_H(\mathcal{C})$.

We have to show that all clash resolvents in $S^{k+1}_H(\mathcal{C}) - S^k_H(\mathcal{C})$ are subsumed by clauses in $S^{k+1}_{Hr}(\mathcal{C})$.
Note that
$$S^{k+1}_{Hr}(\mathcal{C}) = sub(S^k_{Hr}(\mathcal{C}) \cup \rho_H(S^k_{Hr}(\mathcal{C})))$$

So clearly $S^{k+1}_{Hr}(\mathcal{C}) \leq_{ss} S^k_{Hr}(\mathcal{C})$ and, by (IH), $S^{k+1}_{Hr}(\mathcal{C}) \leq_{ss} S^k_H(\mathcal{C})$.

Thus it is sufficient to show $S^{k+1}_{Hr}(\mathcal{C}) \leq_{ss} \rho_H(S^k_H(\mathcal{C}))$. Now let Γ' : $(C'; D'_1, \ldots, D'_n)$ be a clash sequence over the set of clauses $S^k_H(\mathcal{C})$ and let E' be a clash resolvent of Γ'. By (IH) there exist clauses $C, D_1, \ldots, D_n \in S^k_{Hr}(\mathcal{C})$ such that $C \leq_{ss} C'$ and $D_i \leq_{ss} D'_1$ for $i = 1, \ldots, n$.

If C is nonpositive then $\Gamma : (C; D_1, \ldots, D_n)$ is a clash sequence over $S^k_{Hr}(\mathcal{C})$. We distinguish two cases:

a) $C \in \mathcal{C}$ and
b) $C \notin \mathcal{C}$.

We first discuss

case b) : As $C \in S^k_{Hr}(\mathcal{C}) - \mathcal{C}$, C must be a positive clause (in fact all nonpositive clauses in $S^k_{Hr}(\mathcal{C})$ must already be in $\mathcal{C}$). From $C = C_P$ and $C \leq_{ss} C'$ we infer the existence of a substitution ϑ such that $\mathrm{LIT}(C)\vartheta \subseteq \mathrm{LIT}(C'_P)$. Now the clash resolvent E' must contain a subclause of the form $C'_P \eta$ for some substitution η. We conclude $\mathrm{LIT}(C)\vartheta\eta \subseteq \mathrm{LIT}(E')$ and therefore $C \leq_{ss} E'$.

We now turn to

case a) : Because C' must be in $\mathcal{C}$ (otherwise Γ' is not a clash sequence) and $\mathcal{C}$ is subsumption reduced we obtain $C' = C$.

In particular $\Gamma' : (C, D_1, \ldots, D_n)$ is a clash sequence over $S^k_{Hr}(\mathcal{C})$. Γ does not necessarily possess a clash resolvent at all. But we will show the following:

Let R'_i be the i-th intermediary resolvent leading to the clash resolvent E', which means $R'_0 = C'$ and for all $i < n : R'_{i+1}$ is a PRF-resolvent of R'_i and D'_{i+1} and $R'_n = E'$.

Remember that PRF-resolution only cuts out the rightmost (negative) literal of the nonpositive clause. In the positive partner clause factoring is allowed and all of its literals are allowed to be resolved upon. For the inductive argument used below we require a refined subsumption relation $\leq_{ss0}$: Let C, D be clauses in PN-form (Definition 3.6.1). Then $C \leq_{ss0} D$ iff there exists a substitution ϑ such that $\mathrm{LIT}(C_P)\vartheta \subseteq \mathrm{LIT}(D_P)$ and, in case $C_N \neq \square$, $C_N \vartheta = D_N$. Clearly $C \leq_{ss0} D$ implies $C \leq_{ss} D$, but not vice versa.

In the next step we will inductively construct a sequence of clauses R_i with the property $R_i \leq_{ss0} R'_i$ for $i = 1, \ldots, n$. These clauses are the intermediary resolvents of the "subsuming clash".

$i = 0 : R_0 = C$. From $R_0 = R'_0$ we directly get $R_0 \leq_{ss0} R'_0$. Suppose that R_i has already been constructed and $i < n$:

case aa) R_i is positive. We define $R_{i+1} = R_i$. By (IH) we have $R_i \leq_{ss0} R'_i$. Because R_i is positive the last relation is equivalent to $R_i \leq_{ss} R'_i$. As R'_{i+1} is a PRF-resolvent of R'_i and D'_i, R'_i must be of the form: $R'_i : E'_{i+1} \vee \neg Q'$ for some atom Q'.

So, clearly, $R_i \leq_{ss} E'_{i+1}$. But R'_{i+1} contains an instance of E'_{i+1} and we conclude $R_i \leq_{ss} R'_{i+1}$. By definition of R_{i+1} we thus obtain $R_{i+1} \leq_{ss} R'_{i+1}$ and, by positivity of R_{i+1}, $R_{i+1} \leq_{ss0} R'_{i+1}$.

case ab) R_i is nonpositive.

subcase aba) $D_{i+1} \leq_{ss} R'_{i+1}$:

In this case we set $R_{i+1} = D_{i+1}$. Because D_{i+1} is a positive clause we also get $R_{i+1} \leq_{ss0} R'_{i+1}$.

subcase abb) $D_{i+1} \not\leq_{ss} R'_{i+1}$:

By (IH) $D_{i+1} \leq_{ss} D'_{i+1}$ and, by definition of R'_{i+1}, R'_{i+1} is PRF-resolvent of R'_i and D'_{i+1}. By construction of R_i we have $R_i \leq_{ss0} R'_i$. For abbreviation we set $M_i = (R_i)_P$, $M'_i = (R'_i)_P$, $N_i = (R_i)_N$ and $N'_i = (R'_i)_N$. By definition of $\leq_{ss0}$ there must be a substitution ϑ such that

$$\text{LIT}(M_i)\vartheta \subseteq \text{LIT}(M'_i) \text{ and } N_i\vartheta = N'_i.$$

Thus there are negative clauses T_i, T'_i (or $T_i = T'_i = \square$) such that

$$R_i = M_i \vee T_i \vee \neg Q \text{ and } R'_i = M'_i \vee T'_i \vee \neg Q'$$

and

$$T_i\vartheta = T'_i, Q\vartheta = Q'.$$

Let F'_{i+1} be a factor of D'_{i+1} with $F'_{i+1} = G'_1 \vee P' \vee G'_2$ such that P' is the cut literal in the binary resolvent of R'_i and F'_{i+1}. By $D_{i+1} \leq_{ss} D'_{i+1}$ there are two possibilities:

The first is $D_{i+1} \leq_{ss} G'_1 \vee G'_2$. Then clearly $D_{i+1} \leq_{ss} R'_{i+1}$, a case settled in aba) and excluded here.

The second possibility is that there exists a factor F_{i+1} of D_{i+1} such that $F_{i+1} = G_1 \vee P \vee G_2$ and there is a substitution η such that $\text{LIT}(G_1 \vee G_2)\eta \subseteq \text{LIT}(G'_1 \vee G'_2)$ and $P\eta = P'$. Let σ be an m.g.u. of $\{P', Q'\}$. Then $R'_{i+1} = (G'_1 \vee G'_2 \vee M'_i \vee T'_i)\sigma$. But $P' = P\eta$ and $Q' = Q\vartheta$. Assuming, as usual, $dom(\eta) \cap dom(\vartheta) = \emptyset$ the substitution $(\eta \cup \vartheta)\sigma$ is a unifier of $\{P, Q\}$. Let τ be the m.g.u. of $\{P, Q\}$. Then we define

$$R_{i+1} = (G_1 \vee G_2 \vee M_i \vee T_i)\tau.$$

From $\tau \leq_s (\eta \cup \vartheta)\sigma$ we derive $R_{i+1} \leq_{ss0} R'_{i+1}$.

This completes the construction of the R_i. By definition of the R_i we directly obtain $R_n \leq_{ss0} R'_n = E'$. The clause R_n is either D_i for some $i = 1, \ldots, n$ or it is a resolvent of a clash $(C, D_1, \ldots, D_k)$ for some $k \leq n$. Thus either $N_c(R_n) \in S^k_{Hr}(\mathcal{C})$ or $N_c(R_n) \in \rho_H(S^k_{Hr}(\mathcal{C}))$. In any case we may define $E = N_c(R_n)$. Then, clearly

$$E \in S^k_{Hr}(\mathcal{C}) \cup \rho_H(S^k_{Hr}(\mathcal{C})) \text{ and } E \leq_{ss} E'.$$

Therefore there must be a clause $E_0 \in sub(S^k_{Hr}(\mathcal{C}) \cup \rho_H(S^k_{Hr}(\mathcal{C})))$ such that $E_0 \leq_{ss} E$ and so also $E_0 \leq_{ss} E'$.

Putting case a) and case b) together we realize the existence of a clause $E \in S_{Hr}^{k+1}(\mathcal{C})$ such that $E \leq_{ss} E'$. As E' was an arbitrary clash resolvent in $S_H^{k+1}(\mathcal{C})$ we obtain

$$S_{Hr}^{k+1}(\mathcal{C}) \leq_{ss} S_H^{k+1}(\mathcal{C}).$$

$\Diamond$

An argument, quite similar to that in the proof of Lemma 4.2.3 yields the completeness of unrestricted resolution + replacement (Exercise 4.2.14). Note that, like in the case of forward subsumption, we cannot directly use the fact that R_H is a refinement of $R_\emptyset$. Indeed $S_{Hr}^i(\mathcal{C}) \subseteq S_{\emptyset r}^i(\mathcal{C})$ does not hold in general (see Exercise 4.2.18).

Theorem 4.2.3 (Completeness of hyperresolution + replacement).
Let $\mathcal{C}$ be an unsatisfiable set of N_c-clauses, then $(S_{Hr}^i(\mathcal{C}))_{i \in \mathbb{N}}$ is a refutation sequence.

Proof. Hyperresolution is complete according to Theorem 3.6.1. That means there exists a number k such that $\square \in S_H^k(\mathcal{C})$. By Lemma 4.2.3 $S_{Hr}^k(\mathcal{C}) \leq_{ss} S_H^k(\mathcal{C})$ for all $k \in N$. Because only $\square$ subsumes $\square$ we get $\square \in S_{Hr}^k(\mathcal{C})$ and even $S_{Hr}^k(\mathcal{C}) = \{\square\}$. It follows that $(S_{Hr}^k(\mathcal{C}))_{k \in \mathbb{N}}$ is a refutation sequence. $\Diamond$

Another direct consequence of Lemma 4.2.3 is that the minimal refutation depth of hyperresolution is not increased by replacement.

Definition 4.2.11. *Let R_x be a refinement such that R_x + replacement is complete and $\mathcal{C}$ be an unsatisfiable set of clauses (of the appropriate normal form). Then*

$$d_{xr}(\mathcal{C}) : min\{k | S_{xr}^k(\mathcal{C}) = \{\square\}\}$$

is called the minimal refutation depth of $\mathcal{C}$ with respect to R_x + replacement.

Proposition 4.2.9. *The minimal refutation depth of hyperresolution + replacement is less or equal to that of hyperresolution; i.e., $d_{Hr}(\mathcal{C}) \leq d_H(\mathcal{C})$ for all unsatisfiable sets of N_c-clauses $\mathcal{C}$.*

Proof. Let $k = min\{m / \square \in S_H^m(\mathcal{C})\}$. From Lemma 4.2.3 we know that $S_{Hr}^k(\mathcal{C}) \leq_{ss} S_H^k(\mathcal{C})$; clearly $\square \in S_{Hr}^k(\mathcal{C})$ and thus $S_{Hr}^k(\mathcal{C}) = \{\square\}$. So we obtain $d_{Hr}(\mathcal{C}) \leq k$. $\Diamond$

The properties of completeness and preservation of minimal refutation depth hold for unrestricted resolution too. The required arguments are quite similar (for minimal refutation depth see Exercise 4.2.16). Reduction sequences may be considered as normal form computations. If, for some refinement R_x, R_x + replacement is complete then the normal form of every unsatisfiable set of clauses $\mathcal{C}$ is $\{\square\}$; indeed we eventually obtain $\{\square\}$ in producing the sequence $(S_{xr}^i(\mathcal{C}))_{i \in \mathbb{N}}$. In case of satisfiable sets of clauses there may or may not exist normal forms. Choosing

$$C = \{P(x_1) \vee Q(x_1), P(x_1) \vee \neg Q(x_1), Q(x_1) \vee \neg P(x_1)\}$$

(see Example 4.2.4) and S_{Hr} we get

$$S^i_{Hr}(C) = \{P(x_1), Q(x_1)\} \text{ for } i \neq 1;$$

Therefore $\{P(x_1), Q(x_1)\}$ is the "normal form" of C under S_{Hr}. Clearly the existence of such a normal form is equivalent to the convergence of the corresponding replacement sequence.

Choosing S_{Hr} and $\mathcal{D} = \{P(a); \neg P(x) \vee P(f(x))\}$, as in Example 4.2.5, $S^i_{Hr}(\mathcal{D})$ diverges and $\mathcal{D}$ is not normalizable under S_{Hr}. By the undecidability of the satisfiability problem of clause logic there exists no complete replacement method which always terminates; so there must always be non-normalizable sets of clauses. However there exist subclasses of clause logic where S^i_{Hr}-replacement always converges and yields such normal forms. For such subclasses S_{Hr} defines a decision procedure. This problem will appear again in Chapter 5, where we investigate resolution decision methods systematically.

Exercises

Exercise 4.2.1. Show that $=_{ss}$ is an equivalence relation.

Exercise 4.2.2. Consider $=_{ss}$ on $CL[N_s]$ (i.e., on the set of all clauses in standard normal form N_s). Prove that $CL[N_s]/_{=_{ss}}$ contains classes of infinite size (construct an infinite sequence of N_s-clauses $C_1, \ldots, C_n$ such that the C_i are pairwise different and $C_i =_{ss} C_j$ for all i, j).

Remark: If clauses are defined as sets of literals [Rob65], [CL73], [Lov78] then Exercise 4.2.2 proves that there are infinite equivalence classes for subsumption $=_{ss}$ on the set of all clauses.

Exercise 4.2.3. $C <_{ss} D$ is defined by ($C \leq_{ss} D$ and not $D \leq_{ss} C$). Prove that $<_{ss}$ is not Noetherian on $CL[N_c]$, i.e., there exists an infinite sequence of N_c-normalized clauses $C_1, \ldots, C_n, \ldots$ such that for all i, j and $i < j$ $C_j <_s C_i$ (there is a strictly descending sequence of clauses without minimal elements).

Exercise 4.2.4. Modify $<_{ss}$ of Exercise 4.2.3 in order to obtain a Noetherian relation $<_{ss1}$ fulfilling:

(a) For all C there exists a D such that $C <_{ss1} D$.
(b) $C <_{ss1} D \Rightarrow C \leq_{ss} D$ and $D \not\leq_s C$.

Exercise 4.2.5. Let C, D be clauses such that D is ground and not a tautology. Suppose that there exists a ground substitution ϑ such that $C\vartheta \wedge \neg D$ is unsatisfiable. Prove that C subsumes D.

Exercise 4.2.6. Suppose that C, D are clauses and $C \leq_{ss} D$. Let D' be an arbitrary factor of D. Show that $C \leq_{ss} D'$ holds.

Exercise 4.2.7. Let C, D be clauses and N be one of the normalization operators N_v, N_0, N_s, N_c. Show that $C \leq_{ss} D$ iff $N(C) \leq_{ss} N(D)$.

Exercise 4.2.8. Prove that unrestricted resolution + forward subsumption is complete, i.e., show that for all unsatisfiable sets of clauses $C : \square \in R_{\emptyset s}(C)$.

Exercise 4.2.9. Show that, in case of unrestricted resolution, forward subsumption reduces search complexity, i.e., $CS_{\emptyset s}(C) \leq CS_{\emptyset}(C)$ for all unsatisfiable sets of clauses C.

Exercise 4.2.10. Let $\leq_{ssl}$ be the subsumption relation on indexed clauses without unlocking: $C \leq_{ssl} D$ iff C, D are indexed clauses and there exists a ϑ such that $\mathrm{LIT}(C)\vartheta \subseteq \mathrm{LIT}(D)$. Show that locking + forward subsumption is complete under use of $\leq_{ssl}$ instead of $\leq_{ss}$ (Hint: look at the proof of Theorem 4.2.1).

Exercise 4.2.11. Let C, C' be two sets of clauses such that $C \leq_{ss} C'$. Suppose that Γ' is an ULI-deduction of C' from C'. Show that there exists an ULI-deduction Γ of a clause C from C such that $C \leq_{ss} C'$ and $\delta(\Gamma) \leq \delta(\Gamma')$.

Exercise 4.2.12. Define an algorithm transforming ULI-refutations Γ into SULI-refutations Δ (of the same set of clauses and having the same top clause) such that $\delta(\Delta) \leq \delta(\Gamma)$ (Hint: use the proof of Theorem 4.2.2).

Exercise 4.2.13. Prove that replacement is always correct, i.e., if $S^i_{xr}(C)$ is a refutation sequence then C is unsatisfiable.

Exercise 4.2.14. Prove that unrestricted resolution fulfills $S^i_{\emptyset r}(C) \leq_{ss} S^i_{\emptyset}(C)$ for all $i \in \mathbb{N}$.

Exercise 4.2.15. Use Exercise 4.2.14 to prove that unrestricted resolution + replacement is complete.

Exercise 4.2.16. Prove that the minimal refutation depth of unrestricted resolution is not increased by replacement, i.e., $d_{\emptyset r}(C) \leq d_{\emptyset}(C)$ for all unsatisfiable sets of clauses C.

Exercise 4.2.17. Let C_1, C_2, D, E be clauses such that $C_1 \leq_{ss} C_2$ and E is a resolvent of C_2 and D. Then either $C_1 \leq_{ss} E$ or there exists a resolvent F of C_1 and D such that $F \leq_{ss} E$ (note that this result does not follow immediately from Lemma 4.2.1).

Exercise 4.2.18. Give an example of a set of clauses C fulfilling the following two properties:

a) $S^1_{Hr}(C) \nsubseteq S^1_{\emptyset r}(C)$,
b) $S^1_{Hr}(C) \nleq_{ss} S^1_{\emptyset r}(C)$.

4.3 Subsumption Algorithms

In Section 4.2 we have shown that subsumption, by keeping the shortest proof of the corresponding refinement and by reducing the number of derivable clauses, always has a positive effect on search complexity. So we may already be content from the theoretical point of view. But, in practice, the high number of subsumption tests (which are necessary in order to make the method effective) and even the costs of the single subsumption tests themselves may create real problems and slow down programs considerably. Consider, for example, a refinement of type R_{xs} (R_x+ forward subsumption); here we have

$$S_{xs}^{i+1}(\mathcal{C}) = S_{xs}^i(\mathcal{C}) \cup sf(S_{xs}^i(\mathcal{C}), \rho_x(S_{xs}^i(\mathcal{C}))).$$

Thus, on every level, we have to test for all $D \in \rho_x(S_{xs}^i(\mathcal{C}))$ whether $S_{xs}^i(\mathcal{C}) \leq_{ss} \{D\}$ holds. Even more tests are required under replacement. Therefore the total number of required subsumption tests is roughly quadratic in the number of generated clauses. There are, in principle, two ways to improve the situtation: a) to reduce the number of subsumption tests without weakening the deletion principle and b) to improve the subsumption algorithms (i.e., the decision algorithms for the subsumption property). Of course a) and b) do not exclude each other and should both be pursued. Here in this section we only investigate b).

In Proposition 4.2.3 we have shown that $\leq_{ss}$ is decidable. The argument in the proof, however, yields an algorithm which is highly inefficient (it exhausts all terms of a certain depth). However, we cannot really hope to find a polynomial algorithm (unless $P = NP$) as the subsumption problem is NP-complete [GJ79]. By $C \leq_{ss} D$ iff there exists a ϑ such that $LIT(C\vartheta) \subseteq LIT(D)$, we have to search for a substitution that "maps" the literals in C to a subset of the literals in D. By Proposition 4.2.2 we may always assume (without loss of generality) that D is a ground clause. Moreover we always assume that all clauses are reduced (i.e., they do not contain multiple literals) but we don't mention this restriction explicitly for the rest of this section.

Example 4.3.1. Let $C = P(x) \vee P(f(x))$ and $D = P(f(a)) \vee P(f(b))$.
We prove that $C \not\leq_{ss} D$.
Let us assume that there exists a ϑ such that $LIT(C)\vartheta \subseteq LIT(D)$. Without loss of generality we may assume that $dom(\vartheta) = \{x\}$. Taking the first literal $P(x)$ of C we know that $P(x)\vartheta$ must be in $\{P(f(a)), P(f(b))\}$. So we obtain the candidates $\vartheta_1 = \{x \leftarrow f(a)\}$ and $\vartheta_2 = \{x \leftarrow f(b)\}$.

But under ϑ_1 the second literal $P(f(x))$ becomes $P(f(x))\vartheta_1 = P(f(f(a)))$ and $P(f(f(a))) \notin LIT(D)$. Similarly ϑ_2 yields $P(f(x))\vartheta_2 = P(f(f(b))) \notin LIT(D)$.

Therefore, neither ϑ_1 nor ϑ_2 do the job and we obtain $C \not\leq_{ss} D$. Moreover, note that $P(x) \leq_{ss} D$ and $P(f(x)) \leq_{ss} D$ both hold. So we see that subsumption tests cannot be "parallelized". Although very simple, Example 4.3.1 indicates that some kind of backtracking will be required in subsumption algorithms. In general, substitutions must be composed of "pieces" and have to be corrected repeatedly.

Example 4.3.2. $C = P(x) \vee P(f(x,y))$, $D = P(a) \vee P(b) \vee P(f(b,a))$.

Let us proceed as in Example 4.3.1. Then, by mapping the first literal $P(x)$ to D, we obtain the substitution candidates $\vartheta_1 = \{x \leftarrow a\}$, $\vartheta_2 = \{x \leftarrow b\}$, $\vartheta_3 = \{x \leftarrow f(b,a)\}$. Obviously ϑ_2 is the right choice and we may extend it to $\lambda = \{x \leftarrow b, y \leftarrow a\}$. Clearly $C \leq_{ss} D$ by $LIT(C\lambda) \subseteq LIT(D)$.

Thus if we start with ϑ_1 we must eventually correct this choice to ϑ_2. Suppose that we interchange the literals in C such that $P(f(x,y))$ becomes the first one; then clearly λ is the only candidate for a subsuming substitution. So we observe that, in searching for a subsuming substitution algorithmically, the ordering of the literals may play a role.

The idea of searching for the right substitution from left to right forms the basis of the algorithm of Stillman [Sti73] which henceforth will be called ST. Suppose that $C = L_1 \vee \ldots \vee L_n$ and a ϑ_i has already been found with $\{L_1, \ldots, L_i\}\vartheta_i \subseteq LIT(D)$; then ST tries to find an extension ϑ_{i+1} of ϑ_i such that $\{L_1, \ldots, L_{i+1}\}\vartheta_{i+1} \subseteq LIT(D)$. If one succeeds to extend some of the ϑ_is to a ϑ_n then $C \leq_{ss} D$; if there is no such ϑ_n then $C \not\leq_{ss} D$.

In Example 4.3.2 ϑ_2 can be extended to the subsuming substitution λ : $\{x \leftarrow b, y \leftarrow a\}$, while ϑ_1 and ϑ_3 cannot. For the specification of ST we use some auxiliary functions: The Boolean function "unify(L, M)" that returns true iff $\{L, M\}$ is unifiable and the function mgu(L, M) which computes an m.g.u. of $\{L, M\}$ if $\{L, M\}$ is unifiable and is undefined otherwise. We represent ST by a 5-place Boolean function ST_0 on (C, D, i, j, ϑ), where i points to the i-th literal in C, j points to the j-th literal in D and ϑ is a substitution. Initially $i = j = 1$ and $\vartheta = \epsilon$ and the truth value of "$C \leq_{ss} D$" is computed by defining $ST(C, D) = ST_0(C, D, 1, 1, \epsilon)$ for all clauses C, D (D ground). The exact specification of ST_0 can be found in Figure 4.4.

The execution of ST on clauses C, D can be described by a "process tree" $T(C, D)$ in which every matching attempt creates a new node.

Definition 4.3.1 (ST-tree). *Let $C : L_1 \vee \ldots \vee L_n$ and $D : K_1 \vee \ldots \vee K_m$ be clauses. We define the ST-tree $T(C, D)$ recursively by defining a set of nodes $\mathcal{N}_i$ and a set of edges $\mathcal{E}_i$ for every level $i \in \{0, \ldots, n\}$. Moreover, to every node there is a corresponding substitution.*
$i = 0 : \mathcal{N}_0 = \{(0,0)\}$ and $\mathcal{E}_0 = \emptyset$. The substitution corresponding to $(0,0)$ is the empty substitution ϵ.

Suppose now that $\mathcal{N}_i$ and $\mathcal{E}_i$ have already been constructed. If $i = n$ then the construction is completed and $T(C, D) = (\mathcal{N}_n, \mathcal{E}_n)$. Let us assume that $i < n$:
Let N be a node in $\mathcal{N}_i$ and ϑ be the substitution corresponding to N. Now

<u>function</u> $ST_0(C, D, i, j, \vartheta)$;
$\{C = L_1 \vee \ldots \vee L_{|C|}, \ D = K_1 \vee \ldots \vee K_{|D|}\}$
<u>begin</u>
    <u>if</u> $j > |D|$ <u>then</u> <u>return</u> <u>false</u> {the literals of D are "exhausted" }
    <u>else</u> <u>begin</u>
            { search for a matching literal in D}
            $a \leftarrow j$;
            <u>while</u> <u>not</u> unify$(L_i\vartheta, K_a)$ <u>and</u> $a \le |D|$
                <u>do</u> $a \leftarrow a + 1$;
            <u>if</u> $a > |D|$ <u>then</u> <u>return</u> <u>false</u> {no matching literal has been found}
            <u>else</u> <u>begin</u>
                    $\mu_i \leftarrow$ mgu$(L_i\vartheta, K_a)$; {extend the match ϑ}
                    <u>if</u> $i = |C|$ <u>or</u> $ST_0(C, D, i+1, 1, \vartheta\mu_i)$
                            <u>then</u> <u>return</u> <u>true</u> {the extension was successful}
                            <u>else</u> $ST_0(C, D, i, a+1, \vartheta)$ {backtracking}
            <u>end</u>
        <u>end</u>
<u>end</u> {ST}

Fig. 4.4. The algorithm of Stillman

let $j \in \{1, \ldots, m\}$ and μ be a substitution (defined on $V(L_{i+1}\vartheta)$) such that
$L_{i+1}\vartheta\mu = K_j$. *Then the node $(i+1, j)$ is in $\mathcal{N}_{i+1}$ and the edge $(N, (i+1, j))$*
is in $\mathcal{E}_{i+1}$; the substitution corresponding to $(i+1, j)$ is $\vartheta\mu$. If there exists no
K_j *such that $L_{i+1}\vartheta \le_s K_j$ then N is a leaf.*

Example 4.3.3. Let $C = P(x, y) \vee P(y, z) \vee P(x, z)$ and $D = P(a, b) \vee P(b, c) \vee$
$P(b, a)$. The ST-tree $T(C, D)$ is shown in Figure 4.5.

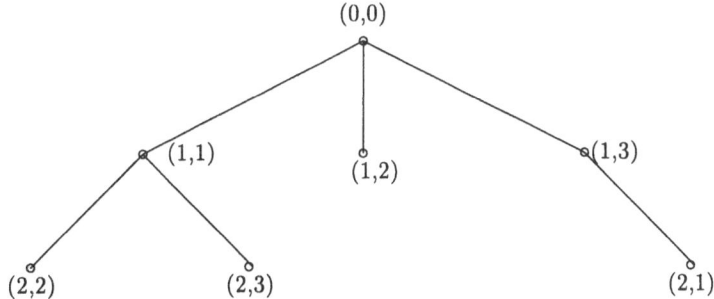

Fig. 4.5. An ST-tree

We write ϑ_{ij} for the substitution corresponding to the node (i, j). So we
obtain $\vartheta_{00} = \epsilon$, $\vartheta_{11} = \{x \leftarrow a, y \leftarrow b\}$ and $L_1\vartheta_{11} = K_1$. Now there exists
a substitution $\mu = \{z \leftarrow c\}$ such that $L_2\vartheta_{11}\mu = K_2$; the corresponding
node is $(2, 2)$ and $\vartheta_{22} = \{x \leftarrow a, y \leftarrow b, z \leftarrow c\}$. After computation of
ϑ_{22} we find no literal K_j such that $L_3\vartheta_{22} \le_s K_j$ and therefore $(2, 2)$ is a

leaf. At this point backtracking is applied and ST produces the substitution $\vartheta_{23} = \{x \leftarrow a, y \leftarrow b, z \leftarrow a\}$. But, here too, we obtain $L_3\vartheta_{23} \not\leq_s K_j$ for $j = 1, 2, 3$ and ST goes back to define ϑ_{12} etc. The sequence of calls of ST_0 is characterized by the sequence of equations:

$$ST_0(C, D, 1, 1, \epsilon) = ST_0(C, D, 2, 2, \vartheta_{11}) = ST_0(C, D, 3, 1, \vartheta_{22}) =$$
$$ST_0(C, D, 3, 2, \vartheta_{22}) = ST_0(C, D, 3, 3, \vartheta_{22}) = false;$$

$$ST_0(C, D, 1, 1, \epsilon) = ST_0(C, D, 1, 2, \epsilon) = ST_0(C, D, 2, 1, \vartheta_{12}) =$$
$$ST_0(C, D, 2, 2, \vartheta_{12}) = ST_0(C, D, 2, 3, \vartheta_{12}) = false; \text{ etc.}$$

Eventually all such call sequences end in $false$ and we obtain $ST_0(C, D, 1, 1, \epsilon) = false$; i.e., $ST(C, D) = false$ and C does not subsume D.

By definition of ST-trees, $ST(C, D)$ is true iff there exists a path of length $|D|$ in $T(C, D)$. The number of nodes in $T(C, D)$ coincides with the number of substitutions generated by ST and thus is an appropriate measure for the computing time of the algorithm.

Theorem 4.3.1. ST *is a decision algorithm for subsumption, i.e., (for $C \neq \Box$ and D ground)* $ST(C, D)$ *always terminates and* $ST(C, D) = true$ *iff* $C \leq_{ss} D$.

Proof. It is easy to see that ST always terminates: The call-depth of ST_0 is bounded by $|C|$, while the maximal number of calls corresponding to a node in the ST-tree is $|D|$. By definition of $T(C, D)$ the total number of calls is always less or equal to $|NOD(T(C, D))||D|$.

We now prove the correctness by induction on $|C|$.

$|C| = 1$: We have to compute $ST(C, D, 1, 1, \epsilon)$. The while-loop of ST_0 is entered with the values $a = 1$ and $\vartheta = \epsilon$.

case a): There exists an a such that $a \leq |D|$ and $\text{unify}(L_1, K_a) = true$. In this case there exists a ϑ such that $L_1\vartheta \in LIT(D)$. Because $C = L_1$ we obtain $C \leq_{ss} D$. But by $|C| = 1$ $ST_0(C, D, 1, 1, \epsilon)$ yields $true$.

case b): For all $a \leq |D|$ we obtain $\text{unify}(L_1, K_a) = false$. By definition of the while-loop in ST_0 we obtain $a = |D| + 1$ and thus $ST(C, D) = ST_0(C, D, 1, 1, \epsilon) = false$. Because all K_a are ground and $\text{unify}(L_1, K_a) = false$ we get $L_1 \not\leq_s K_a$ for all K_a in $LIT(D)$ and so $L_1 \not\leq_{ss} D$. But $C = L_1$.

(IH) Let us suppose that ST is correct on all pairs (C, D) such that $|C| \leq k$. Now let C be a clause such that $|C| = k + 1$ for $k \geq 1$.

case 1): The first execution of the while-loop gives $a = |D| + 1$. By definition of ST_0 we obtain $ST_0(C, D, 1, 1, \epsilon) = ST(C, D) = false$. But if a becomes $|D| + 1$ then there exists no ϑ such that $L_1\vartheta \in LIT(D)$; that means $L_1 \not\leq_{ss} D$. As $L_1 \leq_{ss} D$ is a necessary condition for $C \leq_{ss} D$ we indeed get $C \not\leq_{ss} D$.

case 2): The first execution of the while-loop gives an a with $a \leq |D|$ and mgu μ_1. Because $k + 1 > 1$, the call $ST_0(C, D, 2, 1, \mu_1)$ is performed. But $ST_0(C, D, 2, 1, \mu_1)$ gives the same value as $ST_0(C'\mu_1, D, 1, 1, \epsilon)$ for $C' = L_2 \vee \ldots \vee L_{k+1}$. Because $|C'| = k$ we obtain by (IH) that $ST_0(C'\mu_1, D, 1, 1, \epsilon) = true$ iff $C'\mu_1 \leq_{ss} D$. If $ST_0(C, D, 2, 1, \mu_1) = true$ then $ST_0(C, D, 1, 1, \epsilon) = true$ and $ST(C, D) = true$.
This value is indeed correct as $L\mu_1 \in LIT(D)$ and $LIT(C'\mu_1\vartheta) \subseteq LIT(D)$ for some ϑ. As $L\mu_1$ is already ground $L\mu_1\vartheta = L\mu_1$ and also $L\mu_1\vartheta \in LIT(D)$. Putting things together we obtain

$$LIT(C)\mu_1\vartheta = LIT(L\mu_1\vartheta \vee C'\mu_1\vartheta) \subseteq D$$

and therefore $C \leq_{ss} D$.

If $ST_0(C, D, 2, 1, \mu_1) = false$ then we may search for a new substitution μ_1. This case is "subsumed" under those to be discussed below:

In general we may obtain several substitutions μ_{1i} by the execution of the while-loop corresponding to the nodes of depth 1 in $T(C, D)$.
If $ST_0(C, D, 2, 1, \mu_{1i})$ gives true for such a μ_{1i} the same argument as before yields $C \leq_{ss} D$ and, by definition of ST and ST_0, $ST(C, D) = true$.
So let us suppose that for all μ_{1i} defined on depth 1 in $T(C, D)$ we obtain $ST_0(C, D, 2, 1, \mu_{1i}) = false$. That means for all K in $LIT(D)$ such that $L_1 \leq_s K$ via μ we obtain $C'\mu \not\leq_{ss} D$.

(*) $C \leq_{ss} D$ iff there exists a ϑ such that $dom(\vartheta) = V(L_1), L_1\vartheta \in LIT(D)$ and $(C \setminus L_1)\vartheta \leq_{ss} D$.

Indeed (*) is just the recursive definition of $\leq_{ss}$. But ST_0 produces all substitutions ϑ with $dom(\vartheta) = V(L_1)$ and $L_1\vartheta \in LIT(D)$. Let us call this set of substitutions Θ. By assumption $ST_0(C, D, 2, 1, \vartheta) = false$ for all $\vartheta \in \Theta$ (Θ consists just of the $\mu'_{1i}s$).
But $ST_0(C, D, 2, 1, \vartheta) = ST_0(C'\vartheta, D, 1, 1, \epsilon)$ and so $ST_0(C'\vartheta, D, 1, 1, \epsilon) = false$ for all $\vartheta \in \Theta$. By(IH) we obtain $C'\vartheta \not\leq_{ss} D$ for all $\vartheta \in \Theta$. But, by (*) and by definition of Θ we obtain the following property:
$C \leq_{ss} D$ iff there exists a $\vartheta \in \Theta : L_1\vartheta \in LIT(D)$ and $C'\vartheta \leq_{ss} D$.
As there is no such $\vartheta \in \Theta$ we obtain $C \not\leq_{ss} D$.
By definition of ST_0 we also obtain $ST_0(C, D, 1, 1, \epsilon) = false$ and, consequently, $ST(C, D) = false$. This concludes the analysis of the case $|C| = k + 1$. $\Diamond$

A closer look at the proof of Theorem 4.3.1 reveals that ST is merely an operational interpretation of the following recursive definition (D ground):

$|C| = 0 : C \leq_{ss} D$ (note that $C = \square$)

$|C| > 0 : C \leq_{ss} D$ iff there exists a ϑ such that $L_1\vartheta \in LIT(D)$ and $(C \setminus L_1)\vartheta \leq_{ss} D$.

The main task of ST is to produce all candidates ϑ for the recursion above. ST is easy to specify and to implement, but may be highly inefficient on larger clauses. The reason is that, in the worst case, we may obtain a tree $T(C, D)$ where all paths are of length $|C| - 1$ and the degree of all nodes (except the leaf nodes) is $|D|$. The total number of nodes in such a tree is of the order $|D|^{|C|}$. Henceforth we measure the computational expenses of ST on a pair (C, D) by the number of nodes in $T(C, D)$. Formally we define

$$stn(C, D) = |NOD(T(C, D))|.$$

Example 4.3.4: Let $(C_n, D_n)_{n \geq 2}$ be a sequence of pairs of clauses such that

$$C_n = P(x_1) \vee \ldots \vee P(x_{n-1}) \vee P(f_n(x_1, \ldots, x_n))$$

and

$$D_n = P(a_1) \vee \ldots \vee P(a_{n-1}) \vee P(f_n(b_1, \ldots, b_n))$$

where the f_n are n-place function symbols and the a_i, b_j are constant symbols which are pairwise different. It is easy to verify that $T(C_n, D_n)$ is a tree fulfilling the following property:

a) Every node that is not a leaf is of degree n
b) Every branch (= path from the root to a leaf) is of length $n - 1$.

Therefore

$$stn(C_n, D_n)) = \sum_{i=0}^{n-1} n^i + 1 = \frac{(n^n - 1)}{n - 1} \in O(n^{n-1}).$$

For every leaf-node we have n additional matching attempts. Thus, by counting also failing attempts to extend substitutions, we have about n^n computing steps. Now let us reorder the literals in C_n and arrange them as follows:

$$E_n : P(f_n(x_1, \ldots, x_n)) \vee P(x_1) \vee \ldots \vee P(x_{n-1}).$$

It is easy to see that $stn(E_n, D_n) = 2$, i.e., $T(E_n, D_n)$ consists of 2 nodes only (independent of n): After mapping $P(f_n(x_1, \ldots, x_n))$ to $P(f_n(b_1, \ldots, b_n))$ (the only possibility!) the remaining part of the E_n-clause is:

$$E_n' : P(b_1) \vee \ldots \vee P(b_{n-1}).$$

E_n' is ground and $P(b_1) \notin LIT(D_n)$; so no further substitutions are generated and $E_n' \not\leq_{ss} D_n$. Because ϑ was the only matching candidate for $P(f_n(x_1, \ldots, x_n))$ we eventually obtain $E_n \not\leq_{ss} D_n$. Example 4.3.4 clearly shows that, in testing $\leq_{ss}$ for larger clauses, it may be a wrong choice just to proceed from left to right – as the clause is written down. What we need is some analysis of C, which can easily be performed and defines some preference strategy based on the structure of the variable occurrences in C. The basic idea is to identify cases where subsumption is fast (i.e., polynomial of a low degree) and to try to reduce the subsumption problems to such cases. There are two typical cases:

1) The literals in C are all variable-disjoint and
2) $|V(C)| \leq 1$.

Definition 4.3.2 (the clause type DEC). *Let* $C : L_1 \vee \ldots \vee L_n$ *be a clause. C is called* decomposed *if for all $i, j \leq n$ and $i \neq j : V(L_i) \cap V(L_j) = \emptyset$. The class of all decomposed clauses is denoted by* DEC.

Definition 4.3.3 (the clause type V1C). V1C *is the class of all clauses C with $|V(C)| \leq 1$.*

Every ground clause belongs to $DEC \cap V1C$. Both classes are characterized by weak internal variable connections.

Definition 4.3.4 (simple clauses). *A clause is called* simple *if it is in* DEC $\cup$ V1C.

ST is fast on V1C (the ST-tree is at most quadratic), but may be exponential on DEC due to superfluous backtracking. To obtain a problem in DEC we change the sequence (C_n, D_n) from Example 4.3.4 to

$$C_n : P(x_1) \vee \ldots \vee P(x_{n-1}) \vee P(g(x_n))$$

and

$$D_n : P(a_1) \vee \ldots \vee P(a_{n-1}) \vee P(a_n).$$

Then, as before,

$$|NOD(T(C_n, D_n))| = \frac{n^n - 1}{n - 1}.$$

But it is easy to see that all literals in C_n can be tested independently. For technical reason we introduce a "parallel" version of ST which is correct on DEC but incorrect in general.

Let $C : L_1 \vee \ldots \vee L_n$ and D be clauses (as always we assume D to be ground); we define the Boolean function STP (parallel Stillman algorithm):

$$\text{STP}(C, D) = \bigwedge_{i=1}^{n} \text{ST}(L_i, D).$$

The number of substitutions generated by STP is given as the sum of all $stn(L_i, D)$. So we may assume that, in the computation of $\text{STP}(C, D)$, the $\text{ST}(L_i, D)$ are computed first. The expense created by the computation of $\text{STP}(C, D)$ out of the values $\text{ST}(L_i, D)$ is inessential (at most linear in n); so to facilitate the analysis we don't count these steps at all. Thus we measure the computational expenses of STP by the new measure *stpn* defined by:

$$stpn(C, D) = \sum_{i=1}^{n} stn(L_i, D).$$

Proposition 4.3.1. *Let C be a clause in* DEC. *Then* $STP(C, D) = true$ *iff* $C \leq_{ss} D$ *(STP is correct on DEC).*

Proof. Let $C = L_1 \vee \ldots \vee L_n$. $C \leq_{ss} D$ clearly implies $STP(C, D) = true$ for all clauses C. By Theorem 4.3.1 $ST(L_i, D) = true$ iff $L_i \leq_{ss} D$. But $L_i \leq_{ss} D$ for $i = 1, \ldots, n$ is a necessary condition for $C \leq_{ss} D$. Moreover, by definition of STP, $STP(C, D) = true$ iff for all $i = 1, \ldots, n : ST(L_i, D) = true$. This concludes the proof of the first direction.

By the correctness of ST there exist substitutions $\vartheta_1, \ldots, \vartheta_n$ such that $dom(\vartheta_i) = V(L_i)$ and $L_i\vartheta_i \in LIT(D)$. By $C \in$ DEC we have $V(L_i) \cap V(L_j) = \emptyset$ for $i \neq j$ and therefore $dom(\vartheta_i) \cap dom(\vartheta_j) = \emptyset$ for $i \neq j$. But then the substitution $\vartheta, \vartheta : \vartheta_1 \cup \ldots \cup \vartheta_n$, is well-defined and $L_i\vartheta = L_i\vartheta_i$ for $i = 1, \ldots, n$. So we obtain $L_i\vartheta \in LIT(D)$ for $i = 1, \ldots, n$, i.e., $LIT(C)\vartheta \subseteq LIT(D)$, and therefore $C \leq_{ss} D$.   ◊

Proposition 4.3.2. $stpn(C, D) \leq 2|C|$.

Proof. Let $C = L_1 \vee \ldots \vee L_n$. By definition of *stpn* we have $stpn(C, D) = \Sigma_{i=1}^{n} stn(L_i, D)$. But the trees $T(L_i, D)$ cannot contain more than one edge (and therefore not more than two nodes). Note that in the computation of $ST(L_i, D)$ at most one of the substitutions μ_1 is constructed. So we obtain $stpn(C, D) \leq 2n = 2|C|$.   ◊

Corollary 4.3.1. *The subsumption test $C \leq_{ss} D$ for $C \in$ DEC is decidable in quadratic time.*

Proof. Let $C = L_1 \vee \ldots \vee L_n$ as in Proposition 4.3.2. Then the root of $T(L_i, D)$ corresponds to at most $|D|$ unification attempts. Thus the total number of unification attempts is bounded by $|C||D|$. If $T(L_i, D)$ contains an edge then the leaf node does not correspond to further unification attempts (success has already been achieved).   ◊

Proposition 4.3.3. *Let C be a ground clause. Then $stn(C, D) \leq |C| + 1$.*

Proof. Let $C = L_1 \vee \ldots \vee L_n$ and all L_i be ground. Then all nodes in $T(C, D)$ are of degree ≤ 1, because no literal L_i can be mapped to more than one literal in D (note that all clauses are reduced). The depth of $T(C, D)$ is bounded by $|C|$. Thus (including the root) there can be no more than $|C| + 1$ nodes in $T(C, D)$. In fact, $C \leq_{ss} D$ iff the only path in $T(C, D)$ is of length $|C|$.   ◊

Now we are in the position to estimate the expenses of ST on the class V1C. We will see that the ST-trees for subsumption problems belonging to V1C are of at most quadratic size (in the number of literals in C and D).

Proposition 4.3.4. *Let C be in V1C and D be an arbitrary ground clause, $|C| \geq 1$ and $|D| \geq 1$. Then $stn(C, D) \leq |C||D| + 1$.*

Proof. If C is ground then Proposition 4.3.3 yields $stn(C, D) \leq |C| + 1$. But by $|D| \geq 1$ we have $|C| + 1 \leq |C||D| + 1$.

It remains to analyze the case $|V(C)| = 1$. For this purpose let us assume that $V(C) = \{x\}$. Then C can be written in the form $C_1 \vee L_i \vee C_2$ such that $C_1 : L_1 \vee \ldots \vee L_{i-1}$ is ground (possibly $C_1 = \square$) and L_i is the first literal in C (from the left) containing the variable x.

Then $T(C_1, D)$ is an initial segment of $T(C, D)$. In the proof of Proposition 4.3.3 we have demonstrated that $T(C_1, D)$ can only consist of a single path and that the length of this path is $\leq |C_1| = i - 1$.

If the length of this path is less than $i - 1$ then $ST(C_1, D)$ returns *false* and, by definition of ST, $ST(C, D)$ returns *false*. In this case we have

$$stn(C, D) = stn(C_1, D) < i - 1 < |C|.$$

So let us assume that the length of the branch in $T(C_1, D)$ is $i - 1$. Then

$$stn(C, D) = stn(C_1, D) + stn(L_i \vee C_2, D) - 1.$$

Note that the root node of $T(L_i \vee C_2, D)$ has to be identified with the (only) leaf node in $T(C_1, D)$. By definition of ST_0, $ST_0(L_i \vee C_2, D, 1, 1, \epsilon)$ reduces to $ST_0(L_i \vee C_2, 2, 1, \mu_1)$ for some substitution $\mu_1 : \{x \leftarrow t\}, t$ being a ground term. But

$$ST_0(L_i \vee C_2, 2, 1, \mu_1) = ST_0(C_2\mu_1, 1, 1, \epsilon).$$

There can be at most $|D|$ possible substitutions μ_1. No matter what they really look like, the clause $C_2\mu_1$ is ground (note that $V(C_2) \subseteq \{x\}$). But

$$ST(C_2\mu_1, D) = ST_0(C_2\mu_1, 1, 1, \epsilon)$$

and, by Proposition 4.3.3,

$$stn(C_2\mu_1, D) \leq |C_2\mu_1| + 1 \leq |C_2| + 1 = |C| - |C_1|.$$

Therefore $stn(L_i \vee C_2, D) \leq 1 + |D|(|C| - |C_1|)$. Putting things together we eventually obtain

$$stn(C, D) = stn(C_1, D) + stn(L_i \vee C_2, D) - 1 \leq$$

$$|C_1| + |D|(|C| - |C_1|) + 1 \leq |C||D| + 1.$$

$\Diamond$

On the basis of the propositions derived so far we can define an efficient (polynomial) decision algorithm for "simple" subsumption problems. If C is a simple clause then, by definition either $C \in$ DEC or $C \in$ V1C. For $C \in$ DEC we apply STP and for $C \in$ V1C simply ST itself. The corresponding algorithm is called SSIMP and is shown in Figure 4.6.

SSIMP(C, D):
begin
    if $C \in$ DEC then return STP(C, D)
    else if $C \in$ V1C then return ST(C, D)
        else error (C is not simple)
end.

Fig. 4.6. A subsumption algorithm for simple clauses

It is easy to see that the total number of nodes in the ST-trees generated by SSIMP is at most quadratic in $max\{|C|, |D|\}$ (exercise 4.3.1).

Every clause C can be split into components $C_1, \ldots, C_k$ such that $LIT(C_1) \cup \ldots \cup LIT(C_k) = LIT(C)$, $V(C_i) \cap V(C_j) = \emptyset$ and $LIT(C_i) \cap LIT(C_j) = \emptyset$ for $i \neq j$. If $C \in$ DEC then C defines $|C|$ one-element components. In Propositions 4.3.1 and 4.3.2 we have shown that parallelizing ST to STP turns out to be fruitful on DEC. But "typical" clauses are not in DEC. On the other hand, mapping a literal from C to D in the ST-algorithm introduces constants into the clause C (note that D is ground); the introduction of constant terms by matching may result in a decomposition (where there was none before). The algorithm DC (division into components [GL85]) is based on the idea to find candidates L in C which optimize the decomposition of C. The aim is to iterate splitting until all components are simple and then to apply SSIMP.

Definition 4.3.5. *Let C be a clause and $\sim_C$ be the equivalence relation on $LIT(C)$ induced by the following relation $\sim_{vc}$: $L \sim_{vc} M$ iff $V(L) \cap V(M) \neq \emptyset$. The equivalence classes in $LIT(C)$ under $\sim_C$ are called* connected components *and are denoted by $COMP(C)$.*

Note that $\sim_{vc}$ is reflexive and symmetric, but not transitive. Thus in order to get the equivalence relation $\sim_C$ we must construct the transitive closure of $\sim_{vc}$. In [GL85] the $\sim_{vc}$-relation was represented by a graph such that the pair of literals (L, M) defines an edge iff $V(L) \cap V(M) \neq \emptyset$. Although the graph structure is very appropriate for investigating good decomposition algorithms, it is easier to represent the DC-method via connected components.

Example 4.3.5.
We consider the sequence (C_n, D_n) defined in Example 4.3.4 for $n \geq 2$:

$$C_n \ : \ P(x_1) \vee \ldots \vee P(x_{n-1}) \vee P(f_n(x_1, \ldots, x_n)),$$
$$D_n \ : \ P(a_1) \vee \ldots \vee P(a_{n-1}) \vee P(f_n(b_1, \ldots, b_n)).$$

Then $P(f_n(x_1, \ldots x_n)) \sim_{vc} P(x_i)$ for all $i = 1, \ldots, n-1$. As $\sim_C$ is the transitive closure of $\sim_{vc}$, C_n consists of one connected component only, i.e., $COMP(C_n) = LIT(C_n)$.

Let us investigate the matching substitutions ϑ of the literals in C_n with respect to D. Then either ϑ is of the form $\{x_i \leftarrow t\}$ or $\vartheta = \{x_1 \leftarrow b_1, \ldots, x_n \leftarrow b_n\}$. C_n is not simple and for all ϑ, such that $\vartheta = \{x_i \leftarrow t\}$, and $n \geq 3$ $C_n\vartheta$ is not simple either.

But by selecting $\vartheta = \{x_1 \leftarrow b_1, \ldots, x_n \leftarrow b_n\}$ we achieve that $C_n\vartheta$ is ground and therefore simple. So we are advised to select $P(f_n(x_1, \ldots, x_n))$ as first literal and to reduce the problem under the matching substitution $\{x_1 \leftarrow b_1, \ldots x_n \leftarrow b_n\}$. We see that "grounding" the last literal in C_n maximally decomposes C_n. The graph representation of $\sim vc$ shown in Figure 4.7 provides intuitive evidence for the choice above.

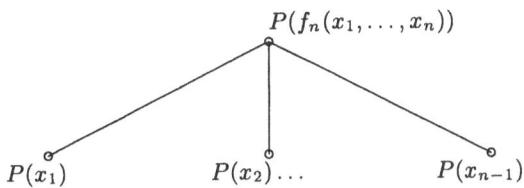

Fig. 4.7. Graph of $\sim_{vc}$

Let $\{L_1, \ldots, L_m\}$ be an element of $COMP(C)$; then we call any reduced disjunction of the $L_1, \ldots, L_m$ (in fact a permutation of $L_1 \vee \ldots \vee L_m$) a clause form of $\{L_1, \ldots, L_m\}$. A component is called simple if a corresponding clause form is simple (note that the property of being simple is independent of the ordering of literals).

We now explain how DC works in general: Let us consider the problem $C \leq_{ss} D$. Initially we compute $COMP(C)$. If all components are simple we test $C_i \leq_{ss} D$ by SSIMP in parallel, where C_i is a clause representation of a $\sim_C$ component $E_i \in COMP(C)$. If there are nonsimple components then we test the simple ones by SSIMP and treat the other components E_i as follows: Let C_i be a clause form of E_i such that C_i is not simple. We select a literal L_{first} from C_i and start an ST-like algorithm on L_{first}. The choice of L_{first} is crucial to the computational behavior of the algorithm. The following choice leads to a strong improvement of the asymptotic worst case behavior with respect to ST:

Let C be a nonsimple clause. Then there are literals containing more than one variable. So we define:

L_{first} = The first literal L such that $|V(L)| \geq 2$ and L contains a maximal number of variables which also occur in other literals of C.

After the selection of L_{first} we try to map L_{first} to D via a matching substitution ϑ. ϑ is a ground substitution and (possibly) decomposes C_i into different components. At this point the algorithm treats the components recursively etc.

Before giving a formal definition of DC we show that different components can indeed be treated independently.

Proposition 4.3.5. *Let $E_1, \ldots, E_n$ be the connected components of a clause C and $C_1, \ldots, C_n$ be their clause forms. Then $C \leq_{ss} D$ iff for all $i = 1, \ldots, n$, $C_i \leq_{ss} D$.*

Proof. Let us suppose that $C_i \leq_{ss} D$ holds for all $i = 1, \ldots, n$. Then $C_0 : C_1 \vee \ldots \vee C_n$ is a permutation of the clause C and, clearly, $C \leq_{ss} D$ iff $C_0 \leq_{ss} D$. Thus it is sufficient to prove $C_0 \leq_{ss} D$.

By $C_i \leq_{ss} D$ there are substitutions ϑ_i such that $dom(\vartheta_i) \subseteq V(C_i)$ and $LIT(C_i)\vartheta_i \subseteq LIT(D)$. By definition of connected components we have $V(C_i) \cap V(C_j) = \emptyset$ for $i \neq j$; so we also obtain $dom(\vartheta_i) \cap dom(\vartheta_j) = \emptyset$ for $i \neq j$. But then $\vartheta : \vartheta_1 \cup \ldots \cup \vartheta_n$ is a (well-defined) substitution with the property: $C_i\vartheta = C_i\vartheta_i$ for $i = 1, \ldots, n$. Therefore $LIT(C_i)\vartheta \subseteq LIT(D)$ for all $i = 1, \ldots, n$ and so $LIT(C_0)\vartheta \subseteq LIT(D)$, i.e., $C_0 \leq_{ss} D$.

For the other direction let us assume $C \leq_{ss} D$. Again, $C_0 \leq_{ss} D$ is an immediate consequence. But $E \vee F \leq_{ss} D$ always implies $E \leq_{ss} D$ and $F \leq_{ss} D$ for arbitrary clauses E and F. Therefore $C_i \leq_{ss} D$ must hold for all $i = 1, \ldots, n$.   ◇

The last proposition provides a justification for the algorithm DC presented in Figures 4.8 and 4.9. If E is a component of a clause we write $\hat{E}$ for a clause form of E.

Theorem 4.3.2. *DC is a decision algorithm for subsumption.*

Proof. (sketch) By Proposition 4.3.5 $C \leq_{ss} D$ iff for all component clauses C_i, $C_i \leq_{ss} D$. In DC the test of the components is either performed by SSIMP or TC. DC return true iff true is obtained for all components. Thus the correctness is reduced to that of SSIMP and TC. The correctness of SSIMP follows from the correctness of ST (Theorem 4.3.1) and that of STP (Proposition 4.3.1). TC proceeds almost like ST, with the exception that TC calls DC instead of itself. The further reduction to the correctness of

```
function DC(C,D) {tests C ≤ₛₛ D};
begin
    ans ← true;
    construct COMP(C);
    for each E ∈ COMP(C) do
        if Ê is simple then ans ← ans and SSIMP (Ê, D)
        else ans ← (ans and TC(Ê, D));
    return ans
end {DC}.
```

Fig. 4.8. The algorithm DC

```
TC(C,D) {TC means "test components"};
begin
    determine L_first in C.
    a ← 1;
    repeat
        while a ≤ |D| and not unify(L_first, Kₐ) do a ← a + 1;
        if a > |D| then sub ← false
            else if |C| = 1 then sub ← true
                else begin
                    μ ← mgu(L_first, Kₐ);
                    if DC(Cμ \ L_first μ, D) then sub ← true
                        else sub ← false
                end;
        a ← a + 1;
    until a > |D| or sub {end repeat};
    return sub
end.
```

Fig. 4.9. The algorithm TC

DC (again) is not cyclic, as the clauses in the call become smaller. Thus the correctness of TC follows from that of ST and an inductive argument reducing correctness to DC again. The formal treatment of this induction is left as an exercise (Exercise 4.3.2).   ◊

The power of DC lies in the recursive use of splitting by a clever selection of the literal L_{first}. Consider, for example, the clause $C_n : P(x_1) \vee \ldots \vee P(x_{n-1}) \vee P(f_n(x_1, \ldots, x_n))$ from Example 4.3.5. $P(f_n(x_1, \ldots, x_n))$ is the only literal containing at least two variables, moreover it contains $n - 1$ variables which also occur in other literals. Therefore the literal $P(f_n(x_1, \ldots, x_n))$ is selected as L_{first} in TC (note that DC calls TC because C_n is not simple). In the next call (of DC from TC) the clause is ground and thus decomposes into simple components. The rest is done by testing $P(b_i) \leq_{ss} D_n$ via ST in parallel.

Note that the number of required subsumption tests in refinements of the type R_{xs} or R_{xr} is roughly quadratic in the number of generated clauses.

Therefore the preprocessing of clauses and the selection of L_{first} do not really matter; even in the worst-case behavior, the computation of the components and of L_{first} is at most quadratic. For practical applications some additional techniques should be added to DC; one such technique is recursive filtering leading to an algorithm DCI ([GL85a]). A detailed empirical analysis and comparison of different algorithms can be found in [Im85]. The problem of optimizing DC (by an adequate choice of L_{first}) is related to the problem of graph decomposition and thus is an interesting problem of combinatorics. In [GL85] it is shown that (using the current choice principle for L_{first}) the worst-case complexity of $O(n^n)$ (of ST) is reduced to about $O(n^{n/2})$. Moreover, there are many cases (see Example 4.3.4) where ST is exponential but DC is polynomial.

Exercises

Exercise 4.3.1. Show that the total number of nodes in ST-trees generated by SSIMP is at most quadratic in $max\{|C|, |D|\}$ (C, D being the clauses for the $\leq_{ss}$-test).

Exercise 4.3.2. Give a mathematical induction proof of the correctness of DC (note that the correctness of DC and TC must be proven "simultaneously").

4.4 The Elimination of Tautologies

Clauses which are subsumed are redundant with respect to other clauses. Thus subsumption is a principle of "relative" redundancy. However, there is a type of clauses which can be considered as "absolutely" redundant; such are the tautological ones. Clearly a clause is a tautology iff it contains a pair of complementary literals. In most cases (of refinements) tautologies can be eliminated without loss of completeness. As in the case of subsumption tautology-elimination always leads to the reduction of search complexity.

First of all it is easy to verify that the elimination of tautologies – applied as preprocessing – preserves completeness. Let $\mathcal{C}$ be a set of clauses and C be a tautological clause in $\mathcal{C}$. Then $F(\mathcal{C} - \{C\}) \sim F(\mathcal{C})$ and therefore $\mathcal{C}$ may be replaced by $\mathcal{C} - \{C\}$. Moreover it is easy to see that R-deductions without tautologies define a complete deduction principle. Recall the proof of completeness of GR-deduction in Section 2.5. In this proof only clauses are resolved which are falsified at failure nodes in a semantic tree. But tautologies cannot be falsified at any node in a semantic tree (simply because there are no falsifying interpretations). Therefore we do not need tautologies in the set of ground instances that has to be refuted. Moreover tautologies can never be created by lifting (if $C\vartheta$ is not a tautology then also C cannot be one). Thus we see that there are always R-refutations that do not contain tautologies, i.e.,

R-deduction without tautologies is complete. The following example shows a close connection between subsumption and tautological clauses with respect to resolution.

Example 4.4.1. Let $C : \{C_1, \ldots, C_5\}$ be the following set of N_c-clauses:

$$C_1 : P(f(x_1)) \vee R(x_1) \vee \neg P(f(x_1)), \quad C_2 : P(x_1) \vee Q(x_1),$$
$$C_3 : R(f(x_1)), \quad C_4 : Q(x_1) \vee \neg R(x_1), \quad C_5 : \neg Q(f(x_1)).$$

C_1 is a tautology, but it is not subsumed by another clause in C. C_1 defines three resolvents in C, the first with C_2, the second with C_4 and the third with itself (note that tautologies always admit self-resolution). The N_c-resolvent with C_2 is:

$$C_6 : P(f(x_1)) \vee Q(f(x_1)) \vee R(x_1).$$

C_6 is not a tautology but it is subsumed by its parent clause C_2. The N_c-resolvent of C_1 and C_4 is:

$$C_7 : P(f(x_1)) \vee Q(x_1) \vee \neg P(f(x_1)).$$

Obviously C_7 is a tautology. There are even two different ways to resolve C_1 with itself, but these are symmmetric; the only N_c-resolvent of C_1 with itself is clearly C_1 itself.

Example 4.4.1 showed that resolving with tautologies either yields tautologies again or clauses which are subsumed by one of the parent clauses. The following proposition shows that this property is not specific to this example only, but is a general principle.

Proposition 4.4.1. *Let C, D be clauses and C be a tautology. Then every resolvent of C and D is either a tautology or is subsumed by D.*

Proof. We may assume without loss of generality that $V(C) \cap V(D) = \emptyset$. As C is a tautology it must be of the form:

$$C : C_1 \vee A \vee C_2 \vee A^d \vee C_3$$

for some literal A. Let ϑ be a G-instance of C defining a factor C'. Then ϑ cannot unify A and A^d; consequently C' must be of the form

$$C' : C_1' \vee A' \vee C_2' \vee A'^d \vee C_3' \text{ or } C' : C_1' \vee A'^d \vee C_2' \vee A' \vee C_3'$$

for $A\vartheta = A'$ (note that the internal ordering of A' with respect to A'^d may be different from that of A with respect to A^d). It suffices to focus on the first form of C'. Now let D' be a factor of D such that

$$D' = D_1' \vee L \vee D_2'$$

and L be the literal in D' selected for cut via binary resolution.

case a) The literal resolved from C' is contained in $C_1' \vee C_2' \vee C_3'$: Let σ be the m.g.u. of the resolution. Then $A'\sigma$ and $A'^d\sigma$ are both contained in $LIT(R)$ for the resolvent R; therefore R is a tautology.

case b) The literal resolved from C' is A': Let σ be an m.g.u. of $\{A', L'^d\}$ and R be the resolvent $R : (C_1' \vee C_2' \vee A'^d \vee C_3' \vee D_1' \vee D_2')\sigma$.
From $A'\sigma = L^d\sigma$ we derive $A'^d\sigma = L\sigma$ and so $L\sigma \in LIT(R)$. Moreover $LIT((D_1' \vee D_2')\sigma) \subseteq LIT(R)$, that means $D' \leq_{ss} R$ via σ.

case c) The literal resolved from C' is A'. This case is completely symmetric to b).

We see that, in case a), R is a tautology; in the cases b) and c) R is subsumed by D.   ◇

Proposition 4.4.1 suggests that, in the presence of a subsumption strategy, tautologies may be redundant and thus can be deleted. If they don't create new tautologies then they produce new clauses which are eliminated by subsumption anyway. In order to turn this observation into an exact result we give a mathematical definition of tautology-elimination within an operator-based refinement.

Definition 4.4.1. *Let C be a set of clauses. By* TAUTEL(C) *we denote the set $C - $ TAUT(C), where* TAUT(C) *denotes the set of all tautological clauses in C. Now let R_x be an arbitrary resolution refinement. We define R_{xT} (R_x under elimination of tautologies) by:*

$$S_{xT}^0(C) = \text{TAUTEL}(C), \ \ S_{xT}^{i+1}(C) = S_{xT}^i(C) \cup \text{TAUTEL}(\rho_x(S_{xT}^i(C))),$$

$$R_{xT}(C) = \bigcup_{i=0}^{\infty} S_{xT}^i(C).$$

If R_x is a refinement then R_{xs} (R_x + forward subsumption) defines a refinement too. Consequently R_{xsT} represents R_x + forward subsumption + elimination of tautologies.

Many refinements can be characterized by subsets of $Res(\{C_1, C_2\})$ for (two) clauses $C_1, C_2 \in C$. In such a case we speak about a binary refinement (for short B-refinement).

Definition 4.4.2 (B-refinements). *A refinement R_x (defined on N-clauses for some normalization operator N) is called a B-refinement if for all sets of N-clauses C: $\rho_x(C) \subseteq N(Res(C))$.*

A-ordering refinements are B-refinements. Lock resolution is a B-refinement on indexed clauses; however it is not a B-refinement on ordinary clauses, because the indices somehow reflect the history of the clauses. Hyperresolution is not a B-refinement. The following theorem shows that, in the presence of forward subsumption, the elimination of tautologies cannot destroy the completeness of B-refinements.

Theorem 4.4.1. *If a B-refinement is complete under forward subsumption then it remains complete under (additional) deletion of tautologies, or more formally: Let R_x be a B-refinement such that R_{xs} is complete; then R_{xsT} is complete too.*

Proof. We have to show that $\square \in R_{xs}(\mathcal{C})$ implies $\square \in R_{xsT}(\mathcal{C})$. It suffices to show (by induction on i):

(*) If $D \in S_{xs}^i(\mathcal{C}) - S_{xsT}^i(\mathcal{C})$ then D is a tautology.

Suppose that (*) has already been proved: Because $\square$ is not a tautology $\square \notin S_{xs}^i(\mathcal{C}) - S_{xsT}^i(\mathcal{C})$ for all i. By the completeness of R_{xs} there exists a k such that $\square \in S_{xsT}^k(\mathcal{C})$ too. It remains to prove (*):

$i = 0$: $S_{xs}^0(\mathcal{C}) - S_{xsT}^0(\mathcal{C}) = \text{TAUT}(\mathcal{C})$ by definition of S_{xs} and S_{xsT}. Clearly (*) is fulfilled for $i = 0$.

(IH) Suppose that $S_{xs}^i(\mathcal{C}) - S_{xsT}^i(\mathcal{C})$ contains tautologies only.

case $i + 1$:
Suppose that $D \in S_{xs}^{i+1}(\mathcal{C}) - S_{xsT}^{i+1}(\mathcal{C})$.
Because R_{xs} is a B-refinement there are clauses $E, F \in S_{xs}^i(\mathcal{C})$ such that $D \in \rho_x(\{E, F\})$.

case a:
Both E and F are in $S_{xsT}^i(\mathcal{C})$: From $D \in S_{xs}^{i+1}(\mathcal{C})$ we conclude that

either $D \in S_{xs}^i(\mathcal{C})$ or $(D \in \rho_x(S_{xsT}^i(\mathcal{C}))$ and $S_{xs}^i(\mathcal{C}) \not\leq_{ss} \{D\})$.

If $D \in S_{xs}^i(\mathcal{C})$ then

$D \in S_{xs}^i(\mathcal{C}) - S_{xsT}^{i+1}(\mathcal{C}) \subseteq S_{xs}^i(\mathcal{C}) - S_{xsT}^i(\mathcal{C})$ and, by (IH), D is a tautology.

So let us assume $S_{xs}^i(\mathcal{C}) \not\leq_{ss} \{D\}$; clearly $D \in \rho_x(S_{xsT}^i(\mathcal{C}))$ by assumption a). Note that $D \notin S_{xsT}^{i+1}(\mathcal{C})$ but $D \in sf(S_{xsT}^i(\mathcal{C}), \rho_x(S_{xsT}^i(\mathcal{C})))$. By definition of R_{xsT}, D must be a tautology.

case b:
$E \notin S_{xsT}^i(\mathcal{C})$, $F \in S_{xsT}^i(\mathcal{C})$: Using $E \in S_{xs}^i(\mathcal{C}) - S_{xsT}^i(\mathcal{C})$ and (IH) we conclude that E must be a tautology. From Proposition 4.4.1 and the fact that D is a resolvent of E and F we derive that either D is a tautology or $F \leq_{ss} D$. If D is a tautology then we have achieved our goal. So let us assume that $F \leq_{ss} D$. Then, clearly, $S_{xs}^i(\mathcal{C}) \leq_{ss} D$ and, by definition of forward subsumption, $D \notin S_{xs}^{i+1}(\mathcal{C}) - S_{xs}^i(\mathcal{C})$.

By assumption we have $D \in S_{xs}^{i+1}(\mathcal{C}) - S_{xsT}^{i+1}(\mathcal{C})$.

We conclude that $D \in S_{xs}^{i}(\mathcal{C}) - S_{xsT}^{i+1}(\mathcal{C})$

and by $S_{xsT}^{i}(\mathcal{C}) \subseteq S_{xsT}^{i+1}(\mathcal{C})$:

$$D \in S_{xs}^{i}(\mathcal{C}) - S_{xsT}^{i}(\mathcal{C}).$$

By (IH) D must be a tautology.

case c:

$E \in S_{xsT}^{i}(\mathcal{C})$, $F \notin S_{xsT}^{i}(\mathcal{C})$: This case is completely symmetric to b).

case d:

$E \notin S_{xsT}^{i}(\mathcal{C})$ and $F \notin S_{xsT}^{i}(\mathcal{C})$: Here we have $E, F \in S_{xs}^{i}(\mathcal{C}) - S_{xsT}^{i}(\mathcal{C})$ and, by (IH), E and F are both tautologies. But D as a resolvent of two tautologies must be a tautology too (note that resolution is a sound inference rule). This concludes the proof of case $i + 1$.   ◊

Corollary 4.4.1. *All A-ordering refinements remain complete under forward subsumption and elimination of tautologies.*

Proof. Immediately from Theorems 4.2.1 and 4.4.1.   ◊

Let R_{xs} be a refinement with forward subsumption. If R_{xs} is complete then we may always add tautology-elimination. To refine R_{xs} to R_{xsT} not only preserves completeness but also reduces the search space. Looking more closely at the proof of theorem 4.4.1 we see that the minimal refutation depth cannot be increased by elimination of tautologies:

$$\text{If } \square \in S_{xs}^{k}(\mathcal{C}) \text{ then also } \square \in S_{xsT}^{k}(\mathcal{C}).$$

Moreover $S_{xsT}^{i}(\mathcal{C}) \subseteq S_{xs}^{i}(\mathcal{C})$ for all $i \in \mathbb{N}$. In particular

$$\square \in S_{xsT}^{d}(\mathcal{C}) \subseteq S_{xs}^{d}(\mathcal{C})$$

where d is the minimal refutation depth of R_{xs} on $\mathcal{C}$ (and thus also of R_{xsT} on $\mathcal{C}$). As an immediate consequence we obtain

$$CS_{xsT}(\mathcal{C}) \leq CS_{xs}(\mathcal{C})$$

for all unsatisfiable sets of clauses $\mathcal{C}$.

As in the case of forward subsumption, locking plays a special role in tautology elimination.

Proposition 4.4.2. *Lock resolution + tautology-elimination is incomplete.*

Proof. Again we may take our old example used to show incompleteness of lock resolution under forward subsumption.

$$C = \{R(\overset{1}{x_1}) \vee P(\overset{5}{x_1}),\ \neg P(\overset{3}{x_1}) \vee R(\overset{6}{x_1}),\ P(\overset{4}{x_1}) \vee \neg R(\overset{7}{x_1}),\ \neg R(\overset{2}{x_1}) \vee \neg P(\overset{8}{x_1})\}$$

The first generation of lock resolvents is

$$S_l^1 = C \cup \{P(\overset{5}{x_1}) \vee \neg P(\overset{8}{x_1}),\ R(\overset{6}{x_1}) \vee \neg R(\overset{7}{x_1})\}.$$

We see that all new clauses on the first level are tautologies (their unlocked forms are tautological clauses). Therefore $S_{lT}^1(C) = C$ and, by definition of R_{xT} for arbitrary refinements R_x, $R_{lT}(C) = C$. But C is unsatisfiable and thus R_{lT} is incomplete.   ◊

We may ask the question whether hyperresolution is complete under elimination of tautologies. In fact the answer is completely trivial: Because $R_H(C) - C$ consists of positive clauses and (in case of unsatisfiability) of □ only, it cannot contain any tautologies. Thus tautology-elimination can only play the role of preprocessing in hyperresolution. The same holds for hyper-resolution with replacement.

As an algorithmic test, tautology-elimination is much faster than subsumption. In order to test the tautology property, we merely order the clause according to their atom formulas and, for equal atoms, put the positive literals in front of the negative ones. If, by this method, C is ordered into a clause C_0 then C_0 is a tautology iff there are clauses D_1, D_2 such that $C_0 = D_1 \vee A \vee \neg A \vee D_2$ for some atom formula A. Therefore the algorithm consists in ordering the atoms first and looking for pairs $A, \neg A$ afterwards. If n denotes the number of symbols in clauses then this procedure is of time complexity $O(n \log n)$ (provided a fast sorting procedure is applied [AHU76]).

Exercises

Exercise 4.4.1. Prove that linear input resolution + subsumption + elimination of tautologies is relatively complete to linear input resolution + subsumption. Define STULI-deductions (subsumption reduced, tautology-free, linear input deductions) out of the concept of SULI-deductions (see Definition 4.2.6) and show that there are STULI-refutations iff there are SULI-refutations.

4.5 Clause Implication

We have seen that, independently of the specific R-deduction methods, tautology elimination and subsumption are admissible as preprocessing (and for some refinements "during" deduction too). Now we might think about further strengthening the preprocessing by deleting clauses that are implied by the remaining set of clauses (this problem is directly connected to the construction of minimal axiom systems). Such a redundancy test, however, is undecidable and therefore is out of range of a computational calculus. A natural restriction of such a general redundancy test could be to check whether one input clause implies another one. Formally this is the problem of finding out whether $F(\{C\}) \to F(\{D\})$ is valid for two clauses C and D; in a more convenient notation we write $\{C\} \to \{D\}$ for $F(\{C\}) \to F(\{D\})$ (this should not be confused with the problem $C \to D$ for open disjunctions C and D). So let us assume that for two clauses $C, D \in \mathcal{C}$ the formula $\{C\} \to \{D\}$ is valid. Then, clearly, $\mathcal{C}$ is logically equivalent to $\mathcal{C} - \{D\}$ and we may work with the smaller set $\mathcal{C} - \{D\}$. Such redundancy reductions, however, are not always as beneficial as subsumption (Exercise 4.5.9). So the investigation of clause implication is mainly of theoretical interest. Most interesting is its behavior relative to subsumption, which will be investigated in this chapter. Concerning subsumption and implication several natural questions arise: Does clause implication coincide with subsumption? If clause implication is different from subsumption what are the mutual relations? Is clause implication decidable? If clause implication would indeed coincide with subsumption, all the questions following the first one would have trivial answers. But we will see that clause implication (essentially) differs from subsumption.

First we find out that there are trivial cases where $\{C\} \to \{D\}$ is valid but $C \leq_{ss} D$ is not: Just take $C = P(x)$ and $D = Q(x) \vee \neg Q(x)$. Because D is a tautology, $\{E\} \to \{D\}$ is valid for all clauses E. On the other hand C does not subsume D. So the next step is to reduce the problem to the case where D is not a tautology; but still $\{C\} \to \{D\}$ does not imply $C \leq_{ss} D$.

Proposition 4.5.1. *There are clauses C, D such that D is not a tautology, $\{C\} \to \{D\}$ is valid, but $C \leq_{ss} D$ does not hold.*

Proof. Let $C = \neg P(x) \vee P(f(x))$ and $D = \neg P(a) \vee P(f(f(a)))$. Clearly C does not subsume D because there is no substitution ϑ such that $\vartheta(x) = a$ and $\vartheta(f(x)) = f(f(a))$ both hold. But C implies D. To establish this fact it suffices to refute the set of clauses $\{\neg P(x) \vee P(f(x)), P(a), \neg P(f(f(a)))\}$, which represents $\{C\} \cup \neg D$.

Indeed resolving C with itself we obtain (a variant of) the clause $\neg P(x) \vee P(f(f(x)))$. Two further resolutions with $P(a)$ and $\neg P(f(f(a)))$ give $\square$. ◊

Looking more closely at the proof of Proposition 4.5.1 we find that C itself does not subsume D, but

$$\neg P(x) \vee P(f(f(x))) \leq_{ss} D.$$

So we see that not C itself but a clause derivable from C (alone) subsumes D. The question remains whether this effect is merely accidental or rather characteristic to the problem. We will show that there is some interesting law behind this phenomenon, namely (a weak form of) the deduction completeness of resolution (note that so far we have dealt with refutational completeness only). The next step will be to prove the important theorem of Lee [Lee67]; afterwards we will show that $\{C\} \to \{D\}$ can be characterized by resolution on $\{C\}$ and by subsumption.

Roughly speaking, the theorem of Lee says that $C \to \{D\}$ is valid iff there exists a $C \in R_\emptyset(\mathcal{C})$ such that $C \leq_{ss} D$ ($\mathcal{C}$ being a set of clauses and D being a clause). That means, if a clause D is implied by C then it can "almost" be derived from C. It is obvious that, in order to make this statement correct, we have to exclude tautological clauses D again. As always, we first analyse the ground case and obtain the general one via lifting.

Lemma 4.5.1. Let $\mathcal{C}$ be a set of ground N_c-clauses and D be a ground N_c-clause such that D is not a tautology and $\mathcal{C} \to \{D\}$ is valid. Then there exists a clause $C \in R_\emptyset(\mathcal{C})$ such that $C \leq_{ss} D$ (i.e., we can derive a clause C from $\mathcal{C}$ which subsumes D).

Proof. We proceed by induction on $occl(\mathcal{C})$, the number of literal occurrences in $\mathcal{C}$ (compare to the proof of Lemma 3.6.1).

$occl(\mathcal{C}) = 0$:
In this case $\mathcal{C} = \emptyset$ or $\mathcal{C} = \{\Box\}$. If $\mathcal{C} = \emptyset$ then $F(\mathcal{C})$ is a tautology and, by the validity of $\mathcal{C} \to \{D\}$, D must be a tautology too. But we have excluded tautological clauses D a priori and so this case cannot occur. If $\mathcal{C} = \{\Box\}$ then, clearly, $\Box \in R_\emptyset(\mathcal{C})$ and $\Box \leq_{ss} D$ (note that $\Box$ subsumes every clause). This concludes the case $occl(\mathcal{C}) = 0$.

(IH) Suppose that, for all sets of ground N_c-clauses $\mathcal{C}$ with $occl(\mathcal{C}) \leq n$, $\mathcal{C} \to \{D\}$ is valid, and D is not a tautology, there exists a clause $C \in R_\emptyset(\mathcal{C})$ such that $C \leq_{ss} D$.

Now let $\mathcal{C}$ be a set of ground N_c-clauses such that $occl(\mathcal{C}) = n + 1$, D is nontautological, and $\mathcal{C} \to \{D\}$ is valid.

case a: $\mathcal{C}$ is unsatisfiable.
By the (refutational) completeness of resolution we obtain $\Box \in R_\emptyset(\mathcal{C})$; clearly $\Box$ subsumes D.

case b: $\mathcal{C}$ is satisfiable.
We show that D must contain a literal L which also occurs in C. Because $\mathcal{C} \to \{D\}$ is valid the set of clauses $\mathcal{C} \cup \neg D$ is unsatisfiable.

By assumption D is not a tautology and thus the set of (ground unit) clauses $\neg D$ is satisfiable. So let $\neg D = \{M_1, \dots, M_k\}$. We assume that for all $i = 1, \dots k$ M_i^d does not occur in $\mathcal{C}$ and derive a contradiction.

Still the literal M_i may occur in $\mathcal{C}$ for some $i \in \{1, \dots, k\}$. But let $\mathcal{C}' = \mathcal{C}$ after application of the pure literal rule on M_i. Then $\mathcal{C}' \sim_{sat} \mathcal{C}$ and $\mathcal{C}'$ contains neither M_i nor M_i^d. Thus $at(M_i)$ does not occur in $\mathcal{C}'$ at all and $\mathcal{C}'$ is satisfiable.

We iterate this procedure until there is no M_i left which occurs in $\mathcal{C}$ (we thus obtain a final set $\mathcal{C}'$). Because $\neg D$ is satisfiable there exists a (propositional) interpretation J on $at(\{M_1, \ldots, M_k\})$ such that $J(M_i) = true$ for all $i = 1, \ldots, k$.

But also $\mathcal{C}'$ is satisfiable and thus there exists some (propositional) model I with $v_I(\mathcal{C}') = true$ and I is defined on $at(\mathcal{C})$. But, by definition of $\mathcal{C}'$, $at(\mathcal{C}') \cap at(\neg D) = \emptyset$ and therefore $J \cup I$ is a model of $\mathcal{C}' \cup \neg D$.

Now note that $v_{I \cup J}(M_i) = true$ for $i = 1, \ldots, k$. As a consequence $J \cup I$ is also a model of $\mathcal{C} \cup \neg D$ ($\mathcal{C}$ may contain some M_i but no M_i^d). But, by assumption, $\mathcal{C} \cup \neg D$ is unsatisfiable and we obtain a contradiction. So we conclude that there must be a literal L such that $L \in LIT(\mathcal{C})$ and $L^d \in LIT'(\neg D)$; but that means $L \in LIT(\mathcal{C}) \cap LIT(D)$.

Now let $\mathcal{D} : \{C_1 \vee L \vee D_1, \ldots, C_m \vee L \vee D_m\}$ be the set of all clauses in $\mathcal{C}$ which contain L and let $D_* = D \setminus L$.

Because $\mathcal{C} \to \{D\}$ is valid, $\mathcal{C} \cup \{L^d\} \to \{D_*\}$ is valid by

$$A \to (B \vee C) \sim (A \wedge \neg B) \to C).$$

By the one-literal rule of Davis and Putnam (applied to L^d) we obtain a set

$$\mathcal{C}_* : (\mathcal{C} - \mathcal{D}) \cup \{C_1 \vee D_1, \ldots, C_m \vee D_m\}$$

which is sat-equivalent to $\mathcal{C} \cup \{L^d\}$.

The next step consists in showing that $\mathcal{C}_* \to \{D_*\}$ is valid. Because D is in N_c-normal form and is not a tautology L^d cannot occur in D; consequently neither L nor L^d occurs in D_*. Thus by applying the one-literal rule to $\mathcal{C} \cup \{L^d\} \cup \neg D_*$ we obtain $\mathcal{C}_* \cup \neg D_*$.

Moreover $\mathcal{C} \cup \{L^d\} \cup \neg D_* \sim_{sat} \mathcal{C}_* \cup \neg D_*$.

Because $\mathcal{C} \cup \{L^d\} \to \{D_*\}$ is valid the set of clauses $\mathcal{C} \cup \{L^d\} \cup \neg D_*$ is unsatisfiable.

We conclude that $\mathcal{C}_* \cup \neg D_*$ is unsatisfiable and, therefore, $\mathcal{C}_* \to \{D_*\}$ is valid. Because L occurs in $\mathcal{C}$, $\mathcal{C}_*$ is strictly smaller than $\mathcal{C}$ and $occl(\mathcal{C}_*) \leq n$.

So we may apply the induction hypothesis (IH) and we obtain a clause $C \in R_\emptyset(\mathcal{C}_*)$ such that $C \leq_{ss} D_*$. It remains to show that there exists a clause $E \in R_\emptyset(\mathcal{C})$ such that $LIT(E) \subseteq LIT(C) \cup \{L\}$. Then, by $LIT(C) \subseteq LIT(D_*)$ (note that C and D_* are ground) we also get

$$LIT(E) \subseteq LIT(C) \cup \{L\} \subseteq LIT(D_*) \cup \{L\} = LIT(D).$$

But that means $E \leq_{ss} D$, which is what we intend to show. In order to construct an appropriate clause E we first observe that the one-literal rule of Davis and Putnam can be simulated by resolution. Particularly the clauses $C_i \vee D_i$ are resolvents of $C_i \vee L \vee D_i$ and L^d. Therefore we get

$$\mathcal{C}_* \subseteq \mathcal{C} \cup Res(\mathcal{C} \cup \{L^d\}) \subseteq R_\emptyset(\mathcal{C} \cup \{L^d\}).$$

By monotonicity and idempotency of $R_\emptyset$ we obtain

$$R_\emptyset(C_*) \subseteq R_\emptyset(C \cup \{L^d\})$$

So we know that $C \in R_\emptyset(C \cup \{L^d\})$ and $C \leq_{ss} D_*$. Let F, G be arbitrary N_c-ground clauses. As in the proof of Lemma 3.6.1 we define:

$$F \leq_L G \quad \text{iff} \quad LIT(F) = LIT(G) \quad \text{or} \quad LIT(F) \cup \{L\} = LIT(G).$$

Cleary $\leq_L$ is reflexive and transitive. We extend $\leq_L$ to sets of clauses via

$$\mathcal{C} \leq_L \mathcal{D} \quad \text{iff for all } C \in \mathcal{C} \text{ there exists a } D \in \mathcal{D} \text{ such that } C \leq_L D.$$

It is easy to show (Exercise 4.5.1) that

$$R_\emptyset(C \cup \{L^d\}) \leq_L R_\emptyset(C) \cup \{L^d\}.$$

From $C \in R_\emptyset(C \cup \{L^d\})$ and the definition of $\leq_L$ we infer that there is a clause $E \in R_\emptyset(C) \cup \{L^d\}$ such that $C \leq_L E$. But C subsumes D_*, a clause which neither contains L nor L^d. Particularly $L^d \notin LIT(C)$. Therefore E cannot be L^d and so $E \in R_\emptyset(C)$ (note that, by satisfiability of C, $C \neq \square$).

But $C \leq_L E$ also implies $LIT(E) \subseteq LIT(C) \cup \{L\}$. Remember we have shown already that

$$LIT(E) \subseteq LIT(C) \cup \{L\} \text{ implies } E \leq_{ss} D.$$

So $E \in R_\emptyset(C)$ and $E \leq_{ss} D$, which concludes the proof of case $n + 1$. $\Diamond$

It remains to show the theorem for the general case.

Theorem 4.5.1 (Theorem of Lee). *Let C be a set of N_c-clauses, let D be a nontautological clause and $C \to \{D\}$ be valid. Then there exists a clause $C \in R_\emptyset(C)$ such that $C \leq_{ss} D$.*

Proof. Let as assume that $C \to \{D\}$ is valid. Then $C \cup \neg D$ is unsatisfiable. As clauses represent universally closed disjunctions $\neg D$ is logically equivalent to the existential closure of a conjunction of literals. By skolemization we obtain $C \cup \neg D_c$, where D_c is D after replacing all variables in $V(D)$ by new distinct constant symbols that do not occur in $C \cup \{D\}$. Clearly $C \cup \neg D \sim_{sat} C \cup \neg D_c$ and D_c is a ground clause.

By Herbrand's theorem there exists a finite set of ground N_c-instances C' of C such that $C' \cup \neg D_c$ is unsatisfiable (note that $\neg D_c$ already consists of unit ground clauses).

But then $C' \to \{D_c\}$ is valid. By Lemma 4.5.1 there exists a clause $C' \in R_\emptyset(C')$ such that $C' \leq_{ss} D_c$. By N_c-ground lifting we obtain a clause $C \in R_\emptyset(C)$ such that there exists a ϑ with $N_c(C\vartheta) = N_c(C') = C'$, i.e., by definition of $\leq_{sc}$ (Definition 3.3.7) $C \leq_{sc} C'$. As $\leq_{ss}$ is strictly stronger than $\leq_{sc}$ we also get $C \leq_{ss} C'$.

Moreover, $\leq_{ss}$ is transitive and we conclude $C \leq_{ss} D_c$. But Proposition 4.2.2 tells us that $C \leq_{ss} D_c$ iff $C \leq_{ss} D$. So we obtain $C \in R_\emptyset(C)$ and $C \leq_{ss} D$. $\Diamond$

Lee's theorem can directly be applied to the case of clause implication. We only have to select $\mathcal{C} = \{C\}$ for a clause C. The following proposition was first proved (using a different method) in [Got87]:

Proposition 4.5.2. *Let C, D be N_c-clauses such that D is not a tautology, $\{C\} \to \{D\}$ is valid, and $R_\emptyset(\{C\}) = \{C\}$; then $C \leq_{ss} D$.*

Proof. By $R_\emptyset(\{C\}) = \{C\}$ and by Lee's theorem C must subsume D.   ◇

Typical clauses fulfilling $R_\emptyset(\{C\}) = \{C\}$ are clauses which are incapable of self-inference. Thus Proposition 4.5.2 holds for all clauses C such that $Res(\{C\}) = \emptyset$. Other candidates for this property are tautologies like $P(x_1) \vee \neg P(x_1)$; but these can be excluded because D may not be a tautology.

Corollary 4.5.1. *If C is not resolvable with itself and D is not a tautology then the validity of $\{C\} \to \{D\}$ implies $C \leq_{ss} D$.*

Proof. Immediate from Proposition 4.5.2.   ◇

Let us assume that D is a nontautological clause. Then, by Theorem 4.5.1, $\{C\} \to \{D\}$ is valid iff there exists an $E \in R_\emptyset(\{C\})$ such that $E \leq_{ss} D$. The question remains, whether there is a general algorithmic method to determine the existence or nonexistence (!) of such a clause E. For such a decision procedure it would suffice to show the existence of a recursive function k on pairs of clauses (C, D) such that the following property holds:

$$R_\emptyset(\{C\}) \leq_{ss} \{D\} \text{ iff } S_\emptyset^{k(C,D)} \leq_{ss} \{D\}.$$

Note that for every $d \in \mathbb{N}$ $S_\emptyset^d(\{C\})$ is finite and $\leq_{ss}$ is decidable. Thus, under the condition that such a bound k exists, we first compute $d = k(C, D)$ and then test $F \leq_{ss} D$ for $F \in S_\emptyset^d(C)$.

But clause implication is undecidable ([SS88]) and thus such a recursive bound k does not exist. It was even shown that the problem remains undecidable when C is restricted to Horn form [MP92] (in fact the problem is already undecidable for Horn clauses C of the form $P \vee \neg P_1 \vee \neg P_2$). As a consequence there exists no algorithmic test for (Horn) clause implication. But, of course, $\{C\} \to \{D\}$ is semidecidable: If $\{C\} \to \{D\}$ is indeed valid then, by the completeness of resolution, we find $\square \in R_\emptyset(\{C\} \cup \neg D)$, although the value of such a test is highly questionable (it is as hard as resolution theorem proving itself!). However, there are relevant cases of decidable clause implication problems (besides the trivial case that D is a tautology or $R_\emptyset(\{C\}) = \{C\}$). In the remaining part of this chapter we present some of the decidable classes and analyze the corresponding decision methods. These methods are based either on Lee's theorem or on resolution decision procedures. In contrast to the Horn clause implication problem, the Krom clause implication problem is decidable. This result was first shown by M. Schmidt-Schauss in [SS88].

Definition 4.5.1 (Krom clause implication KCI). : *Let KCI be the set of all pairs of clauses (C, D) such that C is a Krom clause (i.e., $|C| \leq 2$) and D is arbitrary. Then KCI is called the Krom clause implication problem.*

Theorem 4.5.2. *The Krom clause implication problem is decidable.*

Proof. Let $(C, D) \in KCI$; we have to develop an algorithm which decides $\{C\} \to \{D\}$.

If $C = P_1 \vee P_2$ or $C = \neg P_1 \vee \neg P_2$ for two atoms P_1, P_2 then clearly $R_\emptyset(\{C\}) = \{C\}$ and, by Proposition 4.5.2, $\{C\} \to \{D\}$ iff $C \leq_{ss} D$. In these cases the implication problem can be decided simply by the subsumption test. Even more trivial is the case where D is a tautology. It remains to investigate the case where C contains a positive and a negative literal and D is not a tautology. For technical reasons we base our analysis on N_c-clause forms again.

If $Res(\{C\}) = \{C\}$ then $\{C\} \to \{D\}$ is valid iff $C \leq_{ss} D$.

So we are left with the only nontrivial case that C is capable of self-resolution. We define recursively:

$$\mathcal{C}_1 = \{C\} \quad \text{and} \quad \mathcal{C}_{n+1} = \hat{\rho}_\emptyset(C, \mathcal{C}_n) \cup \mathcal{C}_n \text{ for } n \geq 1$$

For the definition of $\hat{\rho}_x$ see Section 4.1. By Exercise 4.5.5 $\mathcal{C}_{n+1} - \mathcal{C}_n$ consists of at most one element, which – in case of existence – we denote by C^{n+1}; moreover if C^{n+1} exists it is indeed the (only) resolvent of C^n and C (for $n \geq 1$). Thus C^n behaves like the "n-th power" of C. It is trivial that $\mathcal{C}_n \subseteq R_\emptyset(\{C\})$ holds for all $n \geq 1$; but here we even have $\bigcup_{i \in \mathbb{N}} \mathcal{C}_i = R_\emptyset(\{C\})$ (see Exercise 4.5.6). Thus the clauses derivable from C itself are precisely the "clause powers" C^n. Let

$$C^{(1)} = \mathcal{C}_1 \quad \text{and} \quad C^{(n+1)} = \mathcal{C}_{n+1} - \mathcal{C}_n \text{ for } n \geq 1.$$

case a: $C^{(n)} = \emptyset$ for some n.
In this case $R_\emptyset(\{C\}) = S_\emptyset^{(n-1)}(\{C\})$ and, by Lee's theorem, $\{C\} \to \{D\}$ is valid iff $S_\emptyset^{(n-1)}(\{C\}) \leq_{ss} \{D\}$. But $S_\emptyset^{(n-1)}(\{C\})$ is a finite set of clauses and $\leq_{ss}$ is decidable.

case b: $C^{(n)} \neq \emptyset$ for all $n \in \mathbb{N}_+$.
In this case the clauses C^n are defined for all $n \geq 1$. We first show that the term depth of the C^n is monotonically increasing, i.e.,

$$\tau(C^n) \leq \tau(C^{n+1}) \text{ for all } n \geq 1.$$

Let $C^n = Q_1 \vee \neg Q_2$ and $C = P_1 \vee \neg P_2$ and η_1, η_2 be two renaming substitutions such that $V(C^n\eta_1) \cap V(C\eta_2) = \emptyset$. There are two possibilities to define a resolvent of C^n and C.

$D_1 : \neg Q_2 \eta_1 \sigma \vee P_1 \eta_2 \sigma$ for some m.g.u. σ or
$D_2 : Q_1 \eta_1 \tau \vee \neg P_2 \eta_2 \tau$ for an m.g.u. τ.

By the uniqueness of C^{n+1} we have $N_c(D_1) = N_c(D_2)$ and in particular $\tau(D_1) = \tau(D_2) = \tau(C^{n+1})$. Obviously $\tau(\neg Q_2) \leq \tau(D_1)$ and $\tau(Q_1) \leq \tau(D_2)$ and therefore

$$\tau(C^n) = max\{\tau(Q_1), \tau(Q_2)\} \leq \tau(C^{n+1}).$$

We now distinguish two cases:

b1: There exists a $k \geq 1$ such that $\tau(C^k) > \tau(D)$.
Then, by the monotonicity shown above, $\tau(C^l) > \tau(D)$ for all $l \geq k$. It follows that, for all substitutions ϑ and for all $l \geq k, \tau(C^l \vartheta) > \tau(D)$. As a consequence $C^l \leq_{ss} D$ is impossible for $l \geq k$. Thus $R_\emptyset(\{C\}) \leq_{ss} \{D\}$ iff $S_0^{k-1}(C) \leq_{ss} D$. Therefore we obtain a bound on the deduction levels like in case a).

b2: $\tau(C^n) \leq \tau(D)$ for all $n \leq 1$.
As all C^n consist of two literals only and the term depth is uniformly bounded there are only finitely many N_c-normalizations $N_c(C^n)$ (note that N_c performs a standard renaming of variables). Therefore the sequence $(C^n)_{n \in \mathbb{N}_+}$ must contain cycles, which means there are numbers m, r with $r > 0$ and
$$C^m = C^{m+r}.$$
As a consequence we get $C^{m+k} = C^{m+k \bmod r}$ for all $k \in \mathbb{N}$.
Let $d = m + r - 1$; then

$$R_\emptyset(\{C\}) = \bigcup_{i=0}^{\infty} \{C^i\} \subseteq \bigcup_{i=0}^{d} \{C^i\} = S_0^d(\{C\}).$$

Clearly $R_\emptyset(\{C\}) \leq_{ss} \{D\}$ iff $S_0^d(\{C\}) \leq_{ss} \{D\}$.

The case analysis in a), b1), and b2) suggests the following decision algorithm KCIA:

(I) Test "$C \leq_{ss} D$"; if $C \leq_{ss} D$ holds then $\{C\} \to \{D\}$ is valid else continue at point (II) with $n = 1$.

(II) (Suppose that we have already computed $C^1, \ldots, C^n$)
(*) If $\{C^1, \ldots, C^n\} \leq_{ss} \{D\}$ then $\{C\} \to \{D\}$ is valid, else compute C^{n+1}.
a: C^{n+1} does not exist: $\{C\} \to \{D\}$ is not valid.

b: C^{n+1} exists:
b1: $\tau(C^{n+1}) > \tau(D) : \{C\} \to \{D\}$ is not valid.

b2: $C^{n+1} = C^i$ for some $i < n + 1 : \{C\} \rightarrow \{D\}$ is not valid.

b3: If neither b1) nor b2) then set $n \leftarrow n + 1$ and continue at point (*).

By the analyses at the points a), b1), and b2) in the proof we see that there must always be an n such that either $C^{(n)} = \emptyset$ or $\tau(C^n) > \tau(D)$ or $C^n = C^i$ for some $i < n$. But this gives us the guarantee that KCIA always terminates; consequently KCIA is a decision algorithm for the Krom clause implication problem.   ◊

We have seen that Lee's theorem can provide tools to decide clause implication problems. There are also several subclasses of the Horn implication problem which can be decided by this method [Lei88]. We now introduce another method to decide clause implication problems. Instead of investigating the problem $\{C\} \rightarrow \{D\}$ directly, we turn to the satisfiability problem $\{C\} \cup \neg D$.

If $\{C\} \cup \neg D$ is unsatisfiable (and thus $\{C\} \rightarrow \{D\}$ is valid) then $\square \in R_x(\{C\} \cup \neg D)$ for every complete refinement operator R_x. The real problem arises when $\{C\} \cup \neg D$ is satisfiable; because the clause implication problem is undecidable we know that R_x cannot terminate on all satisfiable sets of clauses provided R_x is complete). There are, however, subclasses of the clause implication where appropriate refinements R_x always terminate and thus yield a decision procedure. The idea of finding complete, terminating refinements is crucial to the resolution decision theory to be presented in Chapter 5. For this reason we will refer to the more general results of Chapter 5 in the analyses of some clause implication problems to follow.

Definition 4.5.2 (one-variable clause implication 1VCI).
Let 1VCI be the class of all pairs of clauses (C, D) such that $|V(C)| = 1$ (D may be arbitrary). Then the problem to decide clause implication in 1VCI is called the one-variable clause implication problem.

Theorem 4.5.3. *The one-variable clause implication problem is decidable.*

Proof. This theorem is a specific subcase of the more general Theorem 5.2.1 and we shall not repeat the proof under these more restricted circumstances. Instead we merely mention the main lines of the proof.

Remember our specific A-ordering $<_d$ defined in Section 3.3:

$$A <_d B \quad \text{iff for all } x \in V(A) : \tau_{max}(x, A) < \tau_{max}(x, B) \text{ and } \tau(A) < \tau(B).$$

For $(C, D) \in 1VCI$ the satisfiability problem $\{C\} \cup \neg D$ is contained in the one-variable class (i.e., the class of all sets of clauses containing only clauses with ≤ 1 variables). By Theorem 5.2.1 $R_{<_d}$ terminates on the one-variable class; particularly $R_{<_d}(\{C\} \cup \neg D)$ is finite for all $(C, D) \in 1VCI$ and there exists an i such that

$$S^i_{<_d}(\{C\} \cup \neg D) = S^{i+1}_{<_d}(\{C\} \cup \neg D).$$

Thus there exists an obvious algorithm to compute $R_{<_d}(\{C\}) \cup \neg D$. Therefore the property "$\square \in R_{<_d}(\{C\} \cup \neg D)$" is decidable for $(C, D) \in 1VCI$.

But the validity of $\{C\} \to \{D\}$ is equivalent to $\square \in R_{<_d}(\{C\} \cup \neg D)$. $\lozenge$

Finally, we present a clause implication class that can be decided by hyperresolution (while 1VCI can be decided by A-ordering).

Definition 4.5.3. *Let H0CI be the class of all pairs of clauses (C, D) such that*

$$\tau(C_+) = 0 \text{ and } V(C_+) \subseteq V(C_-) \text{ and } D \text{ is arbitrary.}$$

H0CI represents the clause implication problem for C_+ being function-free and "variable-dominated" by C_-.

The clause implication problem for H0CI can be reduced to the satisfiability problem of the class PVD defined in Chapter 5 (Definition 5.3.3).

Theorem 4.5.4. *The clause implication problem is decidable for $(C, D) \in$ H0CI.*

Proof. If $(C, D) \in H0CI$ then $\{C\} \cup \neg D$ is in PVD. By Theorem 5.3.1 hyperresolution terminates on PVD, i.e., for all $\mathcal{C} \in$ PVD the set $R_H(\mathcal{C})$ is finite.

In particular $R_H(\{C\} \cup \neg D)$ is finite for all $(C, D) \in$ H0CI.
Recall (again) that $\{C\} \to \{D\}$ is valid iff $\{C\} \cup \neg D$ is unsatisfiable and that R_H is complete. Therefore, in order to decide the validity of $\{C\} \to \{D\}$, we just compute the set $R_H(\{C\} \cup \neg D)$. We know that there must be a $k \in \mathbb{N}$ such that

$$S^k_H(\{C\} \cup \neg D) = S^{k+1}_H(\{C\} \cup \neg D).$$

Clearly $\{C\} \to \{D\}$ is valid iff $\square \in R_H(\{C\} \cup \neg D)$. $\lozenge$

More decidable clausal classes (all of them Horn classes) can be found in [Lei88], [Lei90], [LG90], and [Rud92]. The algorithms in the latter three papers are based on hyperresolution (but not all of them on termination); the algorithms in [Lei88] are based on self-inference and on Lee's theorem. It is not sufficiently investigated how clause implication behaves as a redundancy method with respect to resolution. What can be said is that clause implication is too strong to enjoy the "uniform" improvements achieved by subsumption (see Exercises 4.5.8 and 4.5.9). However, there are some trivial cases where clause implication and subsumption coincide; one such case occurs in hyperresolution: As R_H produces positive clauses only which are not capable of self-resolution and which can be implied by positive clauses only (!), implication collapses to ordinary subsumption by Proposition 4.5.2 (note that positive clauses cannot be tautologies!). Linear input deduction in Horn logic shows a similar behavior; in this case all generated clauses are purely negative.

Horn clause implication naturally models rule dependency in logical inference systems; here propositional inference rules (having several formulas in the antecedent and one in the consequent) can be represented as Horn clauses. Then clause implication represents redundancy of rules (i.e., one rule logically implies the other one).

Exercises

Exercise 4.5.1. Let C be a set of ground N_c-clauses and L be a ground literal. Then
$$R_\emptyset(C \cup \{\neg L\}) \leq_L R_\emptyset(C) \cup \{\neg L\}$$
($\leq_L$ is defined in the proof of Lemma 4.5.1).

Exercise 4.5.2. Let C, D be N_c-clauses such that D is not a tautology and let R_H be the operator of hyperresolution defined in Definition 3.6.5. Show that
$$C \leq_{ss} D \text{ iff } \Box \in S_H^2(\{C\} \cup \neg D).$$

In general we cannot replace $R_\emptyset$ by a refinement operator R_x in Lee's theorem (which means the validity of $\{C\} \to \{D\}$ usually does not imply the existence of an $E \in R_x(C)$ such that $E \leq_{ss} D$). But for positive clauses we can obtain a stronger form of Lee's theorem:

Exercise 4.5.3. Let C be a set of N_c-clauses and D be a positive clause such that $C \to \{D\}$ is valid. Then there exists a clause $E \in R_H(C)$ such that $E \leq_{ss} D$ (i.e., there exists a clause derivable by hyperresolution which subsumes D).

Exercise 4.5.4. Let Γ be the class of all clauses pairs (C, D) such that $\tau(C) = 0$ (i.e., C is function free). Show that the clause implication problem for Γ is decidable (hint: construct a ground clause D' such that $\tau(D') = 0$ and $\{C\} \to \{D\}$ is valid iff $\{C\} \to \{D'\}$ is valid).

Exercise 4.5.5 (clause powers). Let C be a Krom clause and
$$C_1 = \{C\} \text{ and } C_{n+1} = \hat{\rho}_\emptyset(C, C_n) \cup C_n$$
for $n \in \mathbb{N}$ (remember that $\hat{\rho}_x(C, D)$ describes the set of all x-resolvents between C and the set of clauses D). Let
$$C^{(1)} = C_1 \text{ and } C^{(n+1)} = C_{n+1} - C_n.$$

Show that, under N_c-normalization, $C^{(n)}$ is either empty or consists of a single clause only (which we may call $C^{(n)}$). Moreover if $C^{(n+1)} \neq \emptyset$ for some $n \geq 1$ then $C^{(n+1)}$ is the (only) resolvent of C and C^n.

Exercise 4.5.6. [SS88] Let C be a Krom clause and C_n be defined as in Exercise 4.5.5. Show that

$$R_\emptyset(\{C\}) = \bigcup_{i \in \mathbb{N}} C_n.$$

Definitions for Exercise 4.5.7:

Let C be a Horn clause such that $C = P \vee \neg Q_1 \vee \ldots \vee \neg Q_n$ and let $C^+ = \{P\}$, $C^- = \{\neg Q_1, \ldots, \neg Q_n\}$. We divide C^- into two groups of literals:

$C_{\mathrm{inf}} = \{L | L \in C^-$ and there exists a renaming substitution η such that $\{\neg P\eta, L\}$ is unifiable $\}$,
$C_{\mathrm{rest}} = C^- - C_{\mathrm{inf}}$.

C_{inf} represents the set of all literals in C^- which lead to self-resolution of C. C_{rest} is the "inference-free" part.

Exercise 4.5.7. [Lei88] Let Γ be the class of all pairs of clauses (C, D) such that C is Horn and $V(C) = V(C_{\mathrm{rest}})$. Show that the clause implication problem for Γ is decidable (Hint: Use hyperresolution and an appropriate ordering of the literals C^-).

Background on Exercise 4.5.8:

The effect of removing implied clauses can be essentially different from that of subsumption. Indeed it may be the case that the minimal refutation depth increases after removal of implied clauses. This phenomenon applies to deletion during deduction and to deletion as preprocessing.

Exercise 4.5.8. Show that there exists a sequence of sets of clauses C_n such that C_n is reduced under implication, $d_\emptyset(C_n) = O(n)$, but the minimal refutation depth of C_n under elimination of implied clauses is $> 2^n$ (for the mathematical treatment use the refinement $R_{\emptyset ci}$ defined like forward subsumption $R_{\emptyset s}$).

Let $\mathrm{IMP}(\mathcal{C})$ be a reduction of $\mathcal{C}$ under implication, i.e., $\mathrm{IMP}(\mathcal{C}) \subseteq \mathcal{C}$ and if $C, D \in \mathrm{IMP}(\mathcal{C})$ such that if $\{C\} \to \{D\}$ is valid then $C = D$.

Exercise 4.5.9. Construct a sentence C_n (similar to that in Exercise 4.5.8) such that $d_\emptyset(C_n) = c$ for some constant c independent of n, but $d_\emptyset(IMP(C_n)) \geq dn$ for a constant d and for all $n \in \mathbb{N}$.

5. Resolution as Decision Procedure

5.1 The Decision Problem

In Section 2.7 we showed that for every unsatisfiable set of clauses there exists a resolution refutation. In the last two chapters we have exhibited some completeness preserving extensions of resolution. Here we will focus on another aspect of refinements, namely termination. Suppose that we start a theorem prover (i.e., a complete resolution refinement R_x) on a set of clauses $\mathcal{C}$, which may be satisfiable or unsatisfiable. Obviously there are three possibilities:

1) R_x terminates on $\mathcal{C}$ and refutes $\mathcal{C}$.
   Because R_x is correct and $\square \in R_x(\mathcal{C})$ we know that $\mathcal{C}$ is unsatisfiable.
2) R_x terminates on $\mathcal{C}$ without producing $\square$.
   By the completeness of R_x $\mathcal{C}$ must be satisfiable.
3) R_x does not terminate on $\mathcal{C}$:
   In this case $R_x(\mathcal{C})$ is infinite and $\square \notin R_x(\mathcal{C})$ (we assume that the production of new clauses is stopped as soon as $\square$ is derived). As in case 2) $\mathcal{C}$ is satisfiable, but we cannot detect this property just by computing $R_x(\mathcal{C})$.

From Proposition 3.1.1 we know that for every complete refinement operator R_x there must exist a (finite) set of clauses $\mathcal{C}$ such that $R_x(\mathcal{C})$ is infinite. That means it is impossible in principle to avoid nontermination on all sets of clauses. Let us trace the logical and historical background of this phenomenon.

By Church's famous result [Chu36] we know that (the validity problem of) predicate logic is undecidable. It is easily verified that the validity and satisfiability problems are recursively equivalent; note that F is valid iff $\neg F$ is unsatisfiable. Therefore, from now on, we may focus on the satisfiability problem only. Let F be a sentence of (first-order) predicate logic. Using the techniques presented in Section 2.2 we can transform F into a sat-equivalent set of clauses $\mathcal{C}$. An immediate consequence of this transformation (which is effectively computable) is the undecidability of the unsatisfiability problem of clause logic (roughly speaking clause logic is undecidable). On the other hand unsatisfiability (but *not* satisfiability) is semi-decidable, i.e., there exists an algorithm producing $\square$ on all unsatisfiable sets of clauses. An algorithm of this type can easily be obtained by computing $R_x()$ for a complete resolution refinement R_x. Clearly R_x cannot be a decision procedure of clause logic and thus it must be nonterminating on some finite, satisfiable sets of clauses; therefore case 3) cannot be avoided.

The area of mathematical logic characterized by the term "decision problem" originated at the beginning of the 20th century. Around 1900 Hilbert formulated the problem of finding an algorithm to decide the validity of first-order predicate logic formulas [Hil01]. He called this decision problem the "fundamental problem of mathematical logic". Indeed, in some informal sense, the problem is even older than modern symbolic logic. In the 17th century Leibniz formulated the vision of a calculus ratiocinator [Leib], which would settle arbitrary problems by purely mechanical computation, once they were translated into an adequate formalism.

At the beginning of the 20th century a positive solution of the decision problem seemed to be merely a question of mathematical invention. Indeed some progress was achieved soon as decidable subclasses of predicate logic were found. The decision algorithms provided for these classes were clearly effective in any plausible intuitive sense of the word (note that before publication of Turing's landmark paper [Tur36] no formal concept of algorithm was available). One of the first results was the decidability of the monadic class [Loew15] (i.e., the class of first-order formulas containing only unary predicate symbols and no function symbols). In the same paper Löwenheim showed that dyadic logic (in which all predicate symbols are binary) is a reduction class, i.e., a class of first-order formulas effectively "encoding" full predicate logic. In the decades between the two world wars many outstanding logicians attacked this problem. The initial strategy (probably) was to enlarge the decidable classes and to "shrink" the reduction classes till they eventually met at some point (the outcome would have been the decidability of first-order logic). We just mention the satisfiability problem of some prefix classes (i.e., classes of closed prenex formulas with function free matrix) proved decidable in this period: $\forall\exists^*$ (a subclass of what is called the Ackermann class today [Ack28]), $\forall\forall\exists^*$ (the Gödel–Kalmar–Schütte class [Göd32]) and $\exists^*\forall^*$ (the Bernays-Schönfinkel class [BS28]). Note that the satisfiability problem is dual to the validity problem and thus the quantifier prefixes become dual as well; e.g., the prefix of the Bernays–Schönfinkel class in the setting of the validity problem is $\forall^*\exists^*$. From now on we use the sentence "*the class K is decidable*" instead of the longer but more precise statement "the satisfiablity problem of K is decidable".

We proved the decidability of the Bernays–Schönfinkel class in Section 2.4 using the finite-model property of this class (i.e., there exists a finite model iff there exists a model at all). A class possessing this property is called *finitely controllable*. The original proofs of decidability for all the classes mentioned above were based on the finite-model property. In fact the set of all PL-formulas having finite models is recursively enumerable. Thus in performing search for a refutation and for a finite model in parallel, we clearly obtain a decision procedure. The algorithms extracted from this method, however, are based on exhaustive search and are hardly applicable in computational logic. Moreover, once we are in possession of the machinery of resolution, we have a proof-theoretic alternative in decision theory which yields not only reasonable algorithms but (very often) also clearer and easier proofs of decidability.

As already outlined in Section 4.5 the following proof-theoretic method proves decidability of the satisfiability problem for a predicate logic class Γ:

Let F be in Γ. Transform F into a sat-equivalent clause form $\mathcal{C}$ (we obtain a clausal class Γ' corresponding to Γ).
Then find a complete resolution refinement which terminates on Γ'.

This principle is quite general and can be applied within other calculi than resolution and other normal forms than clause form. In 1968 S.Y. Maslov proved decidability of the so called K-class (a decision class properly containing the Gödel–Kalmar–Schütte class) using this proof-theoretic paradigm. His approach is based on the inverse method, which is a resolution-type method formulated within the framework of a sequent calculus [Mas68]. More results along this line have been obtained by other representatives of the Russian school [Zam72]. A common feature of resolution and of the inverse method is the use of the unification principle. Note that, in order to prove decidability of classes, we need calculi with a restricted substitution rule (otherwise a formula may define infinitely many successors under the inference principle). A principle yielding a condition of finiteness for substitution is most general unification: Only unifying substitutions are necessary and there is always a single most general unifier which "does the whole job". Thus most general unification is not only a powerful principle in the design of computational algorithms, but also a strong tool in proving decidability of first-order classes.

In the same spirit as Maslov, but on the basis of the resolution calculus, Joyner showed in his thesis [Joy73] that resolution theorem provers can be used as decision procedures for some classical prefix classes (e.g., the Ackermann class and the Gödel-Kalmar-Schütte class). His idea of finding complete resolution refinements R_x that terminate on clause classes corresponding to prefix classes is of central importance to all results in this chapter. The method can easily be extended to clause classes that cannot be obtained from prefix classes (via skolemization). As an example take the class of all formulas of the form $(\forall v)M(v)$ where $M(v)$ is a quantifier-free formula with $V(M(v)) = \{v\}$ which may contain arbitrary terms. Its decidability was proved by Y. Gurevich [Gur73] by a model-theoretic method. In Section 5.2 we will define a resolution decision procedure for this class. More recently decidability results for several functional (nonprefix) classes have been obtained by various resolution methods [FLTZ93]; some of these classes will be discussed in this chapter.

In order to demonstrate the power of resolution as decision procedure we start with a simple "motivating" decision class.

Definition 5.1.1. *The class of all prenex forms* $(Q_1 x_1) \ldots (Q_m x_m) M$, *where M is a function-free conjunction of literals, is called the* Herbrand *class HC [Her31].*

Proposition 5.1.1. *The Herbrand class is decidable (by unrestricted resolution).*

Proof. Let F be a formula of HC and HC′ be the class of all clause forms of formulas in HC. Then HC′ $\subseteq$ Γ where $\Gamma = \{ \mathcal{C} \mid \mathcal{C}$ is a finite set of unit clauses$\}$.

Now let $\mathcal{C}$ be the set of clauses in Γ corresponding to F. Because $\mathcal{C}$ consists of unit clauses only, $Res(\mathcal{C}) = \emptyset$ or $Res(\mathcal{C}) = \{\Box\}$.

Thus we obtain a decision procedure for Γ by computation of $R_\emptyset$; this procedure also decides HC′ and (modulo transformation to clause form) HC.

$\Diamond$

In proving HC decidable Herbrand showed the decidability of the unification problem for atom formulas. Although he did not develop the principle of most general unification his proof first illustrated the importance of unification. Once resolution and unification are already available the decidability of HC is in fact trivial. However, we cannot hope for such an easy game in case of other decidable classes. Indeed the resolution operator $R_\emptyset$ does not terminate on all other classes mentioned above and thus is generally useless as decision procedure. It is the purpose of Sections 5.2 and 5.3 to investigate several types of refinements with respect to their potential to serve as decision methods. Note that by restricting existing refinements under preservation of completeness we always increase the decision potential of the method: If R_x and R_y are both complete refinements and $R_y(\mathcal{C}) \subseteq R_x(\mathcal{C})$ for all sets of clauses $\mathcal{C}$ then the class decided by R_x is a subclass of that decided by R_y.

Besides serving as theoretical tools for the decision problem, resolution decision procedures are valuable as "ordinary" theorem provers too. Because refinements used as decision procedures mostly are very restricted and must terminate, they favor the production of clauses having low complexity; this property makes them quite efficient in practice.

Exercises

A class Γ of (finite) sets of clauses is called "corresponding to Δ", Δ being a class of formulas in predicate logic, if $\Gamma = \{ \text{clf}(F) \mid F \in \Delta \}$, where clf is a function mapping closed predicate logic formulas to clausal form (clf can be computed by the techniques introduced in Section 2.2).

Let VAR1 $= \{ \mathcal{C} \mid \mathcal{C}$ is a finite set of clauses with $|V(C)| \leq 1$ for all $C \in \mathcal{C} \}$.

Exercise 5.1.1. Show that VAR1 does not correspond to a prefix class of predicate logic.

Let PS2 be the class of all finite sets of clauses C such that every predicate symbol occurs at most twice in C.

Exercise 5.1.2. Show that PS2 can be decided by lock resolution, or more exactly: Define an algorithm α which produces an indexed set of clauses $\alpha(C)$ for every $C \in$ PS2 and show that $R_l(\alpha(C))$ is finite for every locking l corresponding to $\alpha(C)$.

Exercise 5.1.3. Show that PS2 cannot be decided by unrestricted resolution, i.e., there exists a $C \in$ PS2 such that $R_\emptyset(C)$ is infinite.

5.2 A-Ordering Refinements as Decision Procedures

Let us consider the class of closed predicate logic formulas of the form

$$F : (\exists x_1)\dots(\exists x_m)(\forall y)(\exists z_1)\dots(\exists z_k)M(x_1,\dots,x_m,z_1,\dots,z_k,y)$$

where $k, m \geq 0$ and M is a function- and constant- free matrix (we always speak about "matrices" when referring to quantifier-free parts of prenex form). This class was shown decidable by Wilhelm Ackermann in 1928 [Ack28] and thus is called the Ackermann class. Because it is characterized by the form of its quantificational prefix it is frequently symbolized by $\exists^*\forall\exists^*$, where Q^* denotes an arbitrary repetition of the quantifier Q. By skolemizing F above we obtain a closed formula

$$F' : (\forall y)M(c_1,\dots,c_m,f_1(y),\dots,f_k(y),y),$$

where $c_1,\dots,c_m$ are (different) constant symbols and $f_1,\dots,f_k$ are (different) one-place function symbols. In transforming the matrix of F' into conjunctive normal form (we may take the straightforward method based on distributivity) we obtain a set of clauses C fulfilling the following properties:

1) All clauses contain at most one variable,
2) all function symbols occuring in C are unary,
3) the term depth of all clauses C in C is ≤ 1.

In particular all sets of clauses obtained from the Ackermann class belong to the *one-variable class* introduced in the following definition:

Definition 5.2.1. *The class* VAR1 *(also called the one-variable class) is the set of all finite sets of clauses C fulfilling the following condition: For all $C \in C$: $|V(C)| \leq 1$.*

We have seen that the clause forms of the formulas of the Ackermann class belong to VAR1; on the other hand there exist sets of clauses in VAR1 that cannot be obtained by transforming Ackermann formulas into clause form (see Exercise 5.1.1).

A decision algorithm α for the class VAR1 can easily be transformed into a decision algorithm β for the Ackermann class: First transform a formula of the Ackermann class into a clause form $\mathcal{C}$ and then apply α to $\mathcal{C}$. If a set of clauses $\mathcal{C}$ in VAR1 is not normalized, then $\mathcal{C}$ (but not its elements) may contain several variables due to variable renaming. But using the N_c-normal form for clauses we even obtain $V(\mathcal{C}) \leq 1$ (every nonground clause can only contain the variable x_1). For technical reasons which will become clear later, we restrict VAR1 to its subset of N_c-normalized sets of clauses.

Definition 5.2.2. *VAR1C is the set of all finite sets of N_c-clauses such that $\mathcal{C} \in$ VAR1.*

The decision problem of VAR1 can easily be reduced to that of VAR1C; we simply have to replace sets of clauses by their normalized forms in a preprocessing step.

Example 5.2.1. The following set of clauses is an element of VAR1 for

$$C_1 = P(a),\ C_2 = \neg P(x) \vee R(f(x)),\ C_3 = \neg R(y) \vee R(f(y)),\ C_4 = \neg R(f(f(b))).$$

By applying condensed normalization N_c we obtain the set

$$\mathcal{D}:\ \{P(a),\ R(f(x_1)) \vee \neg P(x_1),\ R(f(x_1)) \vee \neg R(x_1),\ \neg R(f(f(b)))\}.$$

It is easy to see that $\mathcal{D}$ (and, of course, $\mathcal{C}$ too) is satisfiable. Thus clearly $\square \notin R_\emptyset(\mathcal{D})$ (here we use unrestricted resolution on N_c-clauses). But, unfortunately, $R_\emptyset(\mathcal{D})$ is infinite and $R_\emptyset$ cannot serve as decision algorithm for VAR1C; note that $R(f^{(n)}(a)) \in R_\emptyset(\mathcal{D})$ for all $n \geq 1$. Even hyperresolution does not yield a decision procedure as $R_H(\mathcal{D})$ is infinite too.

We will prove in this chapter that VAR1C can be decided by a specific A-ordering refinement. Remember the A-ordering $<_d$ defined in Section 3.3:

$A <_d B$ iff

(1) $\tau(A) < \tau(B)$ and
(2) For all $x \in V(A)$: $\tau_{\max}(x, A) < \tau_{\max}(x, B)$ (and also $V(A) \subseteq V(B)$).

Applying $R_{<_d}$ to the set of clauses in Example 5.2.1 we obtain

$$R_{<_d}(\mathcal{D}) = \mathcal{D} \cup \{\neg R(f(b)),\ \neg P(f(b)),\ \neg P(b),\ \neg R(b)\}.$$

Thus $R_{<_d}(\mathcal{D})$ is finite and $\square \notin R_{<_d}(\mathcal{D})$ and we have shown that $\mathcal{D}$ is satisfiable by computing $R_{<_d}$ on $\mathcal{D}$.

Our goal is to prove the finiteness of $R_{<_d}(\mathcal{C})$ for all $\mathcal{C} \in$ VAR1C. But in the attempt to show termination of $R_{<_d}$ on VAR1C we face the technical problem that VAR1C is not invariant under $R_{<_d}$, i.e., $\rho_{<_d}(\mathcal{C})$ may contain clauses C with $|VAR(C)| > 1$ for $\mathcal{C} \in$ VAR1C.

Example 5.2.2. Let $C = \{P(a) \vee R(x_1), \; Q(x_1) \vee \neg P(a)\}$; clearly $C \in$ VAR1C. In resolving the two clauses in C we must rename the variables and obtain the N_c-resolvent $Q(x_1) \vee R(x_2)$. Obviously the set of clauses $C \cup \{Q(x_1) \vee R(x_2)\}$ is no longer in VAR1C.

Therefore our first step consists in enlarging the class VAR1C in order to obtain a class which is invariant under $R_{<_d}$. The classes defined below are even invariant under $R_{\emptyset}$ and contain VAR1C.

Definition 5.2.3. *The class K_{∞} is the set of all (possibly infinite!) sets of clauses C in N_c-normal form such that for all $C \in C$ and for all literals L in C: $V(L) \leq 1$ (every literal contains at most one variable). The class K is the subclass of K_{∞} containing only finite sets of clauses.*

By definition the sets of clauses C in K_{∞} and in K possess the following property:

> For all $C \in C$ the connected components $C_1, \ldots, C_k$ of C (see Definition 4.3.5) contain at most one variable.

If D is the set of clauses defined in Example 5.2.1 then $R_{<_d}(D) \in K$ and $R_{\emptyset}(D) \in K_{\infty}$ but $R_{\emptyset}(D) \notin K$. Our final goal is to show that K is invariant under $R_{<_d}$, i.e., $R_{<_d}(C) \in K$ for all $C \in K$ (this property implies termination as K contains only finite sets of clauses). The next step consists in showing that for every $C \in R_{\emptyset}(C)$ and $C \in K$ also $\{C\} \in K$. As indicated above we cannot expect $R_{\emptyset}(C)$ to be contained in K but merely in K_{∞}.

Lemma 5.2.1. *Let C be in K, then $R_{\emptyset}(C) \in K_{\infty}$.*

Proof. By definition of $R_{\emptyset}$ it is sufficient to prove $S^i_{\emptyset}(C) \in K$ for all i. This in turn can be reduced further to

(I) Let $C \in K$ and $C_1, C_2 \in C$ and let C be a resolvent of C_1 and C_2; then $\{C\} \in K$ (and thus $C \cup \{C\} \in K$.

We first show that factoring a clause does not lead outside of K, i.e., if $\{C\} \in K$, C' is a variant of C and $C'\vartheta$ is a factor of C' then $\{N_c(C'\vartheta)\} \in K$.

Let us assume first that we factor inside a connected component. Then ϑ must be a ground substitution (note that corresponding pairs (x,t) such that $x \neq t$ and $x \in V(t)$ cannot be unified). But if ϑ is a ground substitution then clearly $\{N_c(C'\vartheta)\} \in K$ (one of the one-variable component is grounded and added to the ground component).

By factoring over different components we obtain factoring substitutions of the form

$$\vartheta: \quad \lambda \cup \{x_i \leftarrow t[x_j] | i \in A, j \in B\}$$

where $A \cup B \subseteq \{1, \ldots, n\}$ for $V(C') = \{x_1, \ldots, x_n\}$ and $A \cap B = \emptyset$ and λ is a ground substitution; note that $A \cap B$ must be empty by the idempotence of m.g.u.'s (indeed all m.g.u.'s σ fulfill $\sigma\sigma = \sigma$, see Exercise 2.6.5). Clearly $C'\vartheta$ again consists of components with ≤ 1 variables only and $\{N_c(C'\vartheta)\} \in K$.

It remains to show that K is closed under binary resolution.
Let D_1, D_2 be factors of variants of clauses C_1, C_2 such that $\{C_1, C_2\} \in K$.

Assume that $D_1 = E_1 \vee P \vee F_1$ and $D_2 = E_2 \vee \neg Q \vee F_2$ and let σ be an m.g.u. of $\{P, Q\}$. Then

$$R: (E_1 \vee F_1 \vee E_2 \vee F_2)\sigma$$

is a binary resolvent of D_1 and D_2.

We have to show that $\{R\} \in K$.

First of all we observe that $\{N_c(G)\}$, for $G = E_1 \vee F_1 \vee E_2 \vee F_2$ is in K. Because σ is a unifier of $\{P, Q\}$ it must unify all elements in $DIFF(P, Q)$ (see Definition 2.6.9). If σ is a ground substitution then $\{N_c(R)\} \in K$ (the argument is the same as for factoring substitutions). Thus we may assume that $V(P) = \{x_i\}$ and $V(Q) = \{y_j\}$ and σ is a nonground m.g.u. of $\{P, Q\}$. Then either $\sigma = \{x_i \leftarrow t[y_j]\}$ or $\sigma = \{y_j \leftarrow s[x_i]\}$ for some terms s and t. In both cases two components are merged into a single one and thus

$$\{N_c(G\sigma)\} = \{N_c(R)\} \in K.$$

This completes the proof of (I).   ◊

We are now ready to attack the main result of this chapter.

Theorem 5.2.1. *The class K can be decided by the A-ordering $<_d$. More exactly: $R_{<_d}(C)$ is finite for all $C \in K$ and $\square \in R_{<_d}(C)$ iff C is unsatisfiable.*

Proof. By using the results in Section 3.3 it is easy to show that $\square \in R_{<_d}(C)$ iff C is unsatisfiable. First of all $R_{<_d}$ is correct and it is also complete by Theorem 3.3.1.

It remains to show termination.

From Lemma 5.2.1 we know that $R_{<_d}(C) \in K_\infty$ for $C \in K$ (note that $R_{<_d}(C) \subseteq R_\emptyset(C)$ for all sets of N_c-clauses C). What we actually need is the result $R_{<_d}(C) \in K$.

Every set of clauses C contains only finitely many function symbols and constant symbols. Resolution does not change the signature (it can only introduce new variables) and thus the set of function and constant symbols in $R_{<_d}(C)$ is the same as in C itself. We therefore can reduce the proof of termination to

(*) There exist constants $k, l \in \mathbb{N}$ such that for all $C \in R_{<_d}(C)$:
   (*1) $|V(C)| \leq k$ and
   (*2) $\tau(C) \leq l$

It is easy to see that there are only finitely many N_c-normalized clauses of term depth $\leq l$ and containing $\leq k$ different variables over a finite signature (Exercise 5.2.2).

We show now (*2) and, in fact, the more specific result:

(I) Let $C \in K$. Then for all $C \in R_{<_d}(\mathcal{C})$:

$$\Pi(C): \quad \tau(C_g) \leq 2\tau(C) \text{ and } \tau(C \setminus C_g) \leq \tau(C)$$

(where C_g denotes a clause form of the ground component of C).

The property (I) trivially implies $\tau(C) \leq 2\tau(C)$ for all $C \in R_{<_d}(\mathcal{C})$. Formally we have proved (I) for $S^i_{<_d}(\mathcal{C})$ via induction. From Lemma 5.2.1 we know that $R_{<_d}(\mathcal{C}) \in K_\infty$ and therefore all $S^i_{<_d}(\mathcal{C})$ are in K. Thus it is sufficient to prove:

Let C, D be clauses in $R_{<_d}(\mathcal{C})$ fulfilling $\Pi(C)$ and $\Pi(D)$ and E be a resolvent of C and D then E fulfills $\Pi(E)$.

For $C \in S^0_{<_d}(\mathcal{C})$ the property (I) is trivially fulfilled.

Again we split our argument into one part concerning factoring and another for binary resolution. Moreover factoring can be simulated by iterated binary factoring. Thus we have to prove that for $\{C\} \in K$ and $\Pi(C)$ and for a binary factor $C\vartheta$ of C we also get $\Pi(C\vartheta)$. The proof is similar to that for binary resolution shown below and is left as an exercise (Exercise 5.2.3).

We now turn to binary resolution:

Let C_1 and C_2 be factors of renamed variants of two clauses in $R_{<_d}(\mathcal{C})$ such that $\Pi(C_1)$ and $\Pi(C_2)$ both hold. Let R be a binary $R_{<_d}$-resolvent of C_1, C_2, i.e., $R \in \rho_{<_d}(\{C_1, C_2\})$ obtained by binary resolution only. We distinguish the following basic cases:

a) C_1 is ground, C_2 is nonground,

b) C_1 is nonground, C_2 is ground,

c) C_1, C_2 are both nonground,

d) C_1, C_2 are both ground.

The simplest case is d):
Here, by assumption, $\tau(C_1) \leq 2\tau(C)$ and $\tau(C_2) \leq 2\tau(C)$, $(C_1)_g = C_1$, $(C_2)_g = C_2$. Because the m.g.u. of the resolution is ϵ we obtain $\mathrm{LIT}(R) \subseteq \mathrm{LIT}(C_1) \cup \mathrm{LIT}(C_2)$ and thus $\tau(R) \leq 2\tau(C)$; R is ground too and thus $R \setminus R_g = \square$

Let us consider the case a):
Let σ be the m.g.u. of the resolution. If $\sigma = \epsilon$ we again obtain

$$\mathrm{LIT}(R) \subseteq \mathrm{LIT}(C_1) \cup \mathrm{LIT}(C_2)$$

and, by assumption,

$$\tau(R_g) \leq \tau(\text{LIT}(C_1) \cup \text{LIT}(C_2)) \leq 2\tau(\mathcal{C}),$$

$$\tau(R \setminus R_g) \leq \max\{\tau(C_1 \setminus (C_1)_g), \tau(C_2 \setminus (C_2)_g\} \leq \tau(\mathcal{C}).$$

If $\sigma \neq \epsilon$ it must be of the form $\{y \leftarrow s\}$ for $y \in V(C_2)$ and a ground term s. In the resolvent R the x-components for $x \neq y$ are identical to those of $C_1 \vee C_2$.

Moreover R does not contain a y-component and

$$\text{LIT}(R \setminus R_g) \subseteq \text{LIT}(C_1 \vee C_2)$$

Let $C_2[y]$ be the y-component of C_2.

By $\Pi(C_2)$ we have $\tau(C_2[y]) \leq \tau(\mathcal{C})$. Now R is an $R_{<_d}$-resolvent. That means there is no literal $L \in \text{LIT}(R)$ such that $A <_d L$, where A is the resolved atom. Because C_1 is ground either A or $\neg A$ must be a literal of C_1 and thus $\tau(A) \leq \tau(C_1) \leq 2\tau(\mathcal{C})$.

Let us assume that there exists a literal L in R such that $\tau(L) > 2\tau(\mathcal{C})$. Then L must occur in R_g and therefore L is ground. Thus, by definition of $<_d$, we obtain $A <_d L$ and $L \in \text{LIT}(R_g)$. But this contradicts the assumption that R is a $<_d$-resolvent. Therefore $\tau(L) > 2\tau(\mathcal{C})$ is impossible and we obtain

$$\tau(R) \leq 2\tau(\mathcal{C})$$

case b): Completely symmetric to case a).

It remains to investigate case c):

Let $C_1 = E_1 \vee L \vee E_2$ and $C_2 = F_1 \vee M \vee F_2$ such that the resolution operates on the literals L and M. If either L or M belongs to a ground component the argument is exactly as in case a). If L and M are both ground then $R = E_1 \vee E_2 \vee F_1 \vee F_2$ and $\Pi(R)$ holds trivially.

The only interesting case is that both components (containing L and M are nonground. Thus let us assume that $L \in C_1(x)$ and $M \in C_2(y)$, $C_1(x)$ being the x-component of C_1 and $C_2(y)$ the y-component of C_2. Let σ be the m.g.u. of $\{L, M^d\}$; then σ must have one of the following forms:

(α) $\sigma = \{x \leftarrow t, y \leftarrow s\}$ for ground terms s and t, or

(β) $\sigma = \{x \leftarrow t[y]\}$ for some term $t[y]$ with $V(t[y]) = \{y\}$, or

(γ) $\sigma = \{y \leftarrow s[x]\}$ for some term $s[x]$ with $V(s[x]) = \{x\}$.

We have to show that in all three cases $\tau(R_g) \leq 2\tau(\mathcal{C})$ and $\tau(R \setminus R_g) \leq \tau(\mathcal{C})$ Let us assume that A is the resolved atom, i.e., $A = at(L\sigma)$.

case (α):

A is ground. Let us assume that L is some ground literal in R. Then $\tau(L) \leq \tau(A)$, for $\tau(A) < \tau(L)$ would imply $A <_d L$, which is impossible as R is a $<_d$-resolvent. It remains to show that $\tau(A) \leq 2\tau(\mathcal{C})$. By the form of σ it is indeed possible that $\tau(A) > \max\{\tau(C_1(x)), \tau(C_2(y))\}$ (Exercise 5.2.4). But $\tau(A) \leq \tau(L) + \tau(M)$. Note that both L and M belong to variable components and thus, by assumption, $\tau(L), \tau(M) \leq \tau(\mathcal{C})$. Therefore $\tau(A) \leq 2\tau(\mathcal{C})$ and thus also $\tau(R_g) \leq 2\tau(\mathcal{C})$. Moreover we have

$$\mathrm{LIT}(R \setminus R_g) \subseteq \mathrm{LIT}((C_1 \lor C_2) \setminus (C_1 \lor C_2)_g).$$

Note that the x- and y-components become ground and the other components are not changed by σ.

By assumption we have

$$\tau(C_1 \lor C_2) \setminus (C_1 \lor C_2)_g) = \max\{\tau(C_1 \setminus (C_1)_g), \tau(C_2 \setminus (C_2)_g)\} \leq \tau(\mathcal{C}).$$

So we infer $\tau(R \setminus R_g) \leq \tau(\mathcal{C})$.

case (β):
Here the resolved atom is of the form $A:\ A_0\{x \leftarrow t[y]\}$ for $A_0 = at(L)$.

In this case we obtain $\mathrm{LIT}(R_g) \subseteq \mathrm{LIT}((C_1 \lor C_2)_g)$ and, by assumption,

$$\tau(R_g) \leq 2\tau(\mathcal{C}).$$

All z-components for $z \notin \{x, y\}$ remain unchanged and the x-component disappears. Let $R(y)$ be the y-component of R. Then it is enough to show that $\tau(R(y)) \leq \tau(\mathcal{C})$. Because $\{x \leftarrow t[y]\}$ is a matching substitution we have $at(M) = A$. By assumption we have

$$\tau(M) \leq \tau(C_2 \setminus (C_2)_g) \leq \tau(\mathcal{C}).$$

Thus we obtain for the resolved atom $\tau(A) \leq \tau(\mathcal{C})$.

It remains to show that for all $L' \in R(y):\ \tau(L') \leq \tau(\mathcal{C})$.

Because R is a $<_d$-resolvent we have $\tau_{\max}(y, L') \leq \tau_{\max}(y, A)$; note that otherwise we would obtain $\tau(A) < \tau(L')$ and $\tau_{\max}(y, L') \leq \tau_{\max}(y, A)$, what implies $A <_d L'$ contradicting the assumption that R is a $<_d$-resolvent.
    If L' comes from $C_2(y)$ then clearly $\tau(L') \leq \tau(\mathcal{C})$ (by $\tau(C_2 \setminus (C_2)_g) \leq \tau(\mathcal{C})$).
Now assume that $L' = L_0'\{x \leftarrow t[y]\}$ for some $L_0' \in \mathrm{LIT}(C_1(x))$. By assumption we have $\tau(L_0') \leq \tau(\mathcal{C})$. Moreover $\tau_{\max}(x, L_0') \leq \tau_{\max}(x, A_0)$, because $\tau_{\max}(x, L_0') > \tau_{\max}(x, A_0)$ would also imply $\tau_{\max}(y, L') > \tau_{\max}(y, A)$, a property which we have already excluded (because it contradicts the assumption that R is a $<_d$-resolvent).

In any case we obtain the inequality

$$\tau(L') \geq \tau_{\max}(x, L_0') + \tau(t[y]).$$

If $\tau(L') > \tau_{\max}(x, L_0') + \tau(t[y])$ then $\tau(L') = \tau(L_0')$ and, consequently,

$$\tau(L') \leq \tau(\mathcal{C}).$$

So let us assume that $\tau(L') = \tau_{\max}(x, L'_0) + \tau(t[y])$.

Then by $\tau_{\max}(x, L'_0) \leq \tau_{\max}(x, A_0)$ and

$$\tau(A) = \max\{\tau_{\max}(x, A_0) + \tau(t[y]), \ \tau(A_0)\} \leq \tau(\mathcal{C})$$

we also obtain

$$\tau(L') \leq \max\{\tau(L_0), \ \tau_{\max}(x, A_0) + \tau(t[y])\} \leq \tau(\mathcal{C})$$

So we have proved $\tau(R(y)) \leq \tau(\mathcal{C})$ and, to sum up, $\tau(R \setminus R_g) \leq \tau(\mathcal{C})$ and $\tau(R_g) \leq 2\tau(\mathcal{C})$.

case (γ): Symmetric to case (β).

This concludes the proof of (I) and thus of (*2).

It remains to prove (*1):
The depth limit on clauses proved in (I) does not depend on the condensation normal form of the clauses. But note that there are infinitely many clauses with depth $\leq k$ even over the empty signature. The N_c-normal form will give us the guarantee that the size of clauses in $R_{<_d}(\mathcal{C})$ is uniformly bounded.

Let k be the maximal term depth occuring in $R_{<_d}(\mathcal{C})$. Then there exists a number l such that for every literal occurring in a clause $C : C \in R_{<_d}(\mathcal{C})$ the number of different subterms in L is $\leq l$. Because all clauses $C : C \in R_{<_d}(\mathcal{C})$ are condensed there can be no literals M, N in C such that M and N are different and $M \sim_v N$: Within one component of C $M \sim_v N$ trivially implies $M = N$ (simply because $V(M) = V(N) = \{x\}$ for some variable x). As every component contains one variable only the number of literals in the components is uniformly bounded by some number, i.e., there exists a number k such that for all connected components $C(x)$ of a clause $C : C \in R_{<_d}(\mathcal{C}), |C(x)| \leq k$ holds. To sum up, there are numbers k and l such that for all $C : C \in R_{<_d}(\mathcal{C})$ and for all components C' of $C: |C'| \leq k$ and $\tau(C') \leq l$.

It remains to show that there exists a number m such that $comp(C) \leq m$ for all $C : C \in R_{<_d}(\mathcal{C})$, where $comp(\ )$ denotes the number of components.

Let us assume on the contrary that for every m there exists a $C : C \in R_{<_d}(\mathcal{C})$ with $comp(C) > m$. By the uniform bound on the size of components there must be a clause C and components $C(x_i), C(x_j)$ of C such that $C(x_i) \sim_v C(x_j)$. But this contradicts the assumption that all clauses are in N_c-normal form (it is here that we actually need N_c-normal form – N_s-normalization would not give us termination). Note that $\{x_i \leftarrow x_j\}$ defines a factor of C which is a proper subclause of C. So we conclude that there must be a uniform bound m on $comp(C)$ for $C \in R_{<_d}(\mathcal{C})$. But by definition of K_∞ and by $R_{<_d}(\mathcal{C}) \in K_\infty$ we have $|V(C)| = comp(C) - 1$ for all $C : C \in R_{<_d}(\mathcal{C})$. Therefore there exists a number m such that $|V(C)| \leq m$ for all $C : C \in R_{<_d}(\mathcal{C})$. It follows that $R_{<_d}(\mathcal{C})$ is finite and $R_{<_d}(\mathcal{C}) \in K$. $\Diamond$

Corollary 5.2.1. $R_{<_d}$ *decides* VAR1C *and the Ackermann class.*

Proof. The Ackermann class is contained (modulo transformation to clause form) in VAR1C and VAR1C $\subseteq K$. $R_{<_d}$ decides K by Theorem 5.2.1.   ◊

The class K can be generalized further to the class K^* fulfilling the following condition: All literals in a clause contain the same variables or are variable-disjoint, and every functional term containing variables contains all of them. In [FLTZ93] it has been shown that K^* can be decided by an A-ordering refinement (different from $R_{<_d}$) under additional use of *saturation*. Saturation is a method which, after computation of a level $S_x^i(\mathcal{C})$, adds a finite set of instances which are not obtained by most general unification. We will use this technique to decide the Bernays–Schönfinkel class in Section 5.3. Saturation techniques are also required to decide the Gödel class $\forall\forall\exists^*$ and the Skolem class (see [Joy76] and [Fer91]). Saturation is a deviation from the pure resolution paradigm, which is based on most general unification. It is necessary in those cases where termination cannot be obtained by the mere use of most general unifiers. Among the classical decision classes not only the Ackermann class, but also the monadic class can be decided by "pure" A-ordering refinements. As we have seen in Section 5.1, the Herbrand class can even be decided by unrestricted resolution.

Definition 5.2.4 (Monadic Class). *Let* PL_0 *be the set of all closed PL-formulas without constant symbols and function symbols. The subclass of* PL_0 *containing only unary predicate symbols is called the* monadic class *and denoted by* MON.

We show first how MON can be transformed into a sat-equivalent class MON*, i.e., we will define a transformation π such that for all $F \in$ MON*: $\pi(F) \in$ MON* and $F \sim_{sat} \pi(F)$.

For the sake of simplicity we transform F into prenex form, i.e., we obtain a formula $G \in \mathrm{PL}_0$ such that $F \sim G$ and

$$G = (Q_1 x_1) \ldots (Q_m x_m) M(x_1, \ldots, x_m)$$

where $M(x_1, \ldots, x_m)$ is a quantifier-free matrix. Let H be the skolemized form of G obtained by the transformation β defined in Section 2.2. Then H is of the form

$$H : (\forall y_1) \ldots (\forall y_n) M(t_1, \ldots, t_n)$$

where the t_i are either variables, constant symbols or functional terms. According to the prenex form of G a functional term s must fulfill the following property: There exist a number $k \leq n$ and a k-ary function symbol f such that $s = f(y_1, \ldots, y_k)$ (Exercise 5.2.5).

Example 5.2.3. Let F be the formula

$$((\forall x)P(x) \wedge (\forall x)Q(x) \wedge (\exists z)(\neg P(z) \vee R(z)) \rightarrow (\forall z)R(z)$$

F is a satisfiable nonvalid formula in MON.
We first transform F into $\alpha(F)$ (see Section 2.2) and obtain

$$F' : \quad (\exists x)\neg P(x) \vee (\exists y)\neg Q(y) \vee (\forall z)(P(z) \wedge \neg R(z)) \vee (\forall u)R(u).$$

A prenex form of F' is

$$(\forall u)(\exists x)(\forall z)(\exists y)(\neg P(x) \vee \neg Q(y) \vee (P(z) \wedge \neg R(z)) \vee R(u)).$$

The skolemized form $\beta(G)$ is

$$H : \quad (\forall u)(\forall z)(\neg P(f(u)) \vee \neg Q(g(u,z)) \vee (P(z) \wedge \neg R(z)) \vee R(u)).$$

By using the law of distributivity we obtain the set of clauses

$\mathcal{C} : \{\neg P(f(u)) \vee \neg Q(g(u,z)) \vee P(z) \vee R(u),$
$\neg P(f(u)) \vee \neg Q(g(u,z)) \vee \neg R(z) \vee R(u)\}$

In both clauses there exists the "dominating" literal $\neg Q(g(u,z))$ which contains all variables of the clause.

The following line of argument is similar to that of Joyner in [Joy76] but is somewhat more general.

Definition 5.2.5. *Two functional terms s,t are called* similar *if $s = g(r_1,\ldots,r_n)$ and $t = f(w_1,\ldots,w_n)$ such that f and g are two possibly different function symbols of the same arity and $\{r_1,\ldots,r_n\} = \{w_1,\ldots,w_n\}$. A functional term s* dominates *a functional term t if there are function symbols f,g and terms $r_1,\ldots,r_n,w_1,\ldots,w_m$ such that $s = f(r_1,\ldots,r_n)$, $t = g(w_1,\ldots,w_m)$ and $n > m$ and $\{w_1,\ldots,w_m\} \subseteq \{r_1,\ldots,r_n\}$. Every functional term dominates every constant symbol.*

We introduce the following relation $<_1$ on terms: $s <_1 t$ iff either s properly occurs in t or t dominates s or t contains a proper subterm which dominates s or is similar to s.

Proposition 5.2.1. *$<_1$ is irreflexive and transitive and for all substitutions ϑ: $s <_1 t$ implies $s\vartheta <_1 t\vartheta$.*

Proof. Exercise 5.2.6.

Example 5.2.4. $f(x,y)$ and $g(x,y)$ are similar, $h(x,y,z)$ dominates both $f(x,y)$ and $g(y,x)$.

$g(f(x,y),x)$ does not dominate $g(y,x)$ but $g(y,x) <_1 g(f(x,y),x)$ because $g(y,x)$ and $f(x,y)$ are similar and $f(x,y)$ is a proper subterm of $g(f(x,y),x)$. Clearly we also have $x <_1 g(y,x)$.

Definition 5.2.6. *We define the following binary relation $<_2$ on atoms: $A <_2 B$ iff there exists an argument t of B such that for all arguments s of A we have $s <_1 t$.*

Example 5.2.5. $P(x,y) <_2 Q(f(x,y))$ as $x <_1 f(x,y)$ and $y <_1 f(x,y)$. $P(x,z) \not<_2 Q(f(x,y))$ because $z \not<_1 f(x,y)$.

Proposition 5.2.2. $<_2$ *is an A-ordering.*

Proof. $<_2$ is irreflexive because $<_1$ is.

Assume that $A <_2 B$. Then there exists an argument t of B such that for all arguments s in A: $s <_1 t$. Let ϑ be a substitution; then $s\vartheta <_1 t\vartheta$ by Proposition 5.2.1. Clearly $t\vartheta$ is an argument of $B\vartheta$ such that $w <_1 t\vartheta$ for all arguments w of $A\vartheta$ and therefore $A\vartheta <_1 B\vartheta$.

It remains to prove transitivity.
Let $A <_2 B$ and $B <_2 C$. By $A <_2 B$ there exists an argument t of B such that for all arguments s of A: $s <_1 t$. By $B <_2 C$ there exists an argument r of C such that for all arguments w of B: $w <_1 r$. By the transitivity of $<_1$ we obtain $s <_1 r$ for all arguments s of A and thus $A <_2 C$.   ◊

We now define the the the clausal class MON* which (properly) contains the clausal forms of the formulas in MON.

Definition 5.2.7. MON* *is the set of all finite sets of N_c-clauses C such that*

1. *All predicate symbols occurring in C are unary.*
2. *All terms occurring as arguments of atoms in C are either constant symbols, variables, or terms of the form $f(x_1, \ldots, x_n)$ for (distinct) variables $x_1, \ldots, x_n$.*
3. *If C is a clause in C and D is a connected component of C containing the functional terms $t_1, \ldots, t_r$ as arguments of literals then there exist variables $y_1, \ldots, y_k$ and function symbols $g_1, \ldots, g_r$ such that $t_i = g_i(y_1, \ldots, y_{l_i})$ for $i = 1, \ldots, r$ and $l_i \le k$. In particular we obtain $V(t_i) \subseteq V(t_j)$ or $V(t_j) \subseteq V(t_i)$ for all $i, j = 1, \ldots, r$.*

Let Σ be a finite signature, i.e., a finite set of predicate, function, and constant symbols. Then MON(Σ) is the subclass of MON* defined over the signature Σ.*

Note that by the properties 2 and 3 in Definition 5.2.7 every component $D \in \mathrm{COMP}(C)$ such that $|V(D)| > 1$ must contain a literal $L : P(t)$ or $L : \neg P(t)$ respectively such that $t = f(x_1, \ldots, x_n)$ for some function symbol f and variables $x_1, \ldots, x_n$. Every other argument s appearing in D is either similar to t or $s <_1 t$. Thus if $M \in \mathrm{LIT}(D)$ then either L and M contain a similar argument or $M <_2 L$.

We show now that MON* restricted to a fixed finite signature is finite.

Proposition 5.2.3. *Let Σ be a finite signature. Then MON*(Σ) is finite.*

Proof. Let k be the maximal arity of a function symbol in Σ. Then by Definition 5.2.7 points 2 and 3 (and by the fact that the maximal term depth is one) every component of a clause C contains at most k variables. Because the number of variables and the term depth are uniformly bounded and Σ is finite there are only finitely many different components modulo renaming of variables. But MON* is a class of condensed sets of clauses, i.e., all clauses are in N_c-normal form. Thus if $C \in \mathcal{C}$ and $\mathcal{C} \in \text{MON}^*(\Sigma)$ then C cannot contain two different components D_1, D_2 such that $D_1 \sim_v D_2$; note that otherwise C would contain a factor which is a proper subclause of C. Therefore the set $\{C | \{C\} \in \text{MON}^*(\Sigma)\}$ is finite and thus $\text{MON}^*(\Sigma)$ must be finite too.   ◊

Let $\mathcal{C}$ be in MON*; then clearly $\mathcal{C} \in \text{MON}^*(\Sigma)$ where Σ is the signature of $\mathcal{C}$. Because $\text{MON}^*(\Sigma)$ is finite, all we need is a resolution refinement R_x such that $R_x(\mathcal{C}) \in \text{MON}^*(\Sigma)$ for every $\mathcal{C} \in \text{MON}^*(\Sigma)$. Our candidate for R_x is $R_{<_2}$ for the A-ordering $<_2$ in Definition 5.2.6. But, unfortunately, MON* is not invariant under $R_{<_2}$ if we allow arbitrary factors in $<_2$-resolution (see Exercise 5.2.7). Fortunately we may restrict factoring in A-ordering refinements without losing completeness. As we presented Robinson's resolution concept in Section 2.8, we demonstrated that factoring can be restricted to literals which are eventually cut out by binary resolution. We used this principle in lock resolution where factoring was restricted to literals of lowest index. The principle of connecting factoring with binary resolution in this way never leads to incompleteness (unless we deal with quite pathological refinements).

Definition 5.2.8. *Let C be a clause and let $\mathcal{F}$ be a set of literals in C (selected for factorization). Let σ be the factoring substitution unifying $\mathcal{F}$ such that $\mathcal{F}\sigma = \{L'\}$ and $(C\sigma)_r = A \vee L' \vee B$ (D_r denotes the reduced form of the clause D – see Definition 2.4.1). Let E be a binary resolvent of $C' : (C\sigma)_r$ and a clause D by resolution on L'. Then we call every literal $L : L \in \mathcal{F}$ selected for resolution (or more exactly selected for this specific resolution).*

Definition 5.2.9. *A factor is called selected for a resolution Π if it is defined by unifying a set of literals $\mathcal{F}$ which are selected for Π. A resolvent E of two clauses C and D is called obtained under selected factoring if E is obtained by a resolution, where the factors of C and D are selected for Π.*

Definition 5.2.10 (A-ordering with selected factoring). *Let $\rho_{<_A}$ be defined as in Definition 3.3.4. Then $\rho^0_{<_A}(\mathcal{C})$ is a subset of $\rho_{<_A}(\mathcal{C})$ obtained by resolution under selected factoring. $R^0_{<_A}$ is defined like $R_{<_A}$, but on the basis of $\rho^0_{<_A}$.*

It is not hard to show that $R^0_{<_A}$ is complete for every A-ordering $<_A$ (see Exercise 5.2.8). Unlike $R_{<_A}$, $R^0_{<_A}$ does not destroy the structure of MON* and thus gives a decision procedure for this class.

Theorem 5.2.2. *The A-ordering $<_2$ with selected factoring decides* MON*, *i.e., for all $C : C \in$ MON*: $R^0_{<_2}(C)$ is finite, and C is unsatisfiable iff $\square \in R^0_{<_2}(C)$.*

Proof. By the completeness of $R^0_{<_2}$, $\square \in R^0_{<_2}(C)$ iff C is unsatisfiable. Thus it suffices to prove termination of $R^0_{<_2}$ on MON*.

Let C be in MON*; then $C \in$ MON*(Σ) where Σ is the signature of C. Proposition 5.2.3 tells us that MON*(Σ) is finite. Therefore it remains to show that MON*(Σ) is invariant under $R^0_{<_2}$, i.e., $R^0_{<_2}(C) \in$ MON*(Σ).

By the definition of $R^0_{<_2}$ it is sufficient to show that

$$\rho^0_{<_2}(C) \in \text{MON}^*(\Sigma).$$

Moreover, $\rho^0_{<_2}$ is a binary operator, i.e.,

$$\rho^0_{<_2}(C) = \bigcup \{\rho^0_{<_2}(\{C_1, C_2\}) | C_1, C_2 \in C\}.$$

Due to this property of $\rho^0_{<_2}$ it is enough to show that for clauses C, D such that $\{C, D\} \in$ MON*(Σ) and for $E \in \rho^0_{<_2}(\{C, D\})$ we obtain $\{E\} \in$ MON*(Σ).

Let C', D' be variable-disjoint variants of C and D. We limit our attention to the components of C' and D' which are affected by the resolution. Still we face the problem that more than just two components may be involved in the resolution, as even selected factoring may work on several components at once.

Let $F \in \text{COMP}(C')$ or $F \in \text{COMP}(D')$; we call F a *constant component* if $V(F) = \emptyset$ and a *v-component* if $V(F) \neq \emptyset$ and F is function-free. Components containing function symbols are called *functional components*.

Let $\mathcal{L} : \{L_1, \ldots, L_m\}$ be the set of literals in C' selected for resolution and similarly $\mathcal{M} : \{M_1, \ldots, M_k\}$ for the clause D'. If $\mathcal{L}$ and $\mathcal{M}$ are both ground then the resolvent E fulfills

$$\text{LIT}(E) \subseteq \text{LIT}(C') \cup \text{LIT}(D') \text{ and } \{N_c(E)\} \in \text{MON}^*(\Sigma).$$

Let $L \in \mathcal{L}$ such that L belongs to a functional component F. We show that L must be maximal in F with respect to $<_2$. For let us assume on the contrary that L is not maximal in F, i.e., there exists an L_0 in F such that $L <_2 L_0$. Because $<_2$ is an A-ordering we obtain $L\vartheta <_2 L_0\vartheta$ for all substitutions ϑ and thus $L_0 \notin \mathcal{L}$ (note that the set $\mathcal{L}$ must be unifiable!).

Let λ be the factoring substitution for $\mathcal{L}$ and σ be the m.g.u. of the binary resolvent; then we have

$$L\lambda\sigma <_2 L_0\lambda\sigma.$$

Because L_0 is not selected for resolution, $L_0\lambda\sigma$ is a literal of the resolvent. That means the resolvent E contains a literal L' such that $L\lambda\sigma <_2 L'$. But

$L\lambda\sigma$ is the resolved literal and therefore E cannot be $<_2$-resolvent. So we obtain a contradiction to the assumption that $E \in \rho^0_{<_2}(\{C, D\})$. We see that only maximal literals of a functional component can be selected for resolution.

We now distinguish the following cases:

a) All L in $\mathcal{L}$ and all M in $\mathcal{M}$ belong to v-components.
b) One of the L in $\mathcal{L}$ or of the M in $\mathcal{M}$ is a constant component.
c) $\mathcal{L}$ contains literals from v-components and from functional components, $\mathcal{M}$ contains literals from v-components only.
d) $\mathcal{L}$ contains only literals from v-components, $\mathcal{M}$ contains literals from v- and from functional components.
e) $\mathcal{L}$ and $\mathcal{M}$ both contain only literals from functional components.

case a):
Here we have $L_i = P(x_i)$ and $M_j = \neg P(y_j)$ for variables x_i, y_j, the factoring substitutions are $\lambda : \{x_i \leftarrow x_1 | i = 2, \ldots, m\}$ and $\mu : \{y_j \leftarrow y_1 | j = 2, \ldots, k\}$; the m.g.u. of the binary resolution then is of the form $\sigma : \{y_1 \leftarrow x_1\}$ (there are in fact different possibilities to define λ, μ and σ, but all are variants of each other). In the resolvent E the x_i- and y_j-components disappear – with the exception of the x_1-component. Clearly this component is a v-component again and $\{N_c(E)\} \in \mathrm{MON}^*(\Sigma)$.

case b):
$at(L)$ (or $at(M)$ respectively) is of the form $P(c)$ for a constant symbol c. Because $\mathcal{M} \cup \mathcal{L}^d$ must be unifiable all literals in $\mathcal{L} \cup \mathcal{M}$ must either contain the atom $P(c)$ or belong to v-components. Therefore the m.g.u. of the binary resolvent is either ϵ or of the form $\sigma : \{y_1 \leftarrow c\}$. In both cases all involved components become constant and the arguments of all ground literals are constant symbols. Again we obtain $\{N_c(E)\} \in MON^*(\Sigma)$.

case c):
$\mathcal{L}$ contains literals from v-components and from functional components, $\mathcal{M}$ contains literals from v-components only.
Let $L_{i_1}, \ldots, L_{i_r}$ be the functional literals in $\mathcal{L}$. Because $\mathcal{L}$ is unifiable all the L_{i_j} are variants of some $P(f(v_1, \ldots, v_n))$. Let $P(z_1), \ldots, P(z_l)$ be the other literals in $\mathcal{L}$. Clearly

$$\{z_1, \ldots, z_l\} \cap V(\{L_{i_1}, \ldots, L_{i_r}\}) = \emptyset$$

for otherwise $\mathcal{L}$ is not unifiable. Moreover there must exist some variables $y_1, \ldots, y_n$ in $V(\mathcal{L})$ such that the factoring substitution is of the form

$$\lambda : \{z_1 \leftarrow f(y_1, \ldots, y_n), \ldots, z_l \leftarrow f(y_1, \ldots, y_n)\} \cup \pi$$

where π is a variable permutation such that $V(rg(\pi)) = \{y_1, \ldots, y_n\}$. After factoring by λ the components of $\mathcal{L}$ are contracted into a functional component containing one or more dominating literals having atoms of the form $P(f(y_1, \ldots, y_n))$ or $P(g(y_1, \ldots, y_n))$ for some $g \neq f$. Note that the L_{i_j} must be maximal in their components! The m.g.u. of the binary resolution must be of the form

$$\sigma : \{x_1 \leftarrow f(y_1, \ldots, y_n), \ldots, x_r \leftarrow f(y_1, \ldots, y_n)\}$$

the x_i being variables in $\mathcal{M}$. Then either all components belonging to $\mathcal{L} \cup \mathcal{M}$ disappear in the resolvent E or E contains (new) components having maximal literals of the form $P(g(y_1, \ldots, y_m))$. Note that any two functional literals L, M belonging to the same component of C' or of D' have argument lists of the form $(x_1, \ldots, x_s)$ and $(x_1, \ldots, x_q)$ respectively. This form is preserved under the permutation π and thus $\{N_c(E)\} \in \text{MON}^*(\Sigma)$.

case d):
This case is completely symmetric to c).

case e):
We already know that all L in $\mathcal{L}$ and all M in $\mathcal{M}$ must be maximal in their component and thus are all functional. Because $\mathcal{L} \cup \mathcal{M}^d$ must be unifiable all atoms of the literals must be variants of some $P(f(v_1, \ldots, v_n))$. The factoring substitutions λ, μ and also the m.g.u. σ of the binary resolution must all be variable permutations. Consequently the resolved literal itself must be of the form $P(f(v_1, \ldots, v_n))$. Because C' and D' fulfill condition 3) in Definition 5.2.7, all functional literals in E containing variables in $\{v_1, \ldots, v_n\}$ must have an argument list of the form $(v_1, \ldots, v_k)$ for some $k \leq n$. Thus condition 3) in Definition 5.2.7 hols for E too; conditions 1) and 2) are trivially fulfilled. Therefore we obtain $\{N_c(E)\} \in \text{MON}^*(\Sigma)$.   ◇

We have shown that the one-variable class and the monadic class are decidable by A-ordering refinements. It is a natural question, whether we can deal similarly with the two-variable class and with the dyadic class (i.e., the class of all function-free PL-formulas containing only two-place predicate symbols). Unfortunately the answer to both questions is negative. The prefix class $\forall \exists \forall$ has been proved undecidable by Kahr, Moore, and Wang [KMW61]; its skolemization clearly is contained in the 2-variable class. That the dyadic class is a reduction class was already known to Löwenheim [Loew15]; thus its undecidability directly follows from Church's result on the undecidability of predicate logic. However, there are possible extensions of MON* under preservation of decidability. In particular we can get rid of the restriction that all terms must be of depth ≤ 1 ([FLTZ93] Chapter 5).

Exercises

Exercise 5.2.1. Show that the set of all clauses obtained by transforming the Ackermann class into clause form is a proper subset of VAR1.

Exercise 5.2.2. Let $\Gamma_{l,k}$ be the set of all N_c-clauses over a finite signature containing $\leq k$ variables having depth $\leq l$. Show that for all $l, k \in \mathbb{N}$, $\Gamma_{l,k}$ is finite (why is this set infinite without normalization?).

Exercise 5.2.3. Let $\{C\}$ be in K and $C\vartheta$ be a factor of C. Then $\tau((C\vartheta)_G) \leq 2\tau(C)$ and $\tau(C\vartheta \setminus (C\vartheta)_G) \leq \tau(C)$ holds (E_G stands for a clause form of the ground component of the clause E).

Exercise 5.2.4. Let L, M be two literals such that $V(L) = \{x\}$ and $V(M) = \{y\}$ and let σ be the m.g.u. of $\{L, M\}$. Prove that

$$\tau(L\sigma) \leq \tau(L) + \tau(M).$$

Exercise 5.2.5. Let $G \in \mathrm{PL}_0$ be a formula in prenex form and $H : (\forall y_1) \ldots (\forall y_n)M$ be its skolemized form obtained by the transformation β. Then for every functional term s in M there exists a $k \leq n$ and a k-ary function symbol f such that $s = f(y_1, \ldots, y_k)$.

Exercise 5.2.6. Give a proof of Proposition 5.2.1.

Exercise 5.2.7. Show that there are clauses C_1, C_2 such that $\{C_1, C_2\} \in \mathrm{MON}^*$, but there exists a $<_2$-resolvent D of C_1 and C_2 which is not in MON^* (hint: define a factor that destroys the syntax structure of MON^*).

Exercise 5.2.8. Show that A-ordering with selected factoring is complete (hint: use the proof of Theorem 3.3.1). Let MON' be the class of all finite sets of clauses $\mathcal{C}$ such that

1. $\mathcal{C}$ contains only monadic predicate symbols.
2. For all $C \in \mathcal{C}$: $\tau(C) \leq 1$.
3. If $C \in \mathcal{C}$ and $D \in \mathrm{COMP}(C)$ then either $|V(D)| \leq 1$ or D or D contains a literal L such that $V(L) = V(D)$.

Exercise 5.2.9. Show that MON' is not invariant under $R^0_{<_2}$, i.e., there exists a $\mathcal{C} \in \mathrm{MON}'$ such that $R^0_{<_2}(\mathcal{C}) \notin \mathrm{MON}'$.

5.3 Hyperresolution as Decision Procedure

Refinements based on A-orderings or on other types of orderings do not suffice to obtain decision procedures for all relevant decision classes. Particularly in the case of the Bernays–Schönfinkel class as well as of many functional clause classes hyperresolution is superior to ordering refinements. On the other hand hyperresolution is not suited as decision method for the one-variable class or for the Ackermann class. Thus we will discuss some syntactic features, which characterize the applicability of decision methods of different types.

Definition 5.3.1 (Bernays–Schönfinkel class). *Let* BS *be the class of all closed formulas of the form*

$$(\exists x_1) \ldots (\exists x_m)(\forall y_1) \ldots (\forall y_m)M$$

where M is free of quantifiers, constant symbols, and function symbols. BS is called the Bernays–Schönfinkel class. *If M is a conjunction of Horn clauses then we obtain the subclass* BSH.

We have shown in Section 2.4 that BS (and thus also BSH) is decidable; but the argument was model theoretic. Here we are interested in the termination behavior of resolution refinements on BS and BSH. For this purpose we have to define clauses classes corresponding to BS and BSH.

Definition 5.3.2. BS* *is the class of all finite sets of clauses C such that for all $C \in \mathcal{C} : \tau(C) = 0$.* BSH* *is the subclass of* BS* *containing sets of Horn clauses only.*

The condition $\tau(C) = 0$ in Definition 5.3.2 guarantees that there are no function symbols in $\mathcal{C}$. All constant symbols appearing in a set $\mathcal{C} \in$ BS* can be thought to have been introduced by skolemization. Thus BS* is exactly the clause class corresponding to BS; similarly BSH* is the clause class corresponding to BSH. BS* and BSH* can be decided by (total) saturation:

Take a $\mathcal{C} \in$ BSH*, replace $\mathcal{C}$ by the set of all ground instances $\mathcal{C}'$ and then decide $\mathcal{C}'$ by a propositional method (e.g., propositional resolution or the Davis–Putnam method).

This method can be very inefficient due to the fact that $\mathcal{C}'$ may be much larger than $\mathcal{C}$ itself. It is also an interesting fact that BS is of highest computational complexity among the classical prefix classes [DL84].

We now investigate how A-ordering refinements behave on BS and BSH.

Example 5.3.1. Let $\mathcal{C} : \{C_1, C_2, C_3, C_4\}$ be the following set of PN-clauses in BSH*:

$$C_1 = P(a,b), \ C_2 = P(x,y) \vee \neg P(y,x), \ C_3 = P(x,z) \vee \neg P(x,y) \vee \neg P(y,z),$$
$$C_4 = \neg P(b,c).$$

C_2 expresses symmetry and C_3 transitivity. $\mathcal{C}$ is satisfiable because $P(b,c)$ cannot be obtained from $P(a,b)$ using the rules of symmetry and transitivity. We show now that all A-ordering refinements are nonterminating on $\mathcal{C}$, i.e., there exists no A-ordering refinement which decides BSH (note that $\mathcal{C}$ even is a set of Horn clauses).

Indeed $R_{<_A}(\{C_2, C_3\}) = R_\emptyset(\{C_2, C_3\})$ for every A-ordering $<_A$. The reason for this effect is the unifiability of resolved atoms with all atoms in the resolvent (for all R-derivable clauses); therefore the case $L <_A M$ can never occur and no resolvents can be excluded. In particular, $R_{<_A}(\{C_2, C_3\})$ contains the infinite sequence of clauses

$$C_n : P(x_1, x_n) \vee \neg P(x_1, x_2) \ldots \vee \neg P(x_k, x_{k+1}) \ldots \vee \neg P(x_{n-1}, x_n)$$

for $n \geq 2$. These clauses cannot be removed by subsumption, nor do they "collapse" by condensing (the N_c-normal form only changes the names of the variables). Thus even $R_{<_A^s}$ (A-ordering + forward subsumption) and $R_{<_A^r}$ (A-ordering + replacement) are nonterminating on $\mathcal{C}$. On the other hand we obtain

$$R_H(\mathcal{C}) = \mathcal{C} \cup \{P(b,a), P(a,a), P(b,b)\}$$

and thus R_H terminates on $\mathcal{C}$. Moreover we know from Theorem 3.6.2 that

$$\mathcal{M} : \{P(a,b), P(b,a), P(a,a), P(b,b)\}$$

is an atom representation of a Herbrand model of $\mathcal{C}$.

We show now that termination of R_H can be guaranteed on BSH*.

Proposition 5.3.1. *Hyperresolution decides* BSH*, *i.e.,* $R_H(\mathcal{C})$ *is finite for all* $\mathcal{C}$ *in* BSH*.

Proof. Let $\mathcal{C}$ be in BSH*. Then $\mathcal{C}$ is in Horn form and $R_H^+(\mathcal{C})$ consists of unit clauses only. There are no function symbols in $\mathcal{C}$ and no function symbols can be introduced by resolution. Therefore we obtain

$$R_H^+(\mathcal{C}) \subseteq \{A | A \text{ is an atom over } \Sigma(\mathcal{C}), \ \tau(A) = 0\}.$$

Moreover all clauses in $RH^+(\mathcal{C})$ are in N_c-normal form and thus are subjected to variable renaming; so we obtain that the set $N_c(\{A | A \text{ is an atom over } \Sigma(\mathcal{C}), \ \tau(A) = 0\})$ is finite.   ◊

Example 5.3.2.

$$\mathcal{C} := \{P(x_1, x_1, a), \ P(x,z,u) \vee \neg P(x,y,u) \vee \neg P(y,z,u),$$
$$P(x,y,u) \vee P(y,z,u) \vee \neg P(x,z,u), \ \neg P(x,x,b)\}.$$

$\mathcal{C}$ is non-Horn and even "essentially" non-Horn; which means there exists no sign renaming γ such that $\gamma(\mathcal{C})$ is a set of Horn clauses. For $\gamma = \{\neg P\}$ only the roles of a and b are exchanged, otherwise $\mathcal{C}$ remains as it is. R_H neither terminates on $\mathcal{C}$ nor on $\gamma(\mathcal{C})$. R_H produces clauses of arbitrary length on $\mathcal{C}$ – even if we add subsumption (i.e., we replace R_H by R_{Hs}). Thus R_{Hs} + sign renaming does not terminate on $\mathcal{C}$. That means hyperresolution cannot decide the Bernays–Schönfinkel class. Moreover none of the other refinements terminates on $\mathcal{C}$. However, there is general semantic clash resolution over arbitrary models $\mathcal{M}$ as defined in Section 3.6; in such a refinement only clauses which are false in $\mathcal{M}$ are derivable. So, in case $\mathcal{C}$ is satisfiable, we only have to choose a model of $\mathcal{C}$; on such a model all clauses are true and thus semantic clash resolution does not produce any resolvents. But this trick can hardly be recommended as a method in resolution decision theory. Note that models should be the outcome of our procedures, not the starting point!

Of course there is the brute force method to decide BS* by ground saturation. We will see later that, by an appropriate use of hyperresolution, saturation can be reduced considerably.

We show now how hyperresolution can be applied as decision procedure on functional clauses classes. These classes can be considered as generalizations of DATALOG [CGT90]. Formally DATALOG is a subclass of BSH such that all positive clauses are ground and $V(C_P) \subseteq V(C_N)$ for all other clauses (for C_P, C_N see Definition 3.6.1).

Definition 5.3.3. *A set of PN-clauses C belongs to* PVD *(positive variable dominated) if for all $C \in \mathcal{C}$:*
PVD-1) $V(C_P) \subseteq V(C_N)$ (C is ground for $C_N = \square$),
PVD-2) $\tau_{\max}(x, C_P) \leq \tau_{\max}(x, C_N)$ for all $x \in V(C_P)$.

PVD corresponds to a subclass of a class named PVD in [FLTZ93], where the properties above were "relativized" under settings. That means there might be some sign renaming γ such that $\gamma(\mathcal{C}) \in$ PVD even if $\mathcal{C}$ itself is not in PVD. Take for example the set of clauses

$$\mathcal{C} : \{P(x_1) \vee Q(g(x_1, x_1)), \ R(f(x_1), x_2), \ P(a), \ R(x, y) \vee \neg Q(y),$$
$$\neg P(x) \vee \neg P(f(x)), \ \neg R(a, a) \vee \neg R(f(b), a)\}.$$

Obviously $\mathcal{C}$ is not in PVD (there are positive clauses containing variables and $R(x, y) \vee \neg Q(y)$ violates PVD-(1)).

But let γ be the sign renaming $\{P, \neg Q, \neg R\}$. Then

$$\gamma(\mathcal{C}) = \{P(x_1) \vee \neg Q(g(x_1, x_1)), \ R(f(x_1), x_2), \ P(a), \ Q(y) \vee \neg R(x, y),$$
$$\neg P(x) \vee \neg P(f(x)), R(a, a) \vee R(f(b), a)\}.$$

and $\gamma(\mathcal{C}) \in$ PVD.

The idea behind PVD is that the positive parts are always "smaller" than the negative ones. As hyperresolution produces positive clauses only, we may hope that the clauses produced are small too (i.e., small enough to achieve termination). The example above suggests a generalization of PVD by use of renaming; this definition is much clearer than the original definition of PVD given in [Lei93] and [FLTZ93] and was suggested first in [Mat93].

Definition 5.3.4. *A set of clauses C belongs to* PVD$_r$ *if there exists a sign renaming γ such that the PN-form of $\gamma(\mathcal{C})$ belongs to* PVD.

Let us assume that R_H decides PVD (which will indeed be shown later); then there exists the following decision procedure for PVD_r:

1. Search for a renaming γ such that the PN-form of $\gamma(\mathcal{C})$ is in PVD (if there is no such γ then $\mathcal{C} \notin$ PVD$_r$).
2. Apply R_H to $\gamma(\mathcal{C})$.

Note that there are only finitely many sign renamings on a set of clauses and that the properties PVD-(1) and PVD-(2) are decidable. Thus we can always decide whether a set of clauses is in PVD_r. Once we have found the right renaming γ, we replace C by $\gamma(C)$ and apply hyperresolution. Thus in order to show the decidability of PVD_r it is sufficient to show that of PVD.

Theorem 5.3.1. *Hyperresolution decides* PVD, *i.e., for every* $C \in$ PVD $R_H(C)$ *is finite.*

Proof. Let C be in PVD and $d = \max\{\tau(C_P) | C \in C\}$. In order to show that $R_H(C)$ is finite it is sufficient to prove:

a) All clauses in $R_H^+(C)$ are ground.
b) For all $C \in R_H^+(C) : \tau(C) \leq d$.

Note that there are only finitely many ground N_c-clauses of depth $\leq d$ over the signature of C. Moreover it is sufficient to focus on the set of positive derivable clauses because

$$R_H(C) - C \subseteq R_H^+(C).$$

Because of $R_H(C) = \bigcup_{i=0}^{\infty} S_H^i(C)$ we may proceed by induction on i and prove (a) and (b) for the sets $S_H^{i\,+}(C)$. To simplify the notation within the proof we write $M_i(C)$ for the set of all positive clauses in $S_H^i(C)$ (including $\square$ if $\square \in S_H^i(C)$).

case $i = 0$:
$M_0(C) = C_+$ For all $C \in C_+$ we have $C_P = C$ and, by definition of d: $\tau(C_+) \leq d$. By condition PVD-(1) in Definition 5.3.3, C_+ consists of ground clauses only. Thus (a) and (b) are both fulfilled.

(IH) Suppose that

a-(i) All clauses in $M_i(C)$ are ground and
b-(i) $\tau(M_i(C)) \leq d$ holds.

Let C be an arbitrary clause in $M_{i+1}(C)$. Then either $C \in M_i(C)$ in which case C is ground and $\tau(C) \leq d$ by (IH), or $C \in M_{i+1}(C) - M_i(C)$.
In the second case C is a hyperresolvent of a clash $\Gamma : (E; D_1, \ldots, D_n)$ such that $E \in C - C_+$ and $D_j \in M_i(C)$ for $j = 1, \ldots, n$. By (IH) and by definition of $M_i(C)$ the D_j are all ground N_c-clauses (i.e., the ground clauses are ordered and reduced). Thus in order to define a hyperresolvent of Γ we don't need to factor the clauses D_j.

So let

$$E = P_1 \vee \ldots \vee P_m \vee \neg Q_1 \vee \ldots \vee \neg Q_n \text{ and } D_j = F_j \vee A_j \vee G_j$$

for $i = 1, \ldots, n$ such that the atoms A_j are selected for resolution. Let C be the resolvent of Γ; then $C = (P_1 \vee \ldots \vee P_m)\lambda$ for some matching substitution λ such that $dom(\lambda) \subseteq V(E_N)$ and $Q_n\lambda = A_1$, $Q_{n-1}\lambda = A_2, \ldots, Q_1\lambda = A_n$ (note that the A_j are ground!).

Because E is in C and $C \in$ PVD we obtain $V(E_P) \subseteq V(E_N)$. But $C = E_P \lambda$ where λ is a ground substitution with domain $V(E_N)$. Therefore $V(E_P) \subseteq dom(\lambda)$ and $E_P \lambda$, i.e., the clash resolvent C of Γ, is ground. So we obtain a-$(i+1)$.

It remains to show $\tau(C) \le d$.

By PVD-(2) we have

$$\tau_{\max}(x, E_P) \le \tau_{\max}(x, E_N)$$

for all $x \in V(E_P)$. Recall that $C = E_P \lambda$ for the ground substitution λ defined by Γ. We distinguish two cases:

(I) $\tau(E_P \lambda) = \tau(E_P)$ and
(II) $\tau(E_P \lambda) > \tau(E_P)$.

In case (I) $\tau(E_P \lambda) \le d$ because $\tau(E_P) \le d$ by $E \in C$ and by definition of d.

In case (II) there exists a variable $x \in V(E_P)$ such that

$$\tau(E_P \lambda) = \tau_{\max}(x, E_P) + \tau(x\lambda).$$

By PVD-(2) we have $\tau_{\max}(x, E_P) \le \tau_{\max}(x, E_N)$ and therefore

$$\tau_{\max}(x, E_P) + \tau(x\lambda) \le \tau_{\max}(x, E_N) + \tau(x\lambda) \le \tau(E_N \lambda).$$

But we have shown before that, for $E_N = \neg Q_1 \vee \ldots \vee \neg Q_n$,

$$\{Q_1, \ldots, Q_n\}\lambda \subseteq \{A_1, \ldots, A_n\} \subseteq \bigcup_{j=1}^{n} LIT(D_j).$$

Therefore $\tau(E_P \lambda) \le \tau(\bigcup_{j=1}^{n} LIT(D_j))$.

But all the clauses D_j are in $M_i(C)$ and, by (IH) b-(i), $\tau(M_i(C)) \le d$. As a consequence we obtain $\tau(\{D_1, \ldots, D_n\}) \le d$ and therefore $\tau(E_N \lambda) = \tau(C) \le d$.

As C was chosen as arbitrary element in $M_{i+1}(C) - M_i(C)$ we get a-$(i+1)$ and b-$(i+1)$, i.e., all clauses in $M_{i+1}(C)$ are ground and $\tau(M_{i+1}(C)) \le d$. $\Diamond$

The proof of Theorem 5.3.1 reveals that the condensation normal form of clauses is not really necessary to obtain termination; indeed all derived (positive) clauses are ground and thus the N_s normal form or even pure clause reduction would do the job.

The class PVD is relatively "tight" with respect to undecidability: If we add the clause

$$T^- : P(x_1, x_2) \vee P(x_2, x_3) \vee \neg P(x_1, x_3)$$

(i.e., the transitivity of $\neg P$) we can encode the word problem of any equational theory (see [FLTZ93] Chapter 3.3). From the fact that there are equational theories with undecidable word problems (e.g., the theory of combinators [Ste71]) it follows that

$$\Gamma : \{\mathcal{C} \cup \{T^-\} | \mathcal{C} \in \text{PVD}\}$$

is an undecidable class of clause sets.

Let us consider the rough structure of the proof of Theorem 5.3.1. The main point consists in showing that $R_H^+(\mathcal{C})$ is ground and $\tau(R_H^+(\mathcal{C})) \leq d$ for some constant d. While the property PVD-(1) is essential (note that T^- does not fulfil PVD-(1)) PVD-(2) can be replaced by a more general condition (term depth is only a specific complexity measure for literals and clauses). In particular we obtain a more general decision class in replacing term depth by arbitrary atom complexity measures ψ fulfilling the general axioms:

1. $\psi(A) \leq \psi(A\vartheta)$ for all atoms A and substitutions ϑ.
2. For all natural numbers k the set $\{\psi(B) \mid \psi(B) \leq k, B \text{ ground}\}$ is finite.

Moreover ψ has to be extended to literals, clauses and sets of clauses in a straightforward manner (precisely like term depth). For such a measure ψ we have to postulate that there exists a constant d such that for all clauses $C \in \mathcal{C}$ and ground substitutions ϑ either $\psi(C_P\vartheta) \leq d$ or $\psi(C_P\vartheta) \leq \psi(C_N\vartheta)$ holds. Then hyperresolution terminates on $\mathcal{C}$ [Lei93].

BS* is not a subclass of PVD$_r$. But we will define a method to transform BS* into BS* $\cap$ PVD under preservation of sat-equivalence. This method is subtler and more efficient than complete ground saturation. The basic idea is the following:

Let $\mathcal{C}$ be in BS*. Search for a renaming γ such that $\gamma(\mathcal{C}) \in$ PVD. If there is such a γ then apply R_H to $\gamma(\mathcal{C})$ else select some arbitrary γ and transform $\gamma(\mathcal{C})$ into PVD by partial saturation of the variables which violate PVD-1). The exact procedure is shown in Figure 5.1.

```
BSALG (input is a set C ∈ BS*);
    { REN(C) denotes the set of all sign-renamings on C }
    begin
        if there exists a γ ∈ REN(C) such that γ(C) ∈ PVD then C' ← γ(C)
            else begin
                    select a γ ∈ REN(C); C ← γ(C);
                    for all C ∈ C do (compute T(C))
                        if V(CP) ⊆ V(CN) then T(C) ← {C}
                        else T(C) ← {Cλ| dom(λ) ⊆ V(CP) − V(CN), rg(λ) ⊆ H(C)};
                    C' ← ⋃{T(C)|C ∈ C}
                end
        compute RH(C')
    end
```

Fig. 5.1. A decision procedure for BS*

BSALG is indeed a decision algorithm for BS*. First of all $C' \sim_{sat} C$ as the set of all ground instances of C and C' are the same. By definition of $T(C)$ all clauses in $T(C)$ fulfil PVD-(1) (if PVD-(1) holds then PVD-(2) follows trivially as there are no function symbols in C for $C \in$ BS*). So we obtain $C' \in$ PVD and, by Theorem 5.3.1, $R_H(C')$ is finite. We conclude that BSALG is correct and always terminating on BS*. For the actual performance of BSALG the right selection of a renaming is crucial; clearly one should try to select a γ for which C' becomes minimal. In the next example we compare brute force saturation with BSALG. For this purpose we replace R_H by the more restrictive operator R_{Hs}. This leads to a further increase of efficiency, but without loss of correctness and termination (note that R_{Hs} is complete and $R_{Hs}(C) \subseteq R_H(C)$ for all sets of clauses C).

Example 5.3.3. We take the set of clauses from Example 5.3.2, i.e.,

$$C = \{P(x_1, x_1, a), P(x, z, u) \vee \neg P(x, y, u) \vee \neg P(y, z, u),$$
$$P(x, y, u) \vee P(y, z, u) \vee \neg P(x, z, u), \neg P(x, x, b)\}.$$

We already know that R_H does not terminate on C. Clearly $C \notin$ PVD but $C \in$ BS*. We compute the set T (without renaming the predicate symbol P) and obtain

$$C' = \{P(a, a, a), \; P(b, b, a), \; P(x, z, u) \vee \neg P(x, y, u) \vee \neg P(y, z, u),$$
$$P(x, a, u) \vee P(a, z, u) \vee \neg P(x, z, u), P(x, b, u) \vee P(b, z, u) \vee \neg P(x, z, u),$$
$$\neg P(x, x, b)\}.$$

$C' \in$ PVD and $|C'| = |C| + 2 = 6$.

$$R_{Hs}(C') = C' \cup \{P(a, b, a) \vee P(b, a, a)\}.$$

Thus R_{Hs} terminates on C' producing only one additional clause ($|R_{Hs}(C')| = 7$). Note that in $\rho_H(S^1_{Hs}(C'))$ we have the clauses

$$P(b, a, a) \vee P(a, a, a) \vee P(a, b, a), \; P(b, b, a) \vee P(b, a, a) \vee P(a, b, a)$$

which are both subsumed by $S^1_{Hs}(C)$. So we obtain

$$R_{Hs}(C') = S^1_{Hs}(C').$$

Using the brute force saturation method we obtain a set of ground clauses C'' which contains 36 clauses. Moreover C'' has still to be tested for satisfiability. Thus we see that BSALG is much faster than the pure saturation method.

5.4 Hyperresolution and Automated Model Building

The construction of models of abstract structures lies at the very heart of mathematical activity. Although mathematical logic brought forward deep results on the existence and on the structure of models (in the discipline of model theory), very little is known about algorithmic methods for model building. Even in automated theorem proving, considering the importance of constructing counterexamples to the practice of automated reasoning, the body of knowledge about model generation is relatively small. But in more recent times many different methods and techniques have been invented, analyzed and applied. Automated model building (sometimes also called model generation) is becoming a discipline on its own and one of the most fascinating applications of computational logic. In this section our aim is to present automated model building as an application of resolution decision theory; thus it is not the right place to give a substantial survey on this topic (instead we refer to [FL96]). We just mention the approaches of T. Tammet [Tam91], R. Manthey and F. Bry [MB88], R. Caferra and N. Zabel [CZ92], J. Slaney [Sla92] [Sla93], and C. Fermüller and the author [FL96].

Tammet's approach, like that of Fermüller and the author, is based on resolution decision procedures. But while Tammet's method directly yields finite models, that of [FL96] produces symbolic representations of Herbrand models (finite models are extracted from Herbrand models in a postprocessing step). Tammet uses narrowing and works with equations (in the object language), while the approach in [FL96] is based on the use of hyperresolution only; a simplified version of this method will be presented in this section.

Caferra and Zabel define an extension (called RAMC) of the resolution calculus by an equational constraint logic, which is specifically designed for the purpose of model building. Their method is symbolic and yields (like that in [FL96]) complete representations of Herbrand models. Meanwhile Caferra et al. have extended the method to clause logic with equality. A recent paper [?] even shows that the calculus naturally simulates hyperresolution, thus generalizing some results in [FL96].

Manthey and Bry describe a hyperresolution prover called SATCHMO which is based on a model generation paradigm [MB88]. Their method of model building, although similar concerning the use of hyperresolution, differs from that in [FL96] in several aspects. They essentially use splitting of positive clauses and backtracking, features that are both avoided in [FL96]. Moreover [FL96] makes use of subsumption and replacement in an essential way and guarantees termination for specific syntax classes.

Slaney [Sla92] devised the program FINDER that identifies finite models (of reasonable small cardinality) of clause sets whenever they exist. The algorithm is based on a clever variant of exhaustive search through all finite interpretations and does not refer to resolution or another first-order inference system. In [Sla93] Slaney defined SCOTT, a combination of FINDER with the resolution theorem prover OTTER [McC90]. His method was applied successfully to solve open problems in algebra (concerning groupoids). In comparison to the other methods mentioned above, Slaney's method behaves like a "numeric" rather than a symbolic one. Despite this difference the main point – demonstrating the usefulness of model building in automated deduction – coincides with that of the approach presented here.

In Theorem 3.6.2 we have shown that $R_H^+(\mathcal{C})$, the set of positive clauses in $R_H(\mathcal{C})$, defines a Herbrand model if $\mathcal{C}$ is a set of Horn clauses. Indeed, for Horn sets, $R_H^+(\mathcal{C})$ consists of unit clauses only which form the atom representation of a Herbrand model. $R_H^+(\mathcal{C})$ is an atom representation of an H-model even in the more general case that the nonpositive clauses in $\mathcal{C}$ may be arbitrary , but $R_H^+(\mathcal{C})$ consists of unit clauses only (Exercise 5.4.3). By resolution decision theory we possess means to guarantee termination of R_H on certain clause classes. If $\mathcal{C}$ is a set of Horn clauses and R_H terminates on $\mathcal{C}$ then (clearly) we obtain a finite atom representation of a Herbrand model of $\mathcal{C}$. In particular we obtain such finite representations on the class PVD∩Hornlogic by Theorem 5.3.1. The main purpose of this section is to show how we can extract atom representations of models from finite sets $R_{Hr}(\mathcal{C})$, where $R_{Hr}^+(\mathcal{C})$ does not necessarily consist of unit clauses only. In particular we will define a procedure that extracts a Herbrand model from every satisfiable set of clauses in PVD; this procedure is free of backtracking and does not rely on search. Our basic operator, however, will not be R_H but the reduction operator R_{Hr}. The choice of R_{Hr} is based on some particular mathematical properties of subsumption-reduced sets which turn out to be fruitful for model building.

Recall (see Definitions 4.2.9 and 4.2.10) that an R_{Hr}-sequence is of the form $(S_{Hr}^i(\mathcal{C}))_{i\in\mathbb{N}}$ such that $S_{Hr}^0(\mathcal{C}) = \mathrm{sub}(\mathcal{C})$ and

$$S_{Hr}^{i+1}(\mathcal{C}) = \mathrm{sub}(S_{Hr}^i(\mathcal{C}) \cup \rho_H(S_{Hr}^i(\mathcal{C}))).$$

If the sequence converges on a class of clause sets Γ then R_{Hr} is a decision procedure for Γ; in this case we have to compute $S_{Hr}^i(\mathcal{C})$ until we obtain a k such that $S_{Hr}^k(\mathcal{C}) = S_{Hr}^{k+1}(\mathcal{C})$. The final set of clauses obtained that way is "stable", i.e., it remains unchanged under further reductions (it is in fact in some normal form).

Definition 5.4.1. *Let $\mathcal{C}$ be a set of N_c-clauses and let $\hat{R}_{Hr}$ be the operator defined by $\hat{R}_{Hr}(\mathcal{C}) = \mathrm{sub}(\mathcal{C} \cup \rho_H(\mathcal{C}))$. Then $\mathcal{C}$ is called $(R_{Hr}\text{-})$ stable if $\hat{R}_{Hr}(\mathcal{C}) = \mathcal{C}$.*

If a R_{Hr}-sequence $(S_{Hr}^i(\mathcal{C}))_{i\in\mathbb{N}}$ converges to $S_{Hr}^k(\mathcal{C})$ then, by Definition 5.4.1, $S_{Hr}^k(\mathcal{C})$ is stable and a fixed point of the operator $\hat{R}_{Hr}$. Let us assume that an R_{Hr}-sequence converges and yields a (stable) set $\mathcal{C}'$ such that all positive clauses in $\mathcal{C}'$ are unit. Then, by the following lemma, these clauses form an atom representation of a Herbrand model of $\mathcal{C}'$; this lemma is closely related to Theorem 3.6.2. For the remaining part of this section we write "AR" for atom representation and "stable" for R_{Hr}-stable.

Lemma 5.4.1. *Let C be a finite set of nonpositive N_c-clauses and A be a finite set of (N_c-normalized) atoms such that $C \cup A$ is satisfiable and stable. Then A is an AR of a Herbrand model of $C \cup A$ (over the signature of A).*

Proof. Assume on the contrary that A is not an AR of a Herbrand model of $C \cup A$. Then the interpretation I induced by A falsifies $C \cup A$. It follows that the (possibly infinite) set

$$C \cup A \cup \{\neg P | P \in \mathrm{HB}(C) - I\}$$

is unsatisfiable (note that the Herbrand interpretation I is a set of ground atoms).

By the compactness theorem of first-order logic (see for example [BJ74]) there exists a finite set $\mathcal{F}$ such that $\mathcal{F} \subseteq \{\neg P | P \in \mathrm{HB}(C) - I\}$ and $\mathcal{D} : C \cup A \cup \mathcal{F}$ is unsatisfiable.

Because R_{Hr} is complete (i.e., for every unsatisfiable set of PN-clauses the H-reduction sequences converge to $\square$) there exists a $k \in \mathbb{N}$ such that $\square \in S_{Hr}^k(\mathcal{D})$. Because $C \cup A$ is satisfiable and $\mathcal{F}$ consists of negative unit clauses only we obtain $\square \notin \mathcal{D}$. So there must be a number $m \geq 1$ such that

$$\square \in S_{Hr}^m(\mathcal{D}) - S_{Hr}^{m-1}(\mathcal{D}).$$

But $S_{Hr}^m(\mathcal{D}) = \hat{R}_{Hr}(S_{Hr}^{m-1}(\mathcal{D}))$.

Because $C \cup A$ is stable we have

$$S_{Hr}^{m-1}(C \cup A) = C \cup A.$$

As a clash with nonpositive clause from $\mathcal{F}$ can only give the resolvent $\square$ we also obtain

$$\hat{R}_{Hr}(S_{Hr}^{m-1}(\mathcal{D})) = \hat{R}_{Hr}(\mathcal{D}) = \mathrm{sub}(\mathcal{D} \cup \rho_H(\mathcal{D})).$$

By the definition of m we get $\square \in \mathrm{sub}(\mathcal{D} \cup \rho_H(\mathcal{D}))$.

By $\mathrm{sub}(\mathcal{D} \cup \rho_H(\mathcal{D})) \subseteq \mathcal{D} \cup \rho_H(\mathcal{D})$ and $\square \notin \mathcal{D}$ this gives us $\square \in \rho_H(\mathcal{D})$. Clearly $\square \notin \rho_H(C \cup A)$ as $C \cup A$ is satisfiable. Thus $\square$ must have been obtained by a clash of the form $(\neg Q, Q')$ for some $\neg Q$ in $\mathcal{F}$. But, by definition of $\mathcal{F}$, $\neg Q$ must be ground and thus there exists a substitution γ such that $Q = Q'\gamma$.

By the definition of I $Q'\gamma$ must be contained in I; but then it is impossible that $\neg Q'\gamma$ is in $\mathcal{F}$. So we obtain a contradiction and therefore

$$C \cup A \cup \{\neg P | P \in \mathrm{HB}(C) - I\}$$

must be satisfiable, i.e., A is an AR of a Herbrand model of $C \cup A$.   ◊

Lemma 5.4.1 suggests the following strategy for finding a model: Suppose that a R_{Hr}-sequence converges to C such that $\Box \notin C$ (which is equivalent to $C \neq \{\Box\}$). Then search for a finite set of atoms $\mathcal{A}$ such that $(C - C_+) \cup \mathcal{A}$ is finite, satisfiable, and implies C. The resulting set $\mathcal{A}$ is an AR of a Herbrand model of $(C - C_+) \cup \mathcal{A}$ which is also a model of C itself. Before developing a method to obtain such a set of atoms $\mathcal{A}$ we show that R_{Hr}-reduction terminates on PVD. Note that this result is not trivial due to the nonmonotonicity of R_{Hr}. The property $S_{Hr}^k(C) \leq_{ss} S_H^k(C)$ merely yields the completeness of R_{Hr} and not its (relative) termination.

Lemma 5.4.2. R_{Hr} *decides* PVD, *i.e., if* $C \in$ PVD *then the replacement sequence* $(S_{Hr}^i(C))_{i \in \mathbb{N}}$ *converges.*

Proof. In the proof of Theorem 5.3.1 we have shown that for $C \in$ PVD, $R_H^+(C)$ is finite and consists of ground clauses only. From Lemma 4.2.3 we know that

$$S_{Hr}^k(C) \leq_{ss} S_H^k(C) \text{ for } k \in \mathbb{N}.$$

Moreover, by the definition of a reduction sequence, we also have

$$S_{Hr}^{k+1}(C) \leq_{ss} S_{Hr}^k(C) \text{ for } k \in \mathbb{N}.$$

Now let us assume that for all $k \in \mathbb{N}$: $S_{Hr}^k(C) \neq S_{Hr}^{k+1}(C)$, i.e., that $(S_{Hr}^i(C))_{i \in \mathbb{N}}$ diverges.
Because $R_H(C)$ is finite there exists an m such that $R_H(C) = S_H^m(C)$. Moreover, we must have

$$S_{Hr}^k(C) \leq_{ss} S_H^m(C) \text{ for } k \geq m.$$

Let C_0 be the set of all nonpositive clauses in C and C_k be the set of all nonpositive clauses in $S_{Hr}^k(C)$ for $k \geq 1$. Then, by definition of hyperresolution,

$$C_k \subseteq C_0 \text{ and } C_{k+1} \subseteq C_k \text{ for all } k \geq 1.$$

Let C_* be the subset of all $C \in C_0$ such that $C \in S_{Hr}^k(C)$ for all $k \in \mathbb{N}$. If $C_* = \emptyset$ then there exists a k such that for all $p \geq k$ the sets $S_{Hr}^p(C)$ consist of positive clauses only. But in this case there can be no more hyperresolvents and therefore

$$S_{Hr}^k(C) = S_{Hr}^{k+1}(C);$$

but this contradicts our assumption of divergence of the sequence $(S_{Hr}^i(C))_{i \in \mathbb{N}}$.
If $C_* \neq \emptyset$ then, at least, there exists a number k such that for all $p \geq k$ and for all $C \in C_0$:

$$C \notin S_{Hr}^p(C) - S_{Hr}^{p+1}(C).$$

The last property indicates that, after some fixed generation, no clause in C_0 can be deleted any more. But by the assumption of divergence we have

$S_{Hr}^p(\mathcal{C}) \neq S_{Hr}^{p+1}(\mathcal{C})$ for all $p \geq k$. Therefore deletions beyond the k-th level apply to positive clauses only.

Now let $r = \max\{k, m\}$. Then for all $p \geq r$, the sets of positive clauses in $S_{Hr}^p(\mathcal{C})$ (we denote in by $\mathcal{D}_p$) subsume the set $\mathcal{F}_m : (S_H^m(\mathcal{C}))_+$. Let $\mathcal{M}$ be the set of all ground N_c-clauses which subsume $\mathcal{F}_m$. As $\mathcal{F}_m$ is a finite set of N_c-clauses, $\mathcal{M}$ must be finite too. Moreover, we have the property

$$\mathcal{D}_p \subseteq \mathcal{M} \text{ for all } p \geq r.$$

Thus we know that the sets $\mathcal{D}_p$ are uniformly bounded by the set $\mathcal{M}$ and $S_{Hr}^p(\mathcal{C})$ by the set $\mathcal{M} \cup \mathcal{C}_*$ for $p \geq r$. Still divergence may occur by "cycling", i.e., clauses deleted in former generations are again introduced in later ones. We show now that this is impossible too.

Let $D \in \mathcal{D}_p - \mathcal{D}_{p+1}$. Then there exists an N_c-clause E in $\rho_H(S_{Hr}^p(\mathcal{C}))$ such that $E \leq_{ss} D$. Because E and D are both ground clauses $D \leq_{ss} E$ cannot hold; note that $E =_{ss} D$ would imply $\text{LIT}(E) = \text{LIT}(D)$ and, by the N_c-form, $E = D$! Therefore E must properly subsume D, i.e., $\text{LIT}(E) \subset LIT(D)$. Clearly $D \notin \mathcal{D}_s$ for $s \geq p + 1$, as $\mathcal{D}_s \leq_{ss} \mathcal{D}_{p+1}$ and either $E \in \mathcal{D}_s$ or there is an $F \in \mathcal{D}_s$ such that $F \leq_{ss} E$. Thus for all $s \geq p + 1$ there exist an $F \in \mathcal{D}_s$ such that $\text{LIT}(F) \subset \text{LIT}(D)$ and thus F properly subsumes D. But all the sets $SHr^s(\mathcal{C})$ are subsumption-reduced and so $D \notin \mathcal{D}_s$ for $s \geq p + 1$. Therefore also cycling is impossible and there must exist a $p \geq r$ such that $\mathcal{D}_p = \mathcal{D}_{p+1}$. According to the choice of r this yields

$$S_{Hr}^p(\mathcal{C}) = S_{Hr}^{p+1}(\mathcal{C})$$

in contradiction to the assumption of divergence. Therefore the sequence $(S_{Hr}^i(\mathcal{C}))_{i \in \mathbb{N}}$ must converge.   ◊

Lemma 5.4.2 gives us finite R_{Hr}-stable sets and provides the raw material for our model-building method. For every set $\mathcal{C} \in$ PVD we obtain a set $S_{Hr}^k(\mathcal{C})$ such that $S_{Hr}^k(\mathcal{C})$ is stable. If $\mathcal{C}$ is satisfiable then, by the logical equivalence of $\mathcal{C}$ and $S_{Hr}^k(\mathcal{C})$, every model of $S_{Hr}^k(\mathcal{C})$ is also a model of $\mathcal{C}$. If the positive clauses in $S_{Hr}^k(\mathcal{C})$ are all unit then, by Lemma 5.4.1, we already have an AR of a Herbrand model of $\mathcal{C}$.

The following lemma gives us the key technique for the transformation of a stable set into another stable set, where all positive clauses are unit. This lemma holds not only for PVD but for a much larger set of clause sets (see [FL96]); but to make things easier, we concentrate on PVD. Essentially we show that, for stable sets of clauses, positive clauses can be replaced by proper subclauses under preservation of satisfiability.

Lemma 5.4.3. *Let $\mathcal{C} \in$ PVD such that $\mathcal{C}$ is satisfiable and stable and let D be a positive nonunit clause in $\mathcal{C}$. Let P be an (arbitrary) atom in D. Then*

(1) $(\mathcal{C} - \{D\}) \cup \{P\}$ is satisfiable and
(2) $(\mathcal{C} - \{D\}) \cup \{P\} \rightarrow \mathcal{C}$ is valid.

Remark:
(1) and (2) together imply the existence of a model of $(\mathcal{C} - \{D\}) \cup \{P\}$ which is also a model of $\mathcal{C}$.

Proof. (2) is trivial by the validity of $\{P\} \to \{D\}$.
It remains to prove (1).

Let us assume (for a proof by contradiction) that

$$\mathcal{C}_1 : (\mathcal{C} - \{D\}) \cup \{P\} \text{ is unsatisfiable.}$$

Then, by the completeness of hyperresolution, $\Box \in R_H(\mathcal{C}_1)$.
Let $E = N_c(D \setminus P)$; we introduce an auxiliary relation $\leq_E$ on clauses such that

$$F \leq_E G \text{ iff } G \leq_{ss} F \vee E \text{ and } F \leq_{ss} G.$$

If F and G are both ground clauses then $F \leq_E G$ iff

$$\text{LIT}(F) \subseteq \text{LIT}(G) \text{ and } \text{LIT}(G) \subseteq \text{LIT}(F) \cup \text{LIT}(E).$$

We extend the relation $\leq_E$ to sets of clauses by

$$\mathcal{C} \leq_E \mathcal{D} \text{ iff for all } C \in \mathcal{C} \text{ there exists a } D \in \mathcal{D} \text{ such that } C \leq_E D.$$

Note that the way $\leq_E$ is extended to sets of clauses differs from that of $\leq_{ss}$.
  Our next step consists in showing (by induction on i) that

$$S_H^i(\mathcal{C}_1) \leq_E R_H(\mathcal{C}) \text{ for } i \in \mathbb{N}.$$

Note that $R_H(\mathcal{C})$ is finite by Theorem 5.3.1.

$i = 0$:
By choice of E we have $\text{LIT}(P) \cup \text{LIT}(E) = \text{LIT}(D)$ and thus $P \leq_E D$. Note that, by $\mathcal{C} \in \text{PVD}$, all positive clauses in $\mathcal{C}$ are ground. Moreover $\mathcal{C}_1 - \{P\} = \mathcal{C} - \{D\}$ and therefore $\mathcal{C}_1 \leq_E \mathcal{C}$. But R_H is a monotonic operator and so $\mathcal{C} \subseteq R_H(\mathcal{C})$. By definition of $\leq_E$ we get $\mathcal{C}_1 \leq_E R_H(\mathcal{C})$ and therefore

$$S_H^0(\mathcal{C}_1) \leq_E R_H(\mathcal{C}).$$

(IH) Suppose that $S_H^i(\mathcal{C}_1) \leq_E R_H(\mathcal{C})$ holds.

Let F be a clause in $S_H^{i+1}(\mathcal{C}_1)$.
By (IH) we have $S_H^i(\mathcal{C}_1) \leq_E R_H(\mathcal{C})$ and thus the only interesting case is $F \in S_H^{i+1}(\mathcal{C}_1) - S_H^i(\mathcal{C}_1)$.
  By the definition of the operator R_H (and of S_H) there must exist a clash $\Gamma : (C; D_1, \ldots, D_n)$ over $S_H^i(\mathcal{C}_1)$ such that F is clash resolvent of Γ. By (IH) there exist clauses $H_1, \ldots, H_n \in R_H(\mathcal{C})$ such that

$$D_j \leq_E H_j \text{ for } j + 1, \ldots, n.$$

By $\mathcal{C} \in \text{PVD}$ all clauses D_j are ground and, because E is ground, the H_j are ground too.

As all clauses D_j are ground and in N_c-form no factoring is required in resolving the clash Γ (note that by definition of R_H only positive clauses may be subjected to factoring).

Let L_j be the literals cut out from the D_j in the intermediary steps of the clash resolution. Then, by $D_j \leq_E H_j$ we get $L_j \in \text{LIT}(H_j)$. As the D_j are positive and ground and also E is ground, the definition of $\leq_E$ gives us

$$\text{LIT}(D_j) \subseteq \text{LIT}(H_j) \subseteq \text{LIT}(D_j) \cup \text{LIT}(E).$$

Thus $\Gamma' : (C; H_1, \ldots, H_n)$ is a clash over $R_H(C)$. By simulating an analogous clash resolution on the L_j in H_j we obtain a clash resolvent F' such that

$$F \leq_{ss} F' \text{ and } F' \leq_{ss} F \vee E.$$

In fact $N_c(F') = N_c(F \vee E_1 \vee \ldots \vee E_n)$ for clauses $E_1, \ldots, E_n$ fulfilling $\text{LIT}(E_i) \subseteq \text{LIT}(E)$ for $i = 1, \ldots, n$. Therefore

$$\text{LIT}(F \vee E_1 \vee \ldots \vee E_n) \subseteq LIT(F \vee E)$$

and consequently $F \leq_{ss} F' \leq_{ss} F \vee E$.
By the definition of $\leq_E$ we obtain $F \leq_E F'$.

Therefore $S_H^{i+1}(C_1) \leq_E R_H(C)$ and the case $i + 1$ is shown.
From $R_H(C_1) = \bigcup_{i=0}^{\infty} S_H^i(C_1)$ we eventually obtain

$$R_H(C_1) \leq_E R_H(C).$$

Now recall that C_1 is unsatisfiable and thus $\square \in R_H(C_1)$. By the definition of $\leq_E$ there must be a clause I in $R_H(C)$ such that $\square \leq_E I$. Because E is ground and by the definition of $\leq_E$ this implies $\text{LIT}(I) \subseteq \text{LIT}(E)$ and particularly $I \leq_{ss} E$.

By Lemma 4.2.3 we know that $S_{Hr}^k(C) \leq_{ss} S_H^k(C)$ for all $k \in \mathbb{N}$. By assumption C is R_{Hr}-stable, i.e., $\hat{R}_{Hr}(C) = C$ and $S_{Hr}^k(C) = C$ for all $k \in \mathbb{N}$. Consequently we obtain

$$C \leq_{ss} R_H(C)$$

Therefore there must be a clause $G \in C$ such that $G \leq_{ss} I$. Now remember that $E = N_c(D \setminus P)$ and $N_c(D) = D$. In particularly $\text{LIT}(E) \subset \text{LIT}(D)$ and D does not subsume E.

$$\text{But } G \leq_{ss} I \leq_{ss} E \text{ and thus } G \leq_{ss} E.$$

From $E \leq_{ss} D$ we also obtain $G \leq_{ss} D$. Now $D \leq_{ss} G$ is impossible by $G \leq_{ss} E <_{ss} D$. Therefore D and G are two clauses in C such that $G \leq_{ss} D$ and $D \not\leq_{ss} G$. But C is R_{Hr}-stable, i.e.,

$$\text{sub}(C \cup \rho_H(C)) = C.$$

In particular C cannot contain two different clauses G, D such that $G \leq_{ss} D$. So we obtain a contradiction and conclude that the set $(C - \{D\}) \cup \{P\}$ must be satisfiable.   ◊

The validity of Lemma 5.4.3 is essentially based on the stability of the set of clauses $\mathcal{C}$. It is very easy to see that the result becomes wrong for nonstable sets $\mathcal{C}$: Just take

$$\mathcal{C} = \{P(a) \vee P(b),\ \neg P(a)\}.$$

Trivially $\mathcal{C}$ is satisfiable. But if we replace $P(a) \vee P(b)$ by $P(a)$ we obtain the set of clauses $\mathcal{C}_1 : \{P(a), \neg P(a)\}$ which is unsatisfiable. But note that $\mathcal{C}$ is not stable; rather we have $\hat{R}_{Hr}(\mathcal{C}) = \{P(b), \neg P(a)\}$ and the replacement sequence converges to the set $\{P(b), \neg P(a)\}$.

The transformation of $\mathcal{C}$ into $(\mathcal{C} - \{D\}) \cup \{P\}$ can be described by an operator α that (deterministically) selects a clause D and a literal P in D; if $\mathcal{C}_+$ consists of unit clauses only we define $\alpha(\mathcal{C}) = \mathcal{C}$. Then we may iterate the application of α and R_{Hr}-closure on the new sets of clauses. Note that $(\mathcal{C} - \{D\}) \cup \{P\}$ need not be stable, even if $\mathcal{C}$ is stable. Therefore we have to compute a reduction sequence on $\alpha(\mathcal{C})$ in order to obtain the next stable set.

Let us assume that the reduction sequence $(S_{Hr}^i(\mathcal{C}))_{i \in \mathbb{N}}$ converges; then we denote its limit by R_{Hr}. The iterated reduction process can be defined comfortably by an operator on stable sets of clauses.

Definition 5.4.2 (the operator T). T is defined on stable sets of clauses in PVD by $T(\mathcal{C}) = R_{Hr}(\alpha(\mathcal{C}))$. The iteration of T is defined by:

$$T^0(\mathcal{C}) = \mathcal{C} \text{ and } T^{i+1}(\mathcal{C}) = T(T^i(\mathcal{C}))$$

if $T^i(\mathcal{C})$ is a stable set in PVD and $i \in \mathbb{N}$.

It is easy to verify that for all stable sets in PVD all $T^i(\mathcal{C})$ are again stable sets in PVD (Exercise 5.4.2). Therefore $T^i(\mathcal{C})$ is well-defined for all stable sets $\mathcal{C} \in$ PVD and $i \in \mathbb{N}$.

Let $\mathcal{C}$ be a stable set in PVD. Then, by Lemma 5.4.3 we know that $\alpha(\mathcal{C}) \to \mathcal{C}$ is valid and that $\alpha(\mathcal{C})$ is satisfiable, provided $\mathcal{C}$ is satisfiable. Thus also $T(\mathcal{C})$ is satisfiable (by the correctness of R_{Hr}) and $T(\mathcal{C}) \to \mathcal{C}$ is valid. Therefore we already know that $T^i(\mathcal{C}) \to \mathcal{C}$ is valid and that $T^i(\mathcal{C})$ is satisfiable for all $i \in \mathbb{N}$. But, unfortunately, we have not yet reached our goal. We still have to show that the sequence $(T^i(\mathcal{C}))_{i \in \mathbb{N}}$ converges, i.e., that there exists a number k such that $T^k(\mathcal{C}) = T^{k+1}(\mathcal{C})$; we obtain such a k when all positive clauses in $T^k(\mathcal{C})$ are unit clauses and $\alpha(T^k(\mathcal{C})) = T^k(\mathcal{C})$. In this case $T^k(\mathcal{C})_+$ is an AR of a Herbrand model of $\mathcal{C}$.

Example 5.4.1. Consider the following set of N_c-clauses

$$\mathcal{C} = \{E(a) \vee S(a),\ Q(a) \vee R(a),\ P(x_1) \vee Q(x_1) \vee \neg R(x_1) \vee \neg S(x_1),$$
$$\neg P(a) \vee \neg Q(a)\}$$

$\mathcal{C}$ is in PVD but $\mathcal{C}$ is not stable. We compute the reduction sequence $(S_{Hr}^i(\mathcal{C}))_{i \in \mathbb{N}}$ which converges and gives

$$R_{Hr}(\mathcal{C}) = S^1_{Hr}(\mathcal{C}) = \mathcal{C} \cup \{E(a) \vee P(a) \vee Q(a)\}.$$

By writing $\mathcal{C}_1$ for $S^1_{Hr}(\mathcal{C})$ and applying α we obtain

$$\alpha(\mathcal{C}_1) = (\mathcal{C}_1 - \{E(a) \vee S(a)\}) \cup \{S(a)\}.$$

By Lemma 5.4.3 we know that $\alpha(\mathcal{C}_1)$ is satisfiable and that each of its models is a model of $\mathcal{C}$ too.

Again $\alpha(\mathcal{C}_1)$ is not stable and we compute its corresponding R_{Hr}-reduction sequence. Then let

$$\mathcal{C}_2 = T(\mathcal{C}_1) = R_{Hr}(\alpha(\mathcal{C}_1)).$$

So we get

$$\begin{aligned}
\mathcal{C}_2 \;=\; &\{P(a) \vee Q(a),\; S(a),\; Q(a) \vee R(a), \\
&P(x_1) \vee Q(x_1) \vee \neg R(x_1) \vee \neg S(x_1),\; \neg P(a) \vee \neg Q(a)\}.
\end{aligned}$$

Note that in the computation of $\mathcal{C}_2$ we obtain the new clash resolvent $P(a) \vee Q(a)$ which subsumes $E(a) \vee P(a) \vee Q(a)$. On $\mathcal{C}_2$ we define

$$\alpha(\mathcal{C}_2) = (\mathcal{C}_2 - \{P(a) \vee Q(a)\}) \cup \{Q(a)\}.$$

Then $\mathcal{C}_3 = R_{Hr}(\alpha(\mathcal{C}_2)) = T(\mathcal{C}_2) =$

$$\{S(a),\; Q(a),\; P(x_1) \vee Q(x_1) \vee \neg R(x_1) \vee \neg S(x_1),\; \neg P(a) \vee \neg Q(a)\}.$$

Clearly $\alpha(\mathcal{C}_3) = \mathcal{C}_3$ and our procedure stops with $T(\mathcal{C}_3) = \mathcal{C}_3$ and $\mathcal{C}_3 = T^2(\mathcal{C}_1)$ (in fact $\mathcal{C}_3$ is a fixed point of T). By the validity of $T^2(\mathcal{C}_1) \to \mathcal{C}$ and by the satisfiability of $T^2(\mathcal{C}_1)$ via the model $\mathcal{M} : \{S(a), Q(a)\}$ we obtain $\mathcal{M}$ as model of $\mathcal{C}$ itself.

To prove the convergence of the sequence $(T^i(\mathcal{C}))_{i \in \mathbb{N}}$ to a stable set of clauses $\mathcal{D}$ such that $\mathcal{D}_+$ consists of unit clauses only, we introduce an ordering on sets of clauses and show that the $T^i(\mathcal{C})$ are "decreasing" in i.

Definition 5.4.3. *Let $\mathcal{C}$ and $\mathcal{D}$ be two finite sets of N_c-clauses. We define $\mathcal{C} < \mathcal{D}$ if the following conditions are fulfilled:*

a) $\mathcal{C} \leq_{ss} \mathcal{D}$.
b) For all $C \in \mathcal{C}$ there exists a $D \in \mathcal{D}$ such that $C \leq_{ss} D$ and $|C| \leq |D|$.
c) $\mathcal{D} \not\leq_{ss} \mathcal{C}$.

Lemma 5.4.4. *$<$ is irreflexive, transitive, and Noetherian (i.e., there exists no infinite strictly descending chain $\ldots \mathcal{C}_i < \mathcal{C}_{i-1} < \ldots \mathcal{C}_0$ of finite sets of N_c-clauses).*

Proof. $<$ is irreflexive because of Definition 5.4.3 - (c).

We now show transitivity:
Let us assume that $\mathcal{C} < \mathcal{D} < \mathcal{F}$ holds. Then in particular $\mathcal{C} \leq_{ss} \mathcal{D}$ and $\mathcal{D} \leq_{ss} \mathcal{F}$; by transitivity of $\leq_{ss}$ we obtain $\mathcal{C} \leq_{ss} \mathcal{F}$.

For Definition 5.4.3 let C be an arbitrary clause in $\mathcal{C}$. Then there exists a $D \in \mathcal{D}$ such that $C \leq_{ss} D$ and $|C| \leq |D|$. For this D, in turn, there must exist an $F \in \mathcal{F}$ such that $D \leq_{ss} F$ and $|D| \leq |F|$. As both $\leq_{ss}$ and $\leq$ are transitive we obtain $C \leq_{ss} F$ and $|C| \leq |F|$.

By Definition 5.4.3(c) we have $\mathcal{D} \not\leq_{ss} \mathcal{C}$ and $\mathcal{F} \not\leq_{ss} \mathcal{D}$. By assuming $\mathcal{F} \leq_{ss} \mathcal{C}$ we obtain $\mathcal{F} \leq_{ss} \mathcal{D}$ (due to $\mathcal{C} \leq_{ss} \mathcal{D}$ and the transitivity of $\leq_{ss}$); this however contradicts our assumption $\mathcal{D} < \mathcal{F}$. We have thus shown that $\mathcal{C} < \mathcal{F}$ and that $<$ is transitive.

It remains to show that there exists no infinite descending chain with respect to $<$. We assume the existence of such a chain and derive a contradiction.

Let $(\mathcal{C}_i)_{i \in \mathbb{N}}$ be a sequence with $\mathcal{C}_{i+1} < \mathcal{C}_i$ for all $i \in \mathbb{N}$. By Definition 5.4.3(b) and by transitivity of $<$ there exists a constant d such that for all $\mathcal{C}_i$ and for all $C \in \mathcal{C}_i$: $|C| \leq d$ (just choose d as the maximal clause size in $\mathcal{C}_0$).

Let $\mathcal{C}_0 = \{C_1, \ldots, C_m\}$. Then (by transitivity of $<$) $\mathcal{C}_i < \mathcal{C}_0$ and (by Definition 5.4.3(b)) for all $C \in \mathcal{C}_i$ there exists a C_j in $\mathcal{C}_0$ such that $C \leq_{ss} C_j$ and $|C| \leq d$.

We show now that the total number of different N_c-clauses appearing in $(\mathcal{C}_i)_{i \in \mathbb{N}}$ must be infinite, or more precisely:

For every $\mathcal{C}_i$ there exists a $D_i \in \mathcal{C}_i$ such that $D_i \not\in \mathcal{C}_k$ for all $k < i$.

This property follows from the fact that

$$\bigcup_{j=0}^{i-1} \mathcal{C}_j \not\leq_{ss} \mathcal{C}_i$$

(Exercise 5.4.3). Therefore we obtain a sequence $(D_i)_{i \in \mathbb{N}}$ of N_c-clauses which are pairwise different and for all D_i there exist a $C_j \in \mathcal{C}_0$ such that $D_i \leq_{ss} C_j$ and $|D_i| \leq d$. By the (transfinite) pigeonhole principle there exist a $C_k \in \mathcal{C}_0$ such that $D_l \leq_{ss} C_k$ and $|D_l| \leq d$ for infinitely many $l \in \mathbb{N}$ (if you file infinitely many objects in finitely many holes then at least one hole must contain infinitely many objects).

But there can't be infinitely many different clauses in condensed form with a uniform bound on the number of literals and on term depth (note that $D_l \leq_{ss} C_k$ implies $\tau(D_l) \leq \tau(C_k)$). So we obtain a contradiction and $<$ must indeed be Noetherian. $\diamond$

We might expect that $T(\mathcal{C}) < \mathcal{C}$ for all stable sets $\mathcal{C}$ in PVD; this property would directly yield a minimal element $T^i(\mathcal{C})$ in the sequence $(T^j(\mathcal{C}_j))_{j \in \mathbb{N}}$ such that $T^i(\mathcal{C})_+$ consists of unit clauses only. But $T(\mathcal{C}) < \mathcal{C}$ does not hold in general and we have to make a detour.

Example 5.4.2. The following set of clauses is stable and in PVD:

$$\mathcal{C} = \{R(a) \lor S(a),\ R(a) \lor U(a),\ P(a) \lor S(a),\ Q(x) \lor \neg P(x) \lor \neg U(x)\}.$$

Note that the only clash resolvent $Q(a) \lor R(a) \lor S(a)$ is subsumed by $R(a) \lor S(a)$.

For α we select $P(a)$ out of $P(a) \lor S(a)$ and obtain

$$\alpha(\mathcal{C}) = \{R(a) \lor S(a),\ R(a) \lor U(a),\ P(a),\ Q(x) \lor \neg P(x) \lor \neg U(x)\}.$$

$\alpha(\mathcal{C})$ is not stable and

$$R_{Hr}(\alpha(\mathcal{C})) = \alpha(\mathcal{C}) \cup \{Q(a) \lor R(a)\}.$$

Note that $Q(a) \lor R(a)$ is not subsumed by a clause in $\alpha(\mathcal{C})$. Moreover there exists no clause $C \in \mathcal{C}$ such that $Q(a) \lor R(a) \leq_{ss} C$. But $T(\mathcal{C}) = R_{Hr}(\alpha(\mathcal{C}))$ and so

$$T(\mathcal{C}) \nleq \mathcal{C}$$

(in fact condition (b) in Definition 5.4.3 is violated).

Example 5.4.2 tells us that we must be careful in using the (nonmonotonic) operator R_{Hr}; there may be subsumption relations on the less general level which cannot be carried over to the more general one (note that always $T(\mathcal{C}) \leq_{ss} \mathcal{C}$). Fortunately we can save termination via the (monotonic) R_H-closure of the sets of clauses.

Lemma 5.4.5. *Let C be a stable satisfiable set of clauses in PVD such that there exists a positive nonunit clause in $\mathcal{C}$; then $R_H(T(\mathcal{C})) < R_H(\mathcal{C})$.*

Proof. We first show $R_H(\alpha(\mathcal{C})) < R_H(\mathcal{C})$ (note that $T(\mathcal{C}) = R_{Hr}(\alpha(\mathcal{C}))$). Let E be the positive unit clause selected by α and P be the selected atom in E. Then

$$\alpha(\mathcal{C}) = (\mathcal{C} - \{E\}) \cup \{P\}.$$

By the definition of $\alpha(\mathcal{C})$ we have $\alpha(\mathcal{C}) \leq_{ss} \mathcal{C}$ and thus also $R_H(\alpha(\mathcal{C})) \leq_{ss} R_H(\mathcal{C})$ (Exercise 5.4.4). This gives us property (a) of Definition 5.4.3.

In our next step we show point (c) i.e., $R_H(\mathcal{C}) \nleq_{ss} R_H(\alpha(\mathcal{C}))$.

In particular we show that $R_H(\mathcal{C}) \nleq_{ss} \{P\}$; note that $P \in \alpha(\mathcal{C})$ and $\alpha(\mathcal{C}) \subseteq R_H(\alpha(\mathcal{C}))$.

Let us assume $R_H(\mathcal{C}) \leq_{ss} \{P\}$. Then, by $\mathcal{C} \leq_{ss} R_H(\mathcal{C})$, we also get $\mathcal{C} \leq_{ss} \{P\}$ ($\mathcal{C}$ is stable and $R_{Hr}(\mathcal{C}) \leq_{ss} R_H(\mathcal{C})$ by Lemma 4.2.3). Because $\mathcal{C} \in$ PVD the set $R_H(\mathcal{C})$ is in PVD too; in fact PVD is "robust" with respect to hyperresolution (see the proof of Theorem 5.3.1). Thus if $D \in \mathcal{C}$ and $D \leq_{ss} P$ then, as D and P are both ground N_c-clauses and $\square \notin \mathcal{C}$, we obtain $D = P$. But P cannot be in $\mathcal{C}$ because P properly subsumes E and $E \in \mathcal{C}$ (note that $\mathcal{C}$ is stable). So we obtain a contradiction and therefore

$$R_H(\mathcal{C}) \not\leq_{ss} R_H(\alpha(\mathcal{C})).$$

It remains to establish point (b) of Definition 5.4.3. For this purpose we prove by induction on n:

(*) For all $n \in \mathbb{N}$ and for all positive clauses C in $S_H^n(\alpha(\mathcal{C}))$ there exists a positive clause D in $R_H(\mathcal{C})$ such that $C \leq_{ss} D$ and $|C| \leq |D|$.

case $n = 0$:
$P \leq_{ss} E$ and $|P| = 1 < |E|$. Moreover $\alpha(\mathcal{C}) - \{P\} = \mathcal{C} - \{E\}$.

(IH) Suppose that (*) holds for n.

Let C be a positive clause in $S_H^{n+1}(\alpha(\mathcal{C}))$. If $C \in S_H^n(\alpha(\mathcal{C}))$ then we may apply (IH) and obtain the desired result.

Thus it is sufficient to assume that $C \in S_H^{n+1}(\alpha(\mathcal{C})) - S_H^n(\alpha(\mathcal{C}))$.

By definition of R_H, C must be a resolvent of a clash $\Gamma : (G; D_1, \ldots, D_m)$ such that G is a nonpositive clause in $\alpha(\mathcal{C})$ and $D_j \in S_H^n(\alpha(\mathcal{C}))$ for $j = 1, \ldots, m$.

By (IH) there are positive clauses $E_1, \ldots, E_m$ in $R_H(\mathcal{C})$ such that $D_j \leq_{ss} E_j$ and $|D_j| \leq |E_j|$ for $j = 1, \ldots, m$. Because all clauses E_j, D_j must be ground (note that $R_H(\mathcal{C})$ and $S_H^n(\alpha(\mathcal{C}))$ are both in PVD) we obtain

$$\mathrm{LIT}(D_j) \subseteq \mathrm{LIT}(E_j) \text{ for } j = 1, \ldots, m.$$

Thus let Γ' be the clash $(G; E_1, \ldots, E_m)$. Then we can "simulate" the clash resolution of Γ by that of Γ'. The outcome is a clash resolvent F of Γ' such that $\mathrm{LIT}(C) \subseteq \mathrm{LIT}(F)$. Because C and F are N_c-normalized it follows that $|C| \leq |F|$ and $C \leq_{ss} F$. So we have shown (*) for the case $n + 1$ and thus, by induction, for every $C \in R_H(\alpha(C))$ there exists a $D \in R_H(\mathcal{C})$ such that $C \leq_{ss} D$ and $|C| \leq |D|$. This gives us condition (b) of Definition 5.4.3 and eventually

$$R_H(\alpha(\mathcal{C})) < R_H(\mathcal{C}).$$

It remains to prove $R_H(T(\mathcal{C})) < R_H(\mathcal{C})$:

By definition of T we have $T(\mathcal{C}) = R_{Hr}(\alpha(\mathcal{C}))$. By definition of R_{Hr} we obtain $R_{Hr}(\alpha(\mathcal{C})) \subseteq R_H(\alpha(\mathcal{C}))$ and by Lemma 4.2.3 $R_{Hr}(\alpha(\mathcal{C})) \leq_{ss} R_H(\alpha(\mathcal{C}))$; these two properties give

$$R_{Hr}(\alpha(\mathcal{C})) =_{ss} R_H(\alpha(\mathcal{C})).$$

By the monotonicity and idempotency of R_H we obtain

$$R_H(R_{Hr}(\alpha(\mathcal{C}))) \subseteq R_H(\alpha(\mathcal{C})) \text{ and also}$$
$$R_H(R_{Hr}(\alpha(\mathcal{C}))) =_{ss} R_H(\alpha(\mathcal{C})).$$

Therefore

$$R_H(T(\mathcal{C})) \subseteq R_H(\alpha(\mathcal{C})) \text{ and } R_H(T(\mathcal{C})) =_{ss} R_H(\alpha(\mathcal{C})).$$

From the last two properties and from $R_H(\alpha(\mathcal{C})) < R_H(\mathcal{C})$ we obtain

$$R_H(T(\mathcal{C})) < R_H(\mathcal{C}). \qquad \Diamond$$

We are now in a position to show our main result, the convergence of the sequence $(T^i(\mathcal{C}))_{i \in \mathbb{N}}$ on stable sets of clauses $\mathcal{C}$ for $\mathcal{C} \in$ PVD. From this result we will extract an algorithm which, on satisfiable sets of clauses $\mathcal{C} \in$ PVD, always terminates with an atom representation of a Herbrand model of $\mathcal{C}$ (if $\mathcal{C}$ is unsatisfiable then $R_{Hr}(\mathcal{C}) = \{\Box\}$ and the model-building procedure does not start at all).

Theorem 5.4.1. *Let $\mathcal{C}$ be a stable and satisfiable set of clauses in* PVD. *Then the sequence $(T^i(\mathcal{C}))_{i \in \mathbb{N}}$ converges to a set of clauses $\mathcal{D}$ such that $\mathcal{D}_+$ is an atom representation of a Herbrand model of $\mathcal{C}$.*

Proof. If $\mathcal{C}_+$ consists of unit clauses only then $\alpha(\mathcal{C}) = \mathcal{C}$ and $T(\mathcal{C}) = \mathcal{C}$; trivially $(T^i(\mathcal{C}))_{i \in \mathbb{N}}$ converges to $\mathcal{C}$.

If $\mathcal{C}_+$ contains nonunit clauses then $R_H(T(\mathcal{C})) < R_H(\mathcal{C})$ by Lemma 5.4.5. Because also $T(\mathcal{C})$ is stable and satisfiable and $T(\mathcal{C}) \in$ PVD we may iterate the application of T and obtain a chain

$$\Gamma : \ldots < R_H(T^i(\mathcal{C})) < \ldots < R_H(\mathcal{C}).$$

By Lemma 5.4.4 $<$ is Noetherian and thus there must be a minimal element in Γ; let $R_H(T^k(\mathcal{C}))$ be this element. We show now that $T^{k+1}(\mathcal{C}) = T^k(\mathcal{C})$ and that $T^k(\mathcal{C})_+$ consists of unit clauses only.

Let us assume on the contrary that $T^k(\mathcal{C})_+$ contains nonunit clauses. In this case Lemma 5.4.5 yields

$$R_H(T^{k+1}(\mathcal{C})) < R_H(T^k(\mathcal{C})),$$

which contradicts the minimality of $R_H(T^k(\mathcal{C}))$.

Therefore $T^{k+1}(\mathcal{C}) = T^k(\mathcal{C})$ and $T^k(\mathcal{C})_+$ consists of unit clauses only.

By Lemma 5.4.1 $T^k(\mathcal{C})_+$ is an AR of a Herbrand model $\mathcal{M}$ of $T^k(\mathcal{C})$. By Lemma 5.4.3 $\alpha(\mathcal{C}) \to \mathcal{C}$ is valid and so is $T(\mathcal{C}) \to \mathcal{C}$. Therefore $\mathcal{M}$ is also a Herbrand model of $\mathcal{C}$. $\qquad \Diamond$

If the sequence $(T^i(\mathcal{C}))_{i \in \mathbb{N}}$ converges then we denote the limit by $T^*(\mathcal{C})$. By Theorem 5.4.1 we may always apply the following algorithm to sets of clauses in PVD:

MB:
   a) Compute $R_{Hr}(\mathcal{C})$.
   b) If $\Box \in R_{Hr}(\mathcal{C})$ then stop else compute $T^*(R_{Hr}(\mathcal{C}))$.

MB is correct and complete; thus MB yields $\Box$ for unsatisfiable sets of clauses $\mathcal{C}$ in PVD and Herbrand models for satisfiable ones. R_{Hr} is complete and always terminates on PVD (Lemma 5.4.2). Thus if $\mathcal{C}$ is unsatisfiable we obtain a stable set of clauses C' which is in PVD too. But then, by Theorem 5.4.1, $T^*(C')$ is defined and and $T^*(C')_+$ is an AR of a Herbrand model of C'. But $C' \rightarrow C$ is valid and thus $T^*(C')_+$ is also an AR of a Herbrand model of the input clauses set C. Note that MB is free of backtracking and search; indeed the computation of T^* is purely "iterative".

MB can be applied to all decision classes of hyperresolution where $R_{Hr}(C)$ is finite and all positive clauses in $R_{Hr}(C)$ are decomposed (i.e., if L and M are two different literals in a positive clause then $V(L) \cap V(M) = \emptyset$). Such a decision class (besides PVD) is OCC1N (see [FLTZ93]) where the positive clauses may contain variables, but all of them occur only once. Moreover the Herbrand models obtained via MB can be transformed into finite models by a filtration technique [FL93] for PVD and OCC1N.

The whole model building method can be extended to the more general class PVD_r (see Definition 5.3.4): Given a set of clauses C, search for a renaming γ such that $\gamma(C) \in$ PVD and then apply MB to $\gamma(C)$. MB then yields an AR of a Herbrand model of $\gamma(C)$; by changing the signs backwards one obtains an atom representation of a Herbrand model of C.

Example 5.4.3. We define the following satisfiable set of clauses:

$$C = \{P(b),\ P(f(x_1)) \vee \neg P(x_1),\ \neg P(a) \vee \neg P(f(a))\}.$$

C is not in PVD and $(S^i_{Hr}(C))_{i \in \mathbb{N}}$ is divergent. Note that for all $i \geq 1$:

$$P(f^i(b)) \in S^i_{Hr}(C) - S^{i-1}_{Hr}(C).$$

But $C \in PVD_r$ as can be seen by computing $\gamma(C)$ for $\gamma = \{\neg P\}$. Indeed

$$\{\neg P(b),\ P(x_1) \vee \neg P(f(x_1)),\ P(a) \vee P(f(a))\}$$

is in PVD. By setting $C_1 = \gamma(C)$ we obtain

$$S^0_{Hr}(C_1) = C_1 \text{ and } S^1_{Hr}(C_1) = \{\neg P(b), P(x_1) \vee \neg P(f(x_1)), P(a)\}.$$

Clearly $\rho_H(S^1_{Hr}(C_1)) = \emptyset$ and $S^2_{Hr}(C_1) = S^1_{Hr}(C_1)$. We see that $(S^i_{Hr}(C))_{i \in \mathbb{N}}$ converges and

$$R_{Hr}(C_1) = \{\neg P(b),\ P(x_1) \vee \neg P(f(x_1)),\ P(a)\}.$$

By Lemma 5.4.1 (and clearly visible in this case) $\mathcal{A} : \{P(a)\}$ is an atom representation of a Herbrand model of C_1 ($\mathcal{A}$ is also a ground representation of this model).

Therefore $\mathcal{M} : \{P(b)\} \cup \{P(f(t)) | t \in H(C)\}$ is a ground representation of a Herbrand model of C. By Theorem 3.6.2 $R^+_H(C)$ must be an atom representation of a Herbrand model of C. Indeed $R^+_H(C) = \mathcal{M}$, but as $\mathcal{M}$ is infinite the computation of R_H on C does not yield a syntactic model representation. However we can compute a finite AR of $\mathcal{M}$ directly out of $\mathcal{A}$ itself; such a representation is $\mathcal{B} : \{P(b), P(f(x))\}$ (over $H(C)$).

It is not hard to show that the complement set of a finite ground representation always possesses a finite AR (Exercise 5.4.5); moreover this AR can be obtained algorithmically.

Exercises

Exercise 5.4.1. Let C be a set of clauses such that $R_H(C) = C$ and all positive clauses in C are unit clauses. Show that $R_H^+(C)$ is an atom representation of a Herbrand model of C (compare with Theorem 5.4.1).

Exercise 5.4.2. Let T be the operator on stable sets in PVD as in Definition 5.4.2 and let C be a stable set in PVD. Show that the $T^i(C)$ are stable sets in PVD for all $i \in \mathbb{N}$.

Exercise 5.4.3. Let $<$ be as in Definition 5.4.3. Show that for all $k \geq 1$ and for every sequence $C_0, \ldots, C_k$ with $C_k < C_{k-1} < \ldots < C_0 \bigcup_{i=0}^{k-1} C_i \not\leq_{ss} C_k$.

Exercise 5.4.4. Let C and D be two sets of clauses such that $C \leq_{ss} D$. Show that $R_H(C) \leq_{ss} R_H(D)$ (hint: look at the proof of Lemma 4.2.3).

Exercise 5.4.5. Let M be a finite ground representation of a Herbrand model of C. Show that $AT(C) - M$ has a finite atom representation over the signature of C.

Exercise 5.4.6. Let $A : \{P(x, f(x))\}$ be an AR of a Hebrand interpretation M over the signature $\Sigma : \{P, f\}$ (M is in fact the corresponding ground interpretation). Show that $AT(A) - M$ does not possess a finite AR.

Exercise 5.4.7. Show that every satisfiable set of clauses C in PVD has a finite model (hint: construct the finite model out of $T^*(C)$).

6. On the Complexity of Resolution

6.1 Herbrand Complexity and Proof Length

By the undecidability of clause logic there exists no recursive bound on the length of refutations of clause sets C in terms of the length of C, regardless of what logic calculus and what concept of length are chosen. Thus we cannot present a complexity theory similar to that of propositional inference systems [Boe92]. Instead there are basically two different mathematical approaches to proof complexity in predicate logic:

a) Analyze the relative complexity of resolution versus other inference methods.
b) Define some absolute complexity measure for sets of clauses which is independent of deduction concepts but is recursively related to all "reasonable" inference systems.

The book of E. Eder [Ede92] is based on approach (a) and presents many interesting complexity results for first-order calculi (among them also resolution). In [BL92], a paper which mainly focuses on the complexity of resolution, Herbrand complexity is taken as inference-independent basic measure. Most of the results in this section are contained in [BL92] but the concepts and the formalism are different. The Herbrand complexity of a set of clauses C is defined as the minimal cardinality of an unsatisfiable set of ground clauses C' (obtained via ground instantiation from C). By Herbrand's theorem (Theorem 2.3.3) every unsatisfiable set of clauses possesses such a finite unsatisfiable set of ground instances; therefore Herbrand complexity is well-defined on the set of all unsatisfiable sets of clauses.

Definition 6.1.1. *Let C be a (finite) unsatisfiable set of clauses and $\mathrm{GI}(C)$ be the set of all ground instances of clauses from C. Then the Herbrand complexity of C is defined as*
$\mathrm{HC}(C) = \min\{|C'| \mid C' \text{ is unsatisfiable and } C' \subseteq \mathrm{GI}(C)\}$.

Example 6.1.1.
Let C be the set of clauses $\{P(x),\ \neg P(f(y)) \vee \neg P(g(y))\}$.
Then the set

$$C' : \{P(f(a)),\ P(g(a)),\ \neg P(f(a)) \vee \neg P(g(a))\}$$

is a minimal unsatisfiable set of ground instances from C. Therefore $\mathrm{HC}(C) \leq$ 3. It is easy to see that all sets of ground instances D of C with $|D| = 2$ are satisfiable and so $\mathrm{HC}(C) = 3$.

As every unsatisfiable set of clauses defines finite, unsatisfiable sets of ground instances (by Herbrand's theorem), HC can be considered as a measure of the "logical complexity" of a theorem; this measure is independent of particular inference systems and can be taken as a basis to compare different computational calculi. Therefore the problem is to find out how many ground instances of a set of clauses are required in order to achieve unsatisfiability. However, this number is not recursively computable, i.e., there exists no computable function f such that $dom(f) = \mathbb{N}$ and $\mathrm{HC}(C) \leq f(|C|)$ for any unsatisfiable set of clauses C. Even if we consider the undecidability of clause logic as given, this property is not completely trivial because the number of sets of ground clauses D such that $|D| \leq k$ (for some $k \in \mathbb{N}$) may be infinite. Indeed if C is a nonground clause and the Herbrand universe $H(C)$ is infinite then C has infinitely many ground instances. Therefore, even if $\mathrm{HC}(C) = k$, it might be necessary to test arbitrary many sets of ground instances D with $|D| \leq k$. But we will see below that for small Herbrand complexity there exist sets of unsatisfiable ground instances consisting of reasonably "flat" clauses.

Example 6.1.2.
Let C be the set of clauses $\{Q(x),\ \neg Q(y) \vee \neg Q(f(y))\}$.
Then clearly $\mathrm{HC}(C) = 3$ and all of the sets D_t ($t \in H(C)$) of ground clauses are unsatisfiable:

$$D_t:\ \{Q(t),\ Q(f(t)),\ \neg Q(t) \vee \neg Q(f(t))\}$$

In particular there are infinitely many sets of ground clauses with 3 elements that are unsatisfiable. We now give a sketch of the general method which, given an arbitrary set D_t, constructs a term-minimal set of the same structure. For every clause $D \in D_t$ take the clause in C having D as instance and keep all clauses variable-disjoint. So we obtain a set

$$C_0 = \{Q(x_1),\ Q(x_2),\ \neg Q(x_3) \vee \neg Q(f(x_3))\}$$

D_t is an instance of C_0 via the substitution $\vartheta : \{x_1 \leftarrow t, x_2 \leftarrow f(t), x_3 \leftarrow t\}$. ϑ is a simultaneous unifier of the sets of expressions

$$W_1:\ \{Q(x_1), Q(x_3)\},\ W_2:\ \{Q(x_2), Q(f(x_3))\}$$

The corresponding most general unifier is $\sigma : \{x_1 \leftarrow x_3, x_2 \leftarrow f(x_3)\}$. The set of clauses

$$C_0\sigma:\ \{Q(x_3),\ Q(f(x_3)),\ \neg Q(x_3) \vee \neg Q(f(x_3))\}$$

is the "schema" of all the sets D_t (by $C_0\sigma \leq_s D_t$ for all t).

By applying the ground instance $\gamma : \{x_3 \leftarrow a\}$ we obtain the set $\mathcal{D}_a$ which is of minimal term depth among all the unsatisfiable sets of ground clauses with 3 elements.

In Example 6.1.2 the sets of clauses $\mathcal{D}_t$ are all logically isomorphic, i.e., if we replace all ground atoms by propositional variables then (modulo renaming) we obtain the same set of propositional clauses.

Definition 6.1.2 (propositional skeleton). *Let C be a set of clauses and π be a one-one mapping from the sets of all atoms in C to the set of all propositional variables VPROP. π can be extended to C in the following obvious way:*
$\pi(\neg A) = \neg \pi(A)$ *for atoms A in C.*
$\pi(L_1 \vee \ldots \vee L_n) = \pi(L_1) \vee \ldots \vee \pi(L_n)$ *for clauses in C.*
$\pi(\mathcal{C}) = \cup\{\pi(C)|C \in \mathcal{C}\}$.
Then the set $\pi(\mathcal{C})$ is called a propositional skeleton *of $\mathcal{C}$.*

Definition 6.1.3. *A* propositional renaming *is a bijective mapping $\psi :$ VPROP $\rightarrow$ VPROP such that the set $\{X|\psi(X) \neq X\}$ is finite. Renamings can be extended to sets of propositional clauses in the canonical (homomorphic) way.*
Two sets of propositional clauses $\mathcal{B}_1, \mathcal{B}_2$ are called r-equivalent if there exists a propositional renaming ψ such that $\psi(\mathcal{B}_1) = \mathcal{B}_2$.

Example 6.1.3.
Let $\mathcal{D}_t = \{Q(t), Q(f(t)), \neg Q(t) \vee \neg Q(f(t))\}$ for some ground term t (see Example 6.1.2).
We define $\pi(Q(t)) = X$, $\pi(Q(f(t)) = Y$. Then

$$\pi(\mathcal{D}_t) : \ \{X, Y, \neg X \vee \neg Y\}$$

is a propositional skeleton of $\mathcal{D}_t$. Now let

$$C' = \{Q(y), Q(f(y)), \neg Q(y) \vee \neg Q(f(y))\}$$

and $\pi'(Q(y)) = U$, $\pi'(Q(f(y)) = V$. Then we get

$$\pi'(C') = \{U, V, \neg U \vee \neg V\}.$$

$\pi'(C')$ is a propositional skeleton of C' which is r-equivalent to $\pi(\mathcal{D}_t)$. Note that all propositional skeletons of the set

$$C_0 : \{Q(x_1), Q(x_2), \neg Q(x_3) \vee \neg Q(f(x_3))\}$$

are r-equivalent to the set of propositional clauses

$$\mathcal{B} : \{U, V, \neg X \vee \neg Y\}$$

$\mathcal{B}$ is not r-equivalent to $\pi'(C')$.

Definition 6.1.4. *Two sets of ground clauses $\mathcal{D}_1$ and $\mathcal{D}_2$ are called logically isomorphic if there are skeletons $\mathcal{B}_1$ of $\mathcal{D}_1$ and $\mathcal{B}_2$ of $\mathcal{D}_2$ such that $\mathcal{B}_1$ and $\mathcal{B}_2$ are r-equivalent.*

Lemma 6.1.1. *Logically isomorphic sets of clauses are sat-equivalent.*

Proof. Exercise 6.1.1.

Example 6.1.2 suggests the existence of a general method which, given an arbitrary unsatisfiable set of ground instances $\mathcal{D}$ of a set of clauses $\mathcal{C}$, constructs a logically isomorphic set of ground instances $\mathcal{D}'$ with minimal term depth. The minimal term depth we obtain this way can always be recursively bounded in terms of $HC(\mathcal{C})$ and the "length" of $\mathcal{C}$. As concept of length we need the number of symbol occurrences in $\mathcal{C}$. It is not hard to show that the minimal term depth of an unsatisfiable set of ground clauses in $\mathcal{C}$ cannot be bounded in terms of $|\mathcal{C}|$ or $\tau(\mathcal{C})$ (Exercise 6.1.2). We give a formal definition of the symbolic length $\|\ \|$ below:

$\|t\| = 1$ if t is a constant symbol or a variable.

$\|f(t_1,\ldots,t_n)\| = 1 + \sum_{i=1}^{n} \|t_i\|$ for functional terms.

$\|P(t_1,\ldots,t_n)\| = 1 + \sum_{i=1}^{n} \|t_i\|$ for atom formulas.

$\|A \odot B\| = 1 + \|A\| + \|B\|$ for $\odot \in \{\wedge, \vee, \rightarrow\}$.

$\|\neg A\| = 1 + \|A\|$.

$\|(Qx)A\| = 2 + \|A\|$ for $Q \in \{\forall, \exists\}$.

$\|L_1 \vee \ldots \vee L_n\| = \|L_1\| + \ldots + \|L_n\|$ for clauses.

$\|\{C_1,\ldots,C_k\}\| = \|C_1\| + \ldots + \|C_k\|$ for sets of clauses.

Lemma 6.1.2. *Let $\mathcal{C}$ be an unsatisfiable set of clauses; then there exists an unsatisfiable set of ground instances $\mathcal{D}$ from $\mathcal{C}$ with $\|\mathcal{D}\| \leq 2^{3HC(\mathcal{C})\|\mathcal{C}\|}$.*

Proof. We formally develop the method indicated in Example 6.1.2.

Suppose that $\mathcal{C} = \{C_1,\ldots,C_n\}$ and let $\mathcal{D}$ be an arbitrary unsatisfiable set of ground instances from $\mathcal{C}$ such that $|\mathcal{D}| = HC(\mathcal{C})$. Then $\mathcal{D}$ must be of the form

$$\mathcal{D} = \mathcal{D}_1 \cup \ldots \cup \mathcal{D}_n$$

where the $\mathcal{D}_i$ are the sets of ground instances coming from the clause C_i. For every set $\mathcal{D}_i : \{D_{i,1},\ldots,D_{i,k_i}\}$ we define a set of clauses $\mathcal{C}_i : \{C_i\eta_{i,1},\ldots,C_i\eta_{i,k_i}\}$ such that all $\eta_{i,j}$ are variable renamings and $V(C_i\eta_{i,j}) \cap V(C_i\eta_{i,k}) = \emptyset$ for $j \neq k$. Then we define

$$\mathcal{C}_0 = \mathcal{C}_1 \cup \ldots \cup \mathcal{C}_n.$$

Because of the variable renamings $\eta_{i,j}$ the propositional skeletons of $\mathcal{D}$ and $\mathcal{C}_0$ are (in general) different. In any case $\mathcal{C}_0$ is more general as $\mathcal{D}$ and we have $\mathcal{C}_0 \leq_s \mathcal{D}$.

Our aim is to construct the most general instance of $\mathcal{C}_0$ which has the same propositional skeleton as $\mathcal{D}$.

By $\mathcal{C}_0 \leq_s \mathcal{D}$ there exists a substitution ϑ such that $\mathcal{C}_0 \vartheta = \mathcal{D}$. Then ϑ acts as simultaneous unifier for sets of atoms occurring in $\mathcal{C}_0$. To be more precise, there exist sets of atoms $\mathcal{A}_1, \ldots, \mathcal{A}_r$ such that the $\mathcal{A}_i$ are pairwise disjoint and ϑ unifies all of the $\mathcal{A}_i$ for $i = 1, \ldots, r$. We can represent ϑ as a unifier of two terms in the following way. Replace every predicate symbol occurring in the sets $\mathcal{A}_i$ by a new function symbol of the appropriate arity. Then we obtain sets of terms $\mathcal{T}_1, \ldots, \mathcal{T}_r$ having the same leading function symbol. Let $\mathcal{T}_i = \{t_{i,1}, \ldots, t_{i,l_i}\}$ and g be a new function symbol of appropriate arity. Then every simultaneous unifier of the family $\mathcal{A}_1, \ldots, \mathcal{A}_r$ is a unifier of the following two terms (and vice versa):

$$\begin{aligned} w_1 &= g(t_{1,1}, \ldots, t_{1,l_1}, \ldots, t_{r,1}, \ldots, t_{r,l_r}), \\ w_2 &= g(t_{1,1}, \ldots, t_{1,1}, \ldots, t_{r,1}, \ldots, t_{r,1}). \end{aligned}$$

By the unification theorem (Theorem 2.5.1) there exists an m.g.u. of w_1 and w_2 which is also an m.g.u. of the simultaneous unification problem of the $\mathcal{A}_1, \ldots, \mathcal{A}_r$. Then clearly $\mathcal{C}_0 \sigma \leq_s \mathcal{D}$ and $\mathcal{C}_0 \sigma$ and $\mathcal{D}$ have r-equivalent propositional skeletons. By Exercise 6.1.3 we have

$$\|w_1 \sigma\| = \|w_2 \sigma\| \leq 2^{\|w_1\| + \|w_2\|}$$

But by construction of the w_i we have $\|w_i\| \leq |\mathcal{C}_0| c_{max}$ where $c_{max} = \max\{\|C\| \,|\, C \in \mathcal{C}\}$. Clearly $c_{max} \leq \|\mathcal{C}\|$ and therefore

$$\|w_1\| + \|w_2\| \leq 2\|\mathcal{C}\| |\mathcal{C}_0|$$

By construction of the set $\mathcal{C}_0$ and by definition of $\mathcal{D}$ we also have $|\mathcal{C}_0| = |\mathcal{D}| = \mathrm{HC}(\mathcal{C})$ and $\|\mathcal{C}_0\| \leq \mathrm{HC}(\mathcal{C}) \|\mathcal{C}\|$. Moreover

$$\|\mathcal{C}_0 \sigma\| \leq \|\mathcal{C}_0\| \|w_1 \sigma\|$$

By putting things together we obtain the inequality

$$\|\mathcal{C}_0 \sigma\| \leq \mathrm{HC}(\mathcal{C}) \|\mathcal{C}\| 2^{2\|\mathcal{C}\| \mathrm{HC}(\mathcal{C})}$$

A rather rough estimation then gives

$$\|\mathcal{C}_0 \sigma\| \leq 2^{3\|\mathcal{C}\| \mathrm{HC}(\mathcal{C})}$$

By definition, $\mathcal{C}_0 \sigma$ has a propositional skeleton that is r-equivalent to that of $\mathcal{D}$. Therefore every ground instance of $\mathcal{C}_0 \sigma$ is unsatisfiable!

Finally we just define $\gamma = \{v \leftarrow c \mid v \in rg(\sigma)\}$ where c is a constant in the

Herbrand universe of C. Then the sets of ground clauses $\mathcal{D}$ and $\mathcal{D}' : C_0 \sigma \gamma$ are logically isomorphic; consequently $\mathcal{D}'$ is unsatisfiable and $\|\mathcal{D}'\| = \|C_0 \sigma\|$ and we obtain
$$\|D'\| \leq 2^{3\|C\|\mathrm{HC}(C)}.$$

$\Diamond$

Corollary 6.1.1. *Let C be an unsatisfiable set of clauses; then there exists an unsatisfiable set of ground instances $\mathcal{D}$ from C with $\tau(\mathcal{D}) \leq 2^{3\mathrm{HC}(C)\|C\|}$.*

Proof. By Lemma 6.1.2 and $\tau(\mathcal{D}) \leq \|\mathcal{D}\|$. $\Diamond$

Lemma 6.1.2 also yields a method to determine a bound on the size of shortest resolution proofs in terms of Herbrand complexity and the size of the input. Properties of this type hold also for other calculi of first-order logic [KP88].

The following proposition shows that there is no recursive bound on the number of instances of clauses necessary to "realize" Herbrand's theorem.

Proposition 6.1.1. *There exists no computable function f such that $\mathrm{domain}(f) = \mathbb{N}$ and for all unsatisfiable sets of clauses C: $\mathrm{HC}(C) \leq f(|C|)$.*

Proof. Let C be an unsatisfiable set of clauses. By Lemma 6.1.2 there exists an unsatisfiable set of ground clauses $\mathcal{D}$ such that for all $D \in \mathcal{D}$:
$$\|D\| \leq 2^{3\mathrm{HC}(C)\|C\|}$$

Because exponentiation is monotonic we also obtain
$$\|D\| \leq 2^{3f(|C|)\|C\|}$$

Then the following procedure working on the class of all (finite) sets of clauses decides satisfiability:

1. Compute $f(|C|)$ for a given set of clauses C.
2. Test all sets of ground clauses $\mathcal{D}$ with
   $\|D\| \leq 2^{3f(|C|)\|C\|}$ for satisfiability
   (note that there are only *finitely many* $\mathcal{D}$'s having this property).

If one of the $\mathcal{D}$'s is unsatisfiable, which can be tested by the Davis–Putnam procedure, then clearly C is unsatisfiable.

Now assume that all set of clauses $\mathcal{D}$ with
$$\|\mathcal{D}\| \leq 2^{3f(|C|)\|C\|}$$

are satisfiable. Then, by definition of f and HC and by Lemma 6.1.2, C itself must be satisfiable.

Thus the above algorithm defines a decision procedure for the satisfiability problem of clause logic. But clause logic is undecidable (clause logic is a reduction class of predicate logic) and consequently $\mathrm{HC}(C)$ cannot be bounded by a computable function f with domain $\mathbb{N}$. $\Diamond$

Our next step consists in comparing the size of resolution proofs with Herbrand complexity. For this purpose we choose the simple length measure l for resolution proofs. Recall (Definition 2.7.6) that $l(C_1, \ldots, C_n) = n$ for any R-deduction $C_1, \ldots, C_n$. We first analyze the lengths of PR- and GR-deductions (see Definitions 2.5.3 and 2.5.4).

Lemma 6.1.3. *Every instance of a PR-deduction from a set of clauses $\mathcal{C}$ is also a PR-deduction from $\mathcal{C}$.*

Proof. We proceed by induction on the length of PR-deductions Γ.

$l(\Gamma) = 0$:
In this case $\Gamma = C\sigma$ for some $C \in \mathcal{C}$ and for some substitution σ. But then $\Gamma\eta = C\sigma\eta$ for every substitution η and $C\sigma\eta$ is an instance of a clause in $\mathcal{C}$. By Definition 2.5.3 $\Gamma\eta$ is a PR-deduction from $\mathcal{C}$.

(IH) Suppose that for all PR-deductions Γ from $\mathcal{C}$ such that $l(\Gamma) \leq n$ and for all substitutions η $\Gamma\eta$ is a PR-deduction from $\mathcal{C}$.

Now let $\Gamma : C_1, \ldots, C_{n+1}$ be a PR-deduction from $\mathcal{C}$.

case a):
C_{n+1} is an instance of a clause in $\mathcal{C}$, i.e., $C_{n+1} = D\mu$ for some $D \in \mathcal{C}$ and some substitution μ:
By (IH) $(C_1, \ldots, C_n)\lambda$ is a PR-deduction for every λ; moreover $C_{n+1}\lambda$ is an instance of D ($C_{n+1}\lambda = D\mu$). Thus by Definition 2.5.3 $(C_1, \ldots, C_{n+1})\lambda$ is a PR-deduction from $\mathcal{C}$.

case b):
C_{n+1} is a PR-resolvent of clauses D and E where D and E are p-reducts of clauses C_i and C_j respectively for some $i, j < n+1$.
Then $D = A_1 \vee L \vee A_2$ and $E = B_1 \vee L^d \vee B_2$ for some clauses A_1, A_2, B_1, B_2 and some literal L and

$$C_{n+1} = A_1 \vee A_2 \vee B_1 \vee B_2.$$

Let λ be a substitution applied to $\Delta : C_1, \ldots, C_n$. Then $C_i\lambda$ and $C_i\lambda$ are clauses in Δ. Moreover the clauses $D\lambda : A_1\lambda \vee L\lambda \vee A_2\lambda$ and $E\lambda : B_1\lambda \vee L^d\lambda \vee B_2\lambda$ are p-reducts of $C_i\lambda$ and $C_j\lambda$ respectively (this follows directly from the definition of a p-reduct). It follows that

$$C_{n+1}\lambda : (A_1 \vee A_2 \vee B_1 \vee B_2)\lambda$$

is a PR-resolvent of p-reducts of $C_i\lambda$ and $C_j\lambda$.
By (IH) $\Delta\lambda$ is a PR-deduction from $\mathcal{C}$. But then, by Definition 2.5.3(b), $(\Delta, C_{n+1})\lambda$ is a PR-deduction from $\mathcal{C}$ too.   ◊

Proposition 6.1.2. *Let $\mathcal{C}$ be a set of clauses and Γ be a PR-refutation of $\mathcal{C}$; then $\mathrm{HC}(\mathcal{C}) \leq l(\Gamma)$.*

Proof. Let $\Gamma : C_1, \ldots, C_n, \Box$ be a PR-refutation of $\mathcal{C}$. By Lemma 6.1.3 every instance $\Gamma\lambda$ of Γ is a PR-refutation of $\mathcal{C}$. Let λ be a substitution such that $\Gamma\lambda$ is a sequence of ground clauses. Then $\Gamma\lambda$ is even a GR-refutation of $\mathcal{C}$; which means $\Gamma\lambda$ is an R-refutation of the set of ground instances of clauses from $\mathcal{C}$ appearing in $\Gamma\lambda$. Because $\Gamma\lambda$ is an R-refutation and the principle of R-deduction is sound, the set $\mathcal{D}$ of all ground instances of input clauses C_i appearing in Γ must be unsatisfiable. It follows that $\text{HC}(\mathcal{C}) \leq |\mathcal{D}| \leq l(\Gamma)$. $\Diamond$

Corollary 6.1.2. *Let $\mathcal{C}$ be a set of clauses and Γ be a GR-refutation of $\mathcal{C}$; then $\text{HC}(\mathcal{C}) \leq l(\Gamma)$.*

Proof. Every GR-deduction is a PR-deduction. $\qquad\qquad\qquad\qquad\qquad\Diamond$

The proof of Proposition 6.1.2 is based on the fact that every instance of a PR-deduction is a PR-deduction too. Of course, due to the principle of variable renaming and of most general unification, instances of R-deductions typically are not R-deductions. But for the purpose of complexity analysis it is interesting to know, whether to an R-deduction Γ there are substitutions $\lambda_1, \ldots, \lambda_n$ such that $C_1\lambda_1, \ldots, C_n\lambda_n$ is a GR-deduction (from the same set of clauses).

Definition 6.1.5. *Let $\Gamma : C_1, \ldots, C_n$ be an R-deduction from $\mathcal{C}$. A sequence of clauses Γ' is called* propositional projection *of Γ if it holds:*

P1) Γ' is a PR-deduction from $\mathcal{C}$.
P2) There exist substitutions $\lambda_1, \ldots, \lambda_n$ such that $\Gamma' = C_1\lambda_1, \ldots, C_n\lambda_n$.

If Γ' is a propositional projection consisting of ground clauses only then Γ' is called a ground projection *of Γ (in this case Γ' is a GR-deduction from $\mathcal{C}$).*

Ground projection is, in some sense, the inverse operation to lifting. As every GR-deduction can be lifted to a general R-deduction (Theorem 2.7.1) we may ask whether the other direction holds too, i.e. whether every R-deduction possesses a ground projection. The following example shows that the answer is negative.

Example 6.1.4. Let $\mathcal{C} = \{\neg P(x) \vee P(f(x)), P(a), \neg P(f^2(a))\}$. Then

$$\Gamma : \neg P(x) \vee P(f(x)), \neg P(x) \vee P(f^2(x)), P(a), P(f^2(a)), \neg P(f^2(a)), \Box$$

is an R-refutation of $\mathcal{C}$. Note that $\neg P(x) \vee P(f^2(x))$ is obtained from $\neg P(x) \vee P(f(x))$ by self-resolution. Now let s_1, s_2 be arbitrary ground terms and

$$\Gamma' : \neg P(s_1) \vee P(f(s_1)), \neg P(s_2) \vee P(f^2(s_2)), P(a), P(f^2(a)), \neg P(f^2(a)), \Box.$$

Then Γ' cannot be an R-refutation of a set of ground instances $\mathcal{C}'$ of $\mathcal{C}$. Note that the second clause in Γ' is neither a ground instance of a clause in $\mathcal{C}$ nor a resolvent of the first clause with itself (no matter how we choose the ground terms s_1 and s_2). In fact the sequence Γ' is not a GR-deduction at all.

The possibility of resolving a clause with a renamed copy of itself is typical to general resolution. In case of a ground clause C, C can only be self-resolving if it is a tautology; then the self-resolvent D of C is a tautology too (if C and D are both N_c-clauses then even $C = D$). It is trivial that there are always R-refutations of a set of clauses having ground projections: if Γ is obtained from a GR-deduction Γ' via lifting then Γ' is a ground projection of Γ. Moreover, as we will see in this section, there are resolution refinements which always admit ground projections. The following proposition shows that HC is a lower bound to the length of R-deductions admitting ground projections.

Proposition 6.1.3. *Let Γ be an R-refutation of $\mathcal{C}$ which has a ground projection; then $\mathrm{HC}(\mathcal{C}) \leq l(\Gamma)$.*

Proof. Let $\Gamma = C_1, \ldots, C_n, \square$ and $\Gamma' : C_1\lambda_1, \ldots, C_n\lambda_n, \square$ be a ground projection of Γ. Then Γ' is a GR-refutation of $\mathcal{C}$. By Corollary 6.1.2 $\mathrm{HC}(\mathcal{C}) \leq l(\Gamma')$. Moreover we also have $l(\Gamma) = l(\Gamma')$.   ◊

Example 6.1.4 tells us that there are R-deductions without ground projections; thus we cannot conclude that Herbrand complexity is a lower bound to the length of all resolution refutations (which in fact is not the case). However, we will show that linear input refutations (see Definition 3.5.5) always have ground projections. For the proof of this result it is convenient to measure the depth of LI-deductions instead of their length.

Definition 6.1.6. *Let $\Gamma : C_0, E_1, \ldots, C_{n-1}, E_n, C_n$ be a linear deduction. Then the depth of Γ (depth(Γ)) is n.*

According to the definition of LI-deductions Γ we always have

$$\mathrm{depth}(\Gamma) = \frac{(l(\Gamma) - 1)}{2}.$$

Lemma 6.1.4. *Every LI-deduction has a ground projection.*

Proof. We first show by induction on the depth of LI-deductions Γ that Γ has a propositional projection Γ' with the following property: If Γ is a deduction of C then Γ' is a deduction of a variant of C.

depth(Γ) = 0:
$\Gamma : C_0$ is an LI-deduction. Thus $\Gamma' : C_0$ is a propositional projection of Γ which derives a variant of C_0 (namely C_0 itself).

(IH) Suppose that every LI-deduction Γ of a clause C such that depth(Γ) $\leq$ n has a propositional projection Γ' such that Γ' derives a variant C' of C.

Now let depth$(\Gamma) = n + 1$.
Then Γ is of the form $C_0, E_1, \ldots, C_n, E_{n+1}, C_{n+1}$.
By (IH) there exist substitutions $\lambda_0, \ldots, \lambda_n$ and $\mu_1, \ldots, \mu_n$ such that

$$\Delta : C_0\lambda_0, E_1\mu_1, \ldots, E_n\mu_n, C_n\lambda_n$$

is a propositional projection of Γ and λ_n is a renaming substitution.

By the definition of LI-deductions E_{n+1} must be an element of the set of input clauses $\mathcal{C}$ and C_{n+1} is LRM-resolvent of (C_n, E_{n+1}). Let μ be the substitution defining the l-factor of C_n corresponding to the resolution. Then the factor D_n of C_n is a p-reduct of $C_n\mu$. Let $E_{n+1}\eta$ be a renamed version of E_{n+1} such that $V(E_{n+1}\eta) \cap V(C_n\lambda_n) = \emptyset$. Then C_{n+1} is a variant of a propositional resolvent of $D_n\sigma$ and $E_{n+1}\eta\sigma$, where σ is an m.g.u. of a binary resolution (note that λ_n is a renaming too). In particular the sequence

$$C_n\mu\sigma, \quad E_{n+1}\eta\sigma, \quad C_{n+1}\tau$$

is a PR-deduction from $\mathcal{C} \cup \{C_n\mu\sigma\}$ where τ is a renaming substitution. By Lemma 6.1.3 every instance of a PR-deduction from $\mathcal{C}$ is also a PR-deduction from $\mathcal{C}$. Particularly $\Delta\mu\sigma$ is a PR-deduction of $C_n\mu\sigma$ from $\mathcal{C}$. But then

$$\Omega : (C_0\lambda_0, E_1\mu_1, \ldots, E_n\mu_n, C_n\lambda_n)\mu\sigma, \quad E_{n+1}\eta\sigma, \quad C_{n+1}\tau$$

is a propositional projection of Γ which derives a variant of C_{n+1}. This completes the induction proof and we know that all LI-deductions have appropriate propositional projections.

Now let Γ be an LI-deduction of $\mathcal{C}$ and Γ' be a propositional projection of Γ deriving a variant C' of C. Let λ be a ground substitution over the signature of the clauses in Γ such that the domain of λ is the set of all variables occurring in Γ'. Because Γ' is a propositional projection of Γ, $\Gamma'\lambda$ is a ground projection of Γ. $\Diamond$

Theorem 6.1.1. *Let $\mathcal{C}$ be a set of clauses and Γ be an LI-refutation of $\mathcal{C}$. Then* $\mathrm{HC}(\mathcal{C}) \leq l(\Gamma)$.

Proof. By Lemma 6.1.4, Γ has a ground projection. Therefore, by Proposition 6.1.3, $\mathrm{HC}(\mathcal{C}) \leq l(\Gamma)$. $\Diamond$

In the proof of Lemma 6.1.4 we have shown that every LI-deduction has a ground projection. This property is important in the theory of logic programming, as it guarantees the existence of a "total" substitution for such a deduction. These total substitutions applied to the variables of the goal clause (i.e., the negative top clause of a LI-deduction in Horn logic) define the answer substitutions of a logic program [Llo87]. Theorem 6.1.1 might suggest that Herbrand complexity is a lower bound to the length of resolution proofs in general. The following theorem shows that, for general resolution, the situation is quite different.

Theorem 6.1.2. *There exists a sequence of sets of clauses $(C_n)_{n \in \mathbb{N}}$ such that all C_n are refutable by R-deductions of length $n + 5$, but $\mathrm{HC}(C_n) > 2^n$ (Herbrand complexity may be exponential in the length of a shortest resolution refutation).*

Proof. Let $C_n = \{P(a), \neg P(x) \vee P(f(x)), \neg P(f^{2^n}(a))\}$ for all $n \geq 0$. Then the deduction Γ_n:

$$\neg P(x) \vee P(f(x)), \neg P(x) \vee P(f^2(x)), \ldots, \neg P(x) \vee P(f^{2^i}(x)),$$

$$\neg P(x) \vee P(f^{2^{i+1}}(x)), \ldots, \neg P(x) \vee P(f^{2^n}(x)), P(a), P(f^{2^n}(a)),$$

$$\neg P(f^{2^n}(a)), \square$$

is an R-refutation of Γ_n of length $n + 5$. The first n resolvents are defined by iterated self-resolution.

To show the exponentiality of $\mathrm{HC}(C_n)$ we prove that every set of ground instances C'_n of C_n such that $|C'_n| < 2^n + 2$ is satisfiable.

Suppose that $\mathcal{D}$ is a minimal unsatisfiable set of ground instances from C_n and $|\mathcal{D}| < 2^n + 2$.

As $\mathcal{D}$ is unsatisfiable it must contain the clauses $P(a)$ and $\neg P(f^{2^n}(a))$ (otherwise all clauses contain a negative (positive) literal and $\mathcal{D}$ is satisfiable). So let

$$\mathcal{D} = \{P(a), \neg P(f^{2^n}(a))\} \cup \{\neg P(f^i(a)) \vee P(f^{i+1}(a)) | i \in J\}$$

where J is a set of natural numbers such that $|J| < 2^n$. By $|J| < 2^n$ there must exist a number $k \geq 0$ such that $k \notin J$ and $k < 2^n$. In particularly the number

$$p: \quad \min\{k | k \notin J, k < 2^n\}$$

is well-defined.

If $p = 0$ then $\neg P(a) \vee P(f(a))$ is not contained in $\mathcal{D}$; but then there is no clause in $\mathcal{D}$ containing $\neg P(a)$ at all. By the pure-literal-rule (Definition 2.4.2) we conclude that $\mathcal{D} \sim_{sat} \mathcal{D} - \{P(a)\}$. But then also $\mathcal{D} - \{P(a)\}$ is an unsatisfiable set of ground clauses from C_n which contradicts the minimality of $\mathcal{D}$.

If $p > 0$ then $p - 1 \in J$ and

$$D: \quad \neg P(f^{p-1}(a)) \vee P(f^p(a)) \in \mathcal{D}.$$

But by $p \notin J$ and $p < 2^n$, D is the only clause containing the atom $P(f^p(a))$. Using the pure literal rule once more we obtain $\mathcal{D} - \{D\} \sim_{sat} \mathcal{D}$, again contradicting the minimality of $\mathcal{D}$. Therefore $\mathrm{HC}(C_n) < 2^n + 2$ leads to a contradiction. So we obtain $\mathrm{HC}(C_n) > 2^n$.

$\Diamond$

Theorem 6.1.2 tells us that resolution can give an exponential "speed-up" relative to Herbrand complexity. However, in the sequence defined there, the terms in the atoms $P(f^{2^n}(a))$ are of exponential size if we count the number of occurrences of function symbols. This effect is notation dependent (e.g., we may use a binary coding for exponents and obtain a representation of linear size). But an exponential speed-up is possible even when we take the standard notation and the number of symbol occurrences as a measure for the length of deductions (see [Ede92]).

The question remains, whether it is possible to obtain even shorter refutations (in terms of HC). The following theorem shows that this is not the case and that the exponential bound is tight.

Theorem 6.1.3. *Let Γ be an R-refutation of a set of clauses $\mathcal{C}$; then $HC(\mathcal{C}) \leq 2^{2l(\Gamma)}$.*

Proof. We transfer every R-deduction Γ of a clause C from $\mathcal{C}$ into a PR-deduction Γ' of C from $\mathcal{C}$ such that

(*) $l(\Gamma') \leq 2^{2l(\Gamma)}$.

Because PR-deductions always have ground projections we then obtain $HC(\mathcal{C}) \leq 2^{2l(\Gamma)}$.

We prove the inequality (*) by induction on $l(\Gamma)$.

$l(\Gamma) = 1$: $\Gamma' = \Gamma = C$ and Γ' is a PR-deduction of C with $l(\Gamma') \leq 2^{2l(\Gamma)}$.

(IH) Assume that for all R-deductions Γ of a clause C from $\mathcal{C}$ such that $l(\Gamma) \leq n$ there exists a PR-deduction Γ' of C from $\mathcal{C}$ with $l(\Gamma') \leq 2^{2l(\Gamma)}$.

Let $\Gamma : C_1, \ldots, C_{n+1}$ be an R-deduction of length $n + 1$ from $\mathcal{C}$. By the definition of an R-deduction we have to distinguish two cases:

a) C_{n+1} is a variant of a clause in $\mathcal{C}$.
b) C_{n+1} is a resolvent two clauses C_i and C_j for $i, j < n + 1$.

By (IH) there exists a PR-deduction Γ'_k of C_k from $\mathcal{C}$ such that $l(\Gamma'_k) \leq 2^{2k}$ for all $k = 1, \ldots, n$.
In case a) $\Gamma' : \Gamma'_n, C_{n+1}$ is a PR-deduction of C_{n+1} from $\mathcal{C}$ and

$$l(\Gamma') \leq 2^{2n} + 1 < 2^{2(n+1)} = 2^{2l(\Gamma)}.$$

It remains to investigate case (b).
As $i, j < n + 1$ we obtain PR-deductions Γ'_i of C_i and Γ'_j of C_j. Now let $C_i\eta$ and $C_j\vartheta$ be variants of C_i and C_j which are variable-disjoint. Then $\Gamma'_i\eta$ is a PR-deduction of $C_i\eta$ and $\Gamma'_j\vartheta$ one of $C_j\vartheta$. Let us choose η and ϑ in a way that C_{n+1} is resolvent of $C_i\eta$ and $C_j\vartheta$ without additional renaming. Then there are instances (in fact G-instances) $C_i\eta\lambda$ of $C_i\eta$ and $C_j\vartheta\mu$ of $C_j\vartheta$ such that C_{n+1} is a propositional resolvent of p-reducts of $C_i\eta\lambda$ and of $C_j\vartheta\mu$.

Because PR-deduction is closed under substitution (Lemma 6.1.3) $\Gamma'_i\eta\lambda$ is a PR-deduction of $C_i\eta\lambda$ and $\Gamma'_j\vartheta\mu$ is a PR-deduction of $C_j\theta\mu$ (both from $\mathcal{C}$).

By Definition 2.5.3 the sequence

$$\Gamma' : \Gamma'_i\eta\lambda, \ \Gamma'_j\vartheta\mu, \ C_{n+1}$$

is a PR-deduction of C_{n+1} from C. For Γ' we obtain

$$l(\Gamma') \le 2^{2i} + 2^{2j} + 1 \le 2^{2n} + 2^{2n} + 1 < 2^{2(n+1)} = 2^{2l(\Gamma)}.$$

$\Diamond$

The sequence $(C_n)_{n \in \mathbb{N}}$ in the proof of Theorem 6.1.2 is a sequence of sets of Horn clauses. By Theorem 3.5.2, LI-deduction is complete on Horn logic and therefore there exist LI-refutations Γ_n of C_n. By Lemma 6.1.4 the Γ_n have ground projections and therefore $HC(C_n) \le l(\Gamma_n)$ for $n \in \mathbb{N}$. By definition of the sequence $(C_n)_{n \in \mathbb{N}}$ we thus obtain $l(\Gamma_n) > 2^n + 2$ for all n, i.e., all LI-refutations of C_n are of length exponential in n. As resolution complexity is linear on $(C_n)_{n \in \mathbb{N}}$ we see that refinements can lead to an exponential increase in proof complexity. That even holds in refining linear deduction to linear input deduction.

Theorem 6.1.4. *There exists a sequence of clauses $(C_n)_{n \in \mathbb{N}}$ such that*

(1) For every n there exists a linear refutation Γ_n of C_n with $l(\Gamma_n) \le 2n + 5$.
(2) For all n and for all LI-refutations Δ of C_n $l(\Delta) > 2^n + 2$; moreover there exist LI-refutations of C_n for all $n \in \mathbb{N}$.

Proof. We only have to adapt the proof of Theorem 6.1.2.
Let $C_n = \{P(a), \neg P(x) \vee P(f(x)), \neg P(f^{2^n}(a))\}$ for $n \ge 0$ as in the proof of Theorem 6.1.2. By definition of LR-deductions (Definition 3.5.3) the sequence

$\Gamma_n :\ \neg P(x) \vee P(f(x)),\ \neg P(x) \vee P(f(x)),\ \neg P(x) \vee P(f^2(x)),$

$\quad \neg P(x) \vee P(f^{2^i}(x)),\ \neg P(x) \vee P(f^{2^i}(x)),\ \neg P(x) \vee P(f^{2^{i+1}}(x)),$

$\quad \neg P(x) \vee P(f^{2^n}(x)),\ P(a),\ P(f^{2^n}(a)),\ \neg P(f^{2^n}(a)), \square$

is an LR-refutation of C_n. Moreover $l(\Gamma_n) = 2n + 5$.

Note that the length of Γ_n is about double the length of the refutations in Theorem 6.1.2; the reason for this effect is purely notational (in linear deduction self-resolution requires clause copying).

For (2): $HC(C_n) > 2^n + 2$ directly follows from the proof of Theorem 6.1.2. As every LI-refutation Δ of C_n has a ground projection we get $l(\Delta) > 2^n + 2$. Because $(C_n)_{n \in \mathbb{N}}$ is a sequence of Horn clauses there exist LI-refutations of C_n for every n (by Theorem 3.5.2). $\Diamond$

In using the depth measure (Definition 6.1.6) instead of length we obtain LR-refutations of depth $n+2$, but in the example of Theorem 6.1.4 the depth of all LI-refutations is $> 2^{n-1} + 1$. Moreover this theorem also tells us that the proof complexity of LR-deduction strongly depends on the choice of the top clauses. Consider

$$C_n :\ \{P(a),\ \neg P(x) \vee P(f(x)),\ \neg P(f^{2^n}(a))\}$$

and select $\neg P(f^{2^n}(a))$ as top clause of a linear refutation of C_n. As C_n is a set of Horn clauses every LR-refutation with this top clause must be an

LI-refutation too; but then the length of the refutation is greater than $2^n + 2$. Similarly all LR-refutations with top clause $P(a)$ are of exponential length. On $\mathcal{C}_n$ hyperresolution is exponential too. To apply hyperresolution we only have to rewrite the clause $\neg P(x) \vee P(f(x))$ to $P(f(x)) \vee \neg P(x)$. It is easy to see that a hyperrefutation must contain all clauses

$$P(a), \ldots, P(f^i(a)), \ldots, P(f^{2^n}(a)).$$

A similar property holds for any A-ordering with $P(x) < P(f(x))$. Note that the self-resolvent $\neg P(x) \vee P(f^2(x))$ of $\neg P(x) \vee P(f(x))$ is not admissible ($P(f(x))$ is the resolved atom and $P(f(x)) < P(f^2(x))$). As a consequence, a $<$-refutation generates all clauses $\neg P(f^i(a))$ for $i < 2^n$. But this does not mean that linear deduction always yields shorter proofs.

The exponential speed-up of proof complexity versus Herbrand complexity in Theorem 6.1.3 is essentially based on the internal renaming of variables for resolution. Without the possibility of self-renaming, self-inference of clauses becomes impossible in the sequence $(\mathcal{C}_n)_{n \in \mathbb{N}}$. In PR-deductions renaming is not allowed and therefore all PR-deductions are exponential on $\mathcal{C}_n$. It is the property of ground projection which is responsible for the effect that only one instance per clause is used in the deduction.

Resolution is just one among many clausal inference methods based on the unification principle. However, concepts like Herbrand complexity (which is inference-independent) and ground projections can also be used in measuring the complexity of other inference methods. We mention just a few of them: the matrix method of D. Prawitz [Pra69], the mating method of P. Andrews [And81], and W. Bibel's connection method [Bib82]. Without giving a formal mathematical analysis, we illustrate how these methods behave with respect to Herbrand complexity. There are different versions of Prawitz's method, but we refer to the later improved one. The basic idea consists in searching for a model of the set of clauses. Clearly a model exists if one can define a path π containing one literal from each clause such that no pairs of different literals on π can be made complementary (in this case we set all these literals to true). Links between complementary literals indicate that there is no such path containing both of them. If all possible paths can be excluded by links connecting complementary literals then the set of connections is called *spanning*. Because in predicate logic literals need not be complementary in advance they have to be made so by unification. Thus we need a "simultaneous" single substitution in such a matrix which admits the construction of such a spanning set of connections. If the Herbrand complexity exceeds the number of given clauses then such a simultaneous instance of the original matrix cannot exist; in this case we need the additional rule of clause copying. The minimal number of clauses in a matrix needed to show unsatisfiability is just the Herbrand complexity.

Example 6.1.5. Let $\mathcal{C}$ be the set of clauses

$$\{ \neg P(x) \vee P(f(x)),\ P(a),\ \neg P(f^2(a))\}.$$

Because $\mathrm{HC}(\mathcal{C}) = 4$ there exists no single ground instance of the formula

$$(\neg P(x) \vee P(f(x))) \wedge P(a) \wedge \neg P(f^2(a))$$

making it propositionally unsatisfiable. In Prawitz's formalism $\mathcal{C}$ is represented by the matrix $M(x)$ shown in Figure 6.1. Obviously there exists no

$$\neg P(x), P(f(x))$$
$$P(a)$$
$$\neg P(f^2(a))$$

Fig. 6.1. A Prawitz matrix

ground substitution $\lambda : \{x \leftarrow t\}$ such that $M(x)\lambda\ (= M(t))$ is unsatisfiable. In this case one generates a copy $M(y)$ of $M(x)$ such that y is a new variable. Afterwards the matrix $M(x)$ is replaced by the matrix

$$M'(x, y) : \begin{array}{c} M(x) \\ M(y) \end{array}$$

In applying the substitution $\lambda : \{x \leftarrow a, y \leftarrow f(a)\}$ we obtain a ground matrix $M'(a, f(a))$ which is unsatisfiable. In $M(x)$ the sets

$$\{\neg P(x), P(a)\} \text{ and } \{P(f(x)), \neg P(f^2(a))\}$$

cannot be made complementary simultaneously. But in $M'(x, y)$ we may focus on the sets

$$\{\neg P(x), P(a)\},\ \{\neg P(y), P(f(x))\},\ \{P(f(y)), \neg P(f^2(a))\}$$

which all can be made complementary by $\lambda : \{x \leftarrow a, y \leftarrow f(a)\}$.

Note that the general method for obtaining the adequate simultaneous substitutions is the same as that used in Lemma 6.1.2.

It is easy to see that Herbrand complexity is a lower bound to the lines in an unsatisfiable matrix. Every line represents a clause and there must be a ground substitution making the set of clauses in the matrix unsatisfiable. For the sequence

$$\mathcal{C}_n :\ \{P(a), \neg P(x) \vee P(f(x)), \neg P(f^{2^n}(a))\}$$

defined in Theorem 6.1.2 we have $\mathrm{HC}(\mathcal{C}_n) = 2^n + 2$ and thus 2^{n-1} copies of the corresponding matrices $M_n(x)$ are required in the method of Prawitz.

The methods of Andrews [And81] and Bibel [Bib82] can be considered as improvements of Prawitz's method. While the search for a simultaneous substitution in a matrix is characteristic to all these methods, only copying of single clauses (instead of copying whole matrices) is performed in [And81] and [Bib82]. As the methods of Andrews and Bibel are quite similar we select one of them (namely Bibel's) for illustration. Figure 6.2 shows connections in a matrix for the set of clauses

$$\mathcal{C}: \ \{\neg P(x) \lor P(f(x)), P(a), \neg P(f^2(a))\}$$

which are not compatible (simultaneous unifiers don't exist).

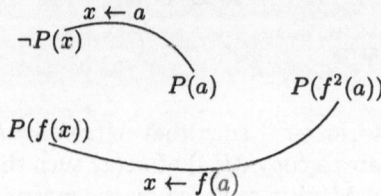

Fig. 6.2. Incompatible connections

Again we produce the copy $\neg P(y) \lor P(f(y))$ of $\neg P(x) \lor P(f(x))$ to obtain spanning compatible set of connections shown in Figure 6.3 (for details see [Bib82]). In fact the term equations $\{x = a, \ y = f(x), \ y = f(a)\}$

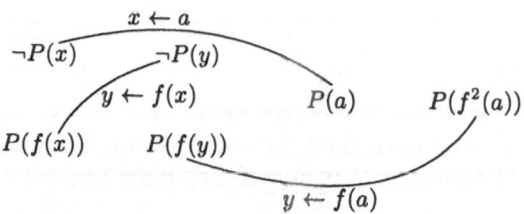

Fig. 6.3. Compatible and spanning connections

have the solution $x = a$, $y = f(a)$. Again Herbrand complexity is a lower bound to the number of clause occurrences in such a matrix. To make the set $\mathcal{C}_n : \ \{P(a), \neg P(x) \lor P(f(x)), \neg P(f^{2^n}(a))\}$ unsatisfiable, 2^{n-1} copies of $\neg P(x) \lor P(f(x))$ are needed.

There is a variant of resolution, called V-resolution [Cha72], where renaming is not allowed and a total substitution must exist (defining a propositional projection). To compensate for the absence of renaming clause copying is required again.

It is significant that all methods mentioned above have ground projections, i.e., there exists a total substitution which can be turned into a ground substitution. In the matrix methods we obtain a ground matrix that is unsatisfiable. V-resolution has a ground projection in the sense of Definition 6.1.5. Iterated "lemmatization" was used to obtain refutations in Theorem 6.1.2. Proofs containing such a mechanism of lemma building do not possess ground projections, i.e., there are no direct ground versions of these proofs.

Exercises

Exercise 6.1.1. Give a proof of Lemma 6.1.1.

Exercise 6.1.2. Let C be an unsatisfiable set of clauses and $\delta(C)$ be the minimal term depth of an unsatisfiable set of ground instances from C. Show that there exists no function $f : \mathbb{N} \to \mathbb{N}$ such that

$$\delta(C) \le f(|C| + \tau(C))$$

for all unsatisfiable sets of clauses C.

Exercise 6.1.3. Let L and M be two unifiable expressions and σ the m.g.u. of L and M. Show that
$$\|L\sigma\| \le 2^{\|L\| + \|M\|}$$

(hint: show that the length can be at most doubled in the single substitution steps of UAL).

6.2 Extension and the Use of Lemmas

In Section 6.1 we investigated the length of resolution proofs relative to Herbrand complexity. The question remains, how resolution behaves relative to other methods of inference. A systematic and thorough treatment of this problem is outside the scope of this book and we refer to [Ede92] for this purpose. However, we will point out in this section that some features of minimality that are characteristic to computational calculi can negatively influence proof complexity.

A substantial weakness of resolution and of most other computational calculi is in their low power to use lemmas in proofs. For this reason resolution proofs usually differ strongly from real mathematical proofs; the latter are typically very structured and expressed over a large base of previously derived theorems. R. Statman [Sta79] has shown that Herbrand complexity may even be nonelementary(!) with respect to the length of a shortest proof in a full logic calculus like LK [Tak75] or Natural Deduction [Pra71]. Because R-refutations maximally give an exponential speed-up with respect to Herbrand complexity it follows that the complexity of resolution relative to that of full logic calculi is nonelementary too.

Example 6.2.1 (Statman's example). Let $(\mathcal{C}_n)_{n\in\mathbb{N}}$ be a sequence of sets of clauses such that

$$\mathcal{C}_n = \mathrm{ST} \cup \mathrm{ID} \cup \{\neg pq = p((\mathbf{T}_n q)q)\},$$

where pq is an abbreviation for $f(p,q)$, p and q are constant symbols, and f is a two-place function symbol. By assuming association to the left, i.e., abc denotes $(ab)c$, we reduce the number of parentheses in the term notation. ST is a set of combinator equations (see [Ste71]) and ID a set of equality axioms.

$$ST = \{\mathbf{S}xyz \doteq (xz)(yz),\ \mathbf{B}xyz \doteq x(yz),\ \mathbf{C}xyz \doteq (xz)y,\ \mathbf{I}x \doteq x,\ px \doteq p(qx)\}.$$

$\mathbf{S}, \mathbf{B}, \mathbf{C}, \mathbf{I}$ are constant symbols, written in capital letters to conform to the standard notation in combinatory logic. We use "$\doteq$" in order to distinguish equality in the object language from that in the metalanguage. The first four equations are the standard definitions of the combinators $\mathbf{S}, \mathbf{B}, \mathbf{C}$, and $\mathbf{I}$ and the fifth is a specific additional axiom. The expression "$\mathbf{T}_n$" appearing in the definition of the set $\mathcal{C}_n$ serves as metatheoretical abbreviation for combinator terms defined by

$$\mathbf{T} \doteq (\mathbf{SB})((\mathbf{CB})\mathbf{I}),\ \mathbf{T}_1 \doteq \mathbf{T},\ \mathbf{T}_{k+1} \doteq \mathbf{T}_k\mathbf{T}\ \text{for } k \in \mathbb{N}_+.$$

Note that () is not associative and thus $\mathbf{T}_k\mathbf{T}$ is not equal to $\mathbf{T}\mathbf{T}_k$.

The set of equality axioms ID is defined as

$$ID = \{x \doteq x, \neg x \doteq y \vee y \doteq x, \neg x \doteq y \vee \neg y \doteq z \vee x \doteq z, \neg x \doteq y \vee \neg u \doteq v \vee xu \doteq yv\}.$$

Now let $s : \mathbb{N} \to \mathbb{N}$ be the following function:

$$s(0) = 1,\ s(n+1) = 2^{s(n)}\ \text{for all } n \in \mathbb{N}.$$

s is not a member of the class of elementary functions [BL74]; in particular there exists no number k such that

$$s(n) \leq 2^{2^{\cdot^{\cdot^{2^n}}}} \quad \}k \text{ times}$$

holds for infinitely many n.

In [Sta79] Statman proved that $\mathrm{HC}(\mathcal{C}_n) \geq \frac{1}{2}s(n)$. The proof requires methods and results from combinatory logic which are outside the scope of our presentation. Using Theorem 6.1.3 we conclude that every resolution refutation of $\mathcal{C}_n$ must be of length $\geq cs(n-1)$ for a constant c which is independent of n.

As clause logic is undecidable the existence of a sequence $(C_n)_{n\in\mathbb{N}}$ with $HC(C_n) \geq \frac{1}{2}s(n)$ comes hardly by surprise (see Proposition 6.1.1). But the importance of Statman's example can be found in the fact that there are short (formal) refutations of C_n. It is not our purpose to formalize such a short proof in a logic calculus, but rather to give a semiformal representation. Readers familiar with Hilbert-type calculi can easily extract a purely formal version of this proof.

First we show that $\mathbf{T}yx \dot= y(yx)$ is derivable in the equational theory defined by ST. Indeed

$$(\mathbf{SB})((\mathbf{CB})\mathbf{I})yx \;=_{(\mathbf{S})}\; (\mathbf{B}y)((\mathbf{CB})\mathbf{I}y)x \;=_{(\mathbf{C})}\; (\mathbf{B}y)((\mathbf{B}y)\mathbf{I})x$$

$$=_{(\mathbf{B})}\; y((\mathbf{B}y\mathbf{I})x) \;=_{(\mathbf{B})}\; y(y(\mathbf{I}x)) \;=_{(\mathbf{I})}\; y(yx).$$

We now define a sequence of PL-formulas H_i:

$H_1(y) = (\forall x_1)px_1 \dot= p(yx_1)$ and

$H_{m+1}(y) = (\forall x_{m+1})(H_m(x_{m+1}) \to H_m(yx_{m+1}))$ for $m = 1, \ldots, n$.

Using some elementary logical rules we can construct the following derivation:

$$\dfrac{\dfrac{(\forall x)px \dot= p(yx) \qquad (\forall x)px \dot= p(yx)}{px \dot= p(yx) \qquad p(yx) \dot= p(y(yx))}}{\dfrac{px \dot= p(y(yx))}{F(y): \; (\forall x)px \dot= p(y(yx))}}$$

But by the property of $\mathbf{T}$ shown above $F(y)$ is logically equivalent to $(\forall x)px \dot= p(\mathbf{T}yx)$, which is exactly $H_1(\mathbf{T}y)$. So we have derived the implication

$$G(y): \; H_1(y) \to H_1(\mathbf{T}y).$$

From this we can obtain the generalized version $(\forall y)G(y)$. But $H_2(\mathbf{T}) = (\forall y)G(y)$.

More generally, for any A, we can derive the implication

$$(\forall x_{m+1})(A(x_{m+1}) \to A(yx_{m+1})) \to (\forall x_{m+1})(A(x_{m+1}) \to A(y(yx_{m+1})))$$

Again by $y(yx_{m+1}) \dot= \mathbf{T}yx_{m+1}$, by setting $A = H_m$ and by definition of H_{m+1} we obtain

$$H_{m+1}(y) \to H_{m+1}(\mathbf{T}y) \text{ and}$$

$$(\forall x_{m+2})(H_{m+1}(x_{m+2}) \to H_{m+1}(\mathbf{T}x_{m+2})).$$

The last formula is just $H_{m+2}(\mathbf{T})$.

In particular we can derive $H_2(\mathbf{T}), \ldots, H_{n+1}(\mathbf{T})$ by short proofs. It is easy to see that (in a Hilbert-type system or in Natural Deduction) the number of steps required to derive $H_i(\mathbf{T})$ is constant and does not depend on i.

Having derived the $H_i(\mathbf{T})$ for $i = 2, \ldots, n+1$ the final refutation is easily obtained (see Figure 6.4).

$$\frac{H_n(\mathbf{T})\quad \dfrac{H_{n+1}(\mathbf{T}):\ (\forall x_{n+1})(H_n(x_{n+1})\to H_n(\mathbf{T}x_{n+1}))}{H_n(\mathbf{T})\to H_n(\mathbf{T}_2)}}{H_n(\mathbf{T}_2):\ (\forall x_n)(H_{n-1}(x_n)\to H_{n-1}(\mathbf{T}_2x_n))}$$

$$\frac{(\forall x_1)px_1\doteq p(qx_1)\quad \dfrac{\dfrac{H_2(\mathbf{T}_n)}{(\forall x_1)px_1\doteq p(qx_1)\to (\forall x_1)px_1\doteq p((\mathbf{T}_nq)x_1)}}{(\forall x_1)px_1\doteq p((\mathbf{T}_nq)x_1)}}{\dfrac{\neg pq\doteq p((\mathbf{T}_nq)q)\qquad\qquad pq\doteq p((\mathbf{T}_nq)q)}{\square}}$$

Fig. 6.4. A refutation of Statman's example

Note that the derivation in Figure 6.4 is essentially based on the substitution rule and on modus ponens.

The essential difference between a resolution refutation and a refutation such as that in Example 6.2.1 (Figure 6.4) consists in the type of formulas used in the deduction. Particularly the formulas $H_m(y)$ contain quantifiers and do not appear in the syntactical representation of the problems $\mathcal{C}_n$; they come from "outside", as it were. By applying resolution on the sets $\mathcal{C}_n$ we can only derive clauses which are defined over the signature of $\mathcal{C}_n$. Not only is it impossible to introduce and use (new) quantificational formulas, but also we are without means to introduce new predicate and function symbols. To obtain short proofs we are forced to somehow introduce new "concepts" into the deduction machinery. In fact we can obtain short proofs by maintaining the clause form and adding extension rules for predicate and function symbols. The introduction of new symbols gives us the means to encode more complex formulas and to use them in deduction. Moreover, extension can also be used in normalization of formulas (note that skolemization is a method of functional extension). Before describing the power of extension principles we first describe their role in formula normalization and then in inference.

6.2.1 Structural Normalization

In constructing propositional normal forms (e.g., conjunctive normal form) we basically have two options:

a) structural transformation and
b) nonstructural (standard) transformation.

In contrast to skolemization (where $\exists$-quantifiers are eliminated) we don't need extension at all to transform formulas in skolemized form into conjunctive normal form (CNF). Clearly it is always possible to transform a quantifier-free formula A into a logically equivalent conjunctive normal form $\text{cnf}(A)$. As logical equivalence is stronger than sat-equivalence and extensions merely preserve sat-equivalence, we might think about preferring nonstructural transformation. However, structural transformation, by keeping more information about the original formula, behaves much better with respect to proof complexity (we will come to this point again). For the sake of simplicity we focus on negation normal forms (NNF, see Section 2.2) and define a transformation γ_{struc} which transforms NNFs into clause form.

In order to define structural CNFs we first have to create names for subformulas of an NNF A. By $\text{subf}(A)$ we denote the set of all subformulas occurring in an NNF A (note that $\text{subf}(A)$ itself consists of formulas in negation normal form).

Definition 6.2.1. *Let A be a formula in NNF. We first define a mapping $\pi_A\colon \text{subf}(A) \to \text{PS}$ with the following properties:*

(1) If $|V(B)| = n$ then the arity of $\pi_A(B)$ is n for $n \geq 1$; if $|V(B)| = 0$ (B is ground) then the arity of $\pi_A(B)$ is 1.
(2) If $B_1, B_2 \in \text{subf}(A)$ and $B_1 \neq B_2$ then $\pi_A(B_1) \neq \pi_A(B_2)$.
(3) For all $B \in \text{subf}(A)$ $\pi_A(B)$ does not occur in A.

π_A can be used to assign (new) literals to the elements of $\text{subf}(A)$. We denote the corresponding mapping by ω_A. ω_A must have the following properties:

(a) If $B \in \text{subf}(A)$ and B is a literal then $\omega_A(B) = B$.
(b) If $B \in \text{subf}(A)$, B is not a literal and $V(B) = \{v_1, \ldots, v_n\}$ (for $n \geq 1$) then $\omega_A(B) = \pi_A(B)(v_1, \ldots, v_n)$.
(c) If $B \in \text{subf}(A)$, B is not a literal and B is ground then $\omega_A(B) = \pi_A(B)(c)$ for some constant symbol c (c need not be new).

We give names to all subformulas of A and literals are named by themselves. Still we have to express in the object language that the (new) literals indeed correspond to some subformulas of A. Note that NNFs are quantifier-free formulas, but semantically they denote universally closed forms. Instead of writing down the whole $\forall$-prefix of a formula A we use the more comfortable notation $\forall(A)$ for the universal closure of A.

Definition 6.2.2. *Let A be a formula in NNF. We define a mapping δ_A assigning universally closed equivalences to the subformulas of A:*

(a) $\delta_A(B) = \text{TAUT}$ for some tautological clause TAUT if $B \in \text{subf}(A)$ and B is a literal.
(b) $\delta_A(B) = \forall(\omega_A(B) \leftrightarrow (\omega_A(B_1) \wedge \omega_A(B_2)))$ if $B \in \text{subf}(A)$ and $B = B_1 \wedge B_2$.
(c) $\delta_A(B) = \forall(\omega_A(B) \leftrightarrow (\omega_A(B_1) \vee \omega_A(B_2)))$ if $B \in \text{subf}(A)$ and $B = B_1 \vee B_2$.

Every $\delta_A(B)$ is called definitional equivalence *induced by A. The set of all definitional equivalences induced by A is denoted by $\mathcal{E}_A$. The formula*

$$\forall(\omega_A(A)) \wedge \bigwedge\{E | E \in \mathcal{E}_A\}$$

is called extension formula *of A and is denoted by $E_A(A)$ (if no confusion can arise we write $E(A)$).*

Example 6.2.2. Let A be the NNF $P(x,y) \vee (Q(x) \wedge \neg R(y))$
and $\pi_A(A) = P_1$, $\pi_A(Q(x) \wedge \neg R(y)) = P_2$.

Then $\omega_A(A) = P_1(x,y)$ and $\omega_A(Q(x) \wedge \neg R(y)) = P_2(x,y)$.

Moreover we have
$\omega_A(P(x,y)) = P(x,y)$, $\omega_A(Q(x)) = Q(x)$ and $\omega_A(\neg R(y)) = \neg R(y)$.

We define

$$\delta_A(P(x,y)) = \delta_A(Q(x)) = \delta_A(\neg R(y)) = R(c) \vee \neg R(c)$$

(note that any tautology would do the job). The other definitional equivalences are:

$$\delta_A(P(x,y) \vee (Q(x) \wedge \neg R(y))) = (\forall x)(\forall y)(P_1(x,y) \leftrightarrow (P(x,y) \vee P_2(x,y))),$$

$$\delta_A(Q(x) \wedge \neg R(y)) = (\forall x)(\forall y)(P_2(x,y) \leftrightarrow (Q(x) \wedge \neg R(y))).$$

Under deletion of tautologies (which is always possible in a conjunction of formulas) the extension formula of A is

$$(\forall x)(\forall y)P_1(x,y) \wedge (\forall x)(\forall y)(P_1(x,y) \leftrightarrow (P(x,y) \vee P_2(x,y)))\wedge$$
$$(\forall x)(\forall y)(P_2(x,y) \leftrightarrow (Q(x) \wedge \neg R(y))).$$

Extension formulas are the cornerstones of the structural normal form transformation. However we already see that (in general) A is not logically equivalent to its extension formula $E(A)$. Fortunately logical equivalence is not really necessary (it gets lost by skolemization anyway); it suffices to preserve sat-equivalence.

Lemma 6.2.1. *Let A be a formula in NNF and $E_A(A)$ be the extension formula of A. Then $\forall(A)$ is sat-equivalent to $E_A(A)$.*

Proof. For every formula F in predicate logic we write F^0 for its open form, i.e., F^0 is F without quantifiers.
First we prove that $E(A)^0 \to A$ is a valid formula. Then clearly the formula $\forall(E(A)^0 \to A)$ is valid too. By distributing the universal quantifiers we eventually obtain the validity of $E(A) \to \forall(A)$. As a consequence the satisfiability of $E(A)$ implies that of $\forall(A)$.

We prove the validity of $E(A)^0 \to A$ by induction on the complexity $comp$ of the formula A, where $comp(A)$ is the number of connectives from $\{\wedge, \vee\}$ occurring in A.

$comp(A) = 0$:

Then A is a literal and (by definition) $E(A) = \forall(A)$. Thus $E(A)^0 \to A = A \to A$ and the validity is obvious.

(IH) Assume that $E(A)^0 \to A$ is valid for all A with $comp(A) \leq k$.

Let A be an NNF-formula with $comp(A) = k + 1$.

Then either $A = A_1 \vee A_2$ or $A = A_1 \wedge A_2$ for two formulas A_1, A_2 in NNF such that $comp(A_1), comp(A_2) \leq k$.

case a) $A = A_1 \wedge A_2$:

By the definition of the extension formula we obtain:

$$E(A_1)^0 = P_1(\bar{y}_1) \wedge \mathcal{E}^0_{A_1},$$
$$E(A_2)^0 = P_2(\bar{y}_2) \wedge \mathcal{E}^0_{A_2},$$
$$E(A)^0 = P(\bar{x}) \wedge \mathcal{E}^0_A \text{ and}$$

$$(*) \; \mathcal{E}^0_A = \mathcal{E}^0_{A_1} \wedge \mathcal{E}^0_{A_2} \wedge (P(\bar{x}) \leftrightarrow (P_1(\bar{y}_1) \wedge P_2(\bar{y}_2))).$$

By (IH) the formulas

$$P_1(\bar{y}_1) \wedge \mathcal{E}^0_{A_1} \to A_1 \text{ and}$$
$$P_2(\bar{y}_2) \wedge \mathcal{E}^0_{A_2} \to A_2$$

are valid.

By the definition of $E(A)$ and $(*)$ we obtain that

$$E(A)^0 \to P_1(\bar{y}_1) \wedge \mathcal{E}^0_{A_1}, \; E(A)^0 \to P_2(\bar{y}_2) \wedge \mathcal{E}^0_{A_2}$$

are both valid.

By transitivity of "$\to$" we obtain the validity of the formulas $E(A)^0 \to A_1$, $E(A)^0 \to A_2$ and eventually that of the formula $E(A)^0 \to (A_1 \wedge A_2)$.

case b) $A = A_1 \vee A_2$:

Let $E(A_1)$, $E(A_2)$, $E(A)$ be as in case a) with the exception that

$$(+) \; \mathcal{E}^0_A = \mathcal{E}^0_{A_1} \wedge \mathcal{E}^0_{A_2} \wedge (P(\bar{x}) \leftrightarrow (P_1(\bar{y}_1) \vee P_2(\bar{y}_2))).$$

Again by (IH) we obtain the validity of

$$P(\bar{y}_1) \wedge \mathcal{E}^0_{A_1} \to A_1,$$
$$P(\bar{y}_2) \wedge \mathcal{E}^0_{A_2} \to A_2.$$

This implies the validity of the formulas

$$P(\bar{y}_1) \wedge \mathcal{E}^0_{A_1} \wedge \mathcal{E}^0_{A_2} \to A_1 \vee A_2,$$
$$P(\bar{y}_2) \wedge \mathcal{E}^0_{A_1} \wedge \mathcal{E}^0_{A_2} \to A_1 \vee A_2.$$

Using the valid propositional schema

$$((\mathcal{A} \wedge \mathcal{B}) \to \mathcal{C}) \wedge ((\mathcal{F} \wedge \mathcal{B}) \to \mathcal{C}) \;\to\; (((\mathcal{A} \vee \mathcal{F}) \wedge \mathcal{B}) \to \mathcal{C})$$

we obtain the validity of the formula

$$(P_1(\bar{y}_1) \vee P_2(\bar{y}_2)) \wedge \mathcal{E}^0_{A_1} \wedge \mathcal{E}^0_{A_2} \;\to\; A_1 \vee A_2.$$

But by definition of $E(A)$ and by (+) the formula

$$E(A)^0 \to \mathcal{E}^0_{A_1} \wedge \mathcal{E}^0_{A_2} \wedge (P_1(\bar{y}_1) \vee P_2(\bar{y}_2))$$

is valid. By transitivity of $\to$ we eventually obtain the validity of $E(A)^0 \to A_1 \vee A_2$. This concludes the first part of the proof.

We show now that the satisfiability of $\forall(A)$ implies that of $E(A)$. This direction is a little more difficult than the first one, due to the fact that $\forall(A) \to E(A)$ need not be valid. Here we can only prove satisfiability of $E(A)$ by extending models of $\forall(A)$. To simplify the notation we write $E(B)$ instead of $E_A(B)$ for $B \in \mathrm{subf}(A)$. As in the first part of the proof we analyze the open formulas A and $E(A)^0$. We reduce the proof to

(I) Suppose that the interpretations $\mathcal{M}_J : (D, \Phi, J)$ are models of A (for some environments J). Then there exists a Φ' such that $\Phi \subseteq \Phi'$ and all interpretations $\mathcal{M}'_J : (D, \Phi', J)$ are models of $E(A)^0$.

Suppose that (I) has already been proved. If $\forall(A)$ is satisfiable then there exist models $\mathcal{M}_I : (D, \Phi, I)$ of A such that $v_{\mathcal{M}_I}(A) = \mathbf{T}$ for all environments I (note that $\forall(A)$ is closed). By (I) every model $\mathcal{M}'_I : (D, \Phi', I)$ is a model of $E(A)^0$, i.e.,

$$v_{\mathcal{M}'_I}(E(A)^0) = \mathbf{T} \text{ for all } I.$$

By the definition of the extension formula we have $V(E(A)^0) = V(A)$ and so $v_{\mathcal{M}'_I}(\forall(E(A)^0)) = \mathbf{T}$, i.e., $\mathcal{M}'_I$ (for any environment I) is a model of $\forall(E(A)^0))$. But $E(A)$ is logically equivalent to $\forall(E(A)^0))$ and therefore $\mathcal{M}'_I$ is a model of $E(A)$.

As in the first part of the proof we show (I) by induction on $comp(A)$.

$comp(A) = 0$:
$E(A)^0 = A$. Thus if $v_{\mathcal{M}}(A) = \mathbf{T}$ then $v_{\mathcal{M}}(E(A)^0) = T$ and $\Phi' = \Phi$.

(IH) Assume that (I) holds for all A with $comp(A) \leq k$.

If $comp(A) = k + 1$ then $A = A_1 \land A_2$ or $A = A_1 \lor A_2$ for A_1, A_2 such that $comp(A_1)$ and $comp(A_2)$ are both $\leq k$. As in the first part of the proof we use the definitions $E(A_1)^0, E(A_2)^0$, and $E(A)^0$:

$$E(A_1)^0 = P_1(\bar{y}_1) \land \mathcal{E}^0_{A_1},$$
$$E(A_2)^0 = P_2(\bar{y}_2) \land \mathcal{E}^0_{A_2},$$
$$E(A)^0 = P(\bar{x}) \land \mathcal{E}^0_A, \quad \text{and}$$

$$(*) \quad \mathcal{E}^0_A = \mathcal{E}^0_{A_1} \land \mathcal{E}^0_{A_2} \land (P(\bar{x}) \leftrightarrow (P_1(\bar{y}_1) \circ P_2(\bar{y}_2))).$$

for $\circ \in \{\land, \lor\}$.

case a) $A = A_1 \land A_2$.

Suppose that $v_{\mathcal{M}}(A) = \mathbf{T}$. Let $\mathcal{M}_1$ be the restriction of $\mathcal{M}$ to $\Sigma(A_1)$ and $\mathcal{M}_2$ be the restriction of $\mathcal{M}$ to $\Sigma(A_2)$ (remember that $\Sigma(F)$ denotes the signature of F). Then clearly

$$v_{\mathcal{M}_1}(A_1) = v_{\mathcal{M}_2}(A_2) = \mathbf{T}.$$

By (IH) there exist extensions $\mathcal{N}_1$ of $\mathcal{M}_1$ and $\mathcal{N}_2$ of $\mathcal{M}_2$ such that

$$\mathcal{N}_i = (D, \Phi'_i, I) \text{ for } \mathcal{M}_i = (D, \Phi_i, I) \text{ and } \Phi_i \subseteq \Phi'_i \text{ for } i = 1, 2 \text{ }^{\bullet}$$

Note that the evaluation functions Φ'_1, Φ'_2 don't depend on the environment I.

By Exercise 6.2.1 $\mathcal{N}_1$ and $\mathcal{N}_2$ must coincide on $\Sigma(E(A_1))^0) \cap \Sigma(E(A_2)^0)$. Therefore there exists a common extension $\mathcal{N}$ of $\mathcal{N}_1$ and $\mathcal{N}_2$ such that

$$\mathcal{N} = (D, \Psi, I) \text{ and } v_{\mathcal{N}}(E(A_1)^0) = v_{\mathcal{N}}(E(A_2)^0) = \mathbf{T}.$$

In particular we obtain

$$v_{\mathcal{N}}(P_1(\bar{y}_1)) = v_{\mathcal{N}}(P_2(\bar{y}_2)) = v_{\mathcal{N}}(\mathcal{E}^0_{A_1}) = v_{\mathcal{N}}(\mathcal{E}^0_{A_2}) = \mathbf{T}.$$

By the definition of the extension formula $E_A(A)$ the predicate symbol P neither occurs in $E(A_1)$ nor in $E(A_2)$. So we may extend $\mathcal{N}$ to $\mathcal{P} : (D, \Xi, I)$ such that $\Xi(P)(\bar{d}) = \mathbf{T}$ for all $\bar{d} \in D^l$ where l is the arity of P. Then clearly

$$v_{\mathcal{P}}(P(\bar{x}) \leftrightarrow (P_1(\bar{y}_1) \land P_2(\bar{y}_2))) = \mathbf{T}, \quad v_{\mathcal{P}}(P(\bar{x})) = \mathbf{T}$$

and

$$v_{\mathcal{P}}(\mathcal{E}^0_{A_1}) = v_{\mathcal{P}}(\mathcal{E}^0_{A_2}) = \mathbf{T}.$$

Moreover $\mathcal{P}$ is an extension of $\mathcal{M}$ and thus fulfills all requirements of $\mathcal{M}'$ in (I).

case b) $A = A_1 \lor A_2$

Let $\mathcal{M} = (D, \Phi, I)$ and $v_{\mathcal{M}}(A) = \mathbf{T}$.

Then either $v_{\mathcal{M}_1}(A_1) = \mathbf{T}$ or $v_{\mathcal{M}_2}(A_2) = \mathbf{T}$ for the restrictions $\mathcal{M}_i$ to $\Sigma(A_i)$. It suffices to investigate the case $v_{\mathcal{M}_1}(A_1) = \mathbf{T}$ (the other case is symmetric). But note that the truth values may "oscillate" under the change of the environments. It is only necessary that for every model (D, Φ, J) of A either the restriction (D, Φ_1, J) verifies A_1 or the restriction (D, Φ_2, J) verifies A_2. For different environments J the roles of A_1 and A_2 may change.

By (IH) there exists an extension $\mathcal{N}_1$ of $\mathcal{M}_1$ such that $v_{\mathcal{N}_1}(E(A_1)^0) = \mathbf{T}$. In particular we obtain

$$v_{\mathcal{N}_1}(P_1(\bar{y}_1)) = v_{\mathcal{N}_1}(\mathcal{E}_{A_1}^0) = \mathbf{T}.$$

$\mathcal{N}_1$ can be extended to an interpretation $\mathcal{N}$ such that $v_{\mathcal{N}}(\mathcal{E}_{A_2}^0) = \mathbf{T}$. Note that $\mathcal{E}_{A_2}^0$ consists of equivalences only and (for the specific environment I!) we can assign the appropriate truth values to the extension predicates. Again these assignments may be different for different environments I. Particularly $\mathcal{N}$ may falsify $P_2(\bar{y}_2)$ and the formula $E(A_2)^0$. We then extend $\mathcal{N}$ further to $\mathcal{P} : (D, \Xi, I)$ such that $\Xi(P)(\bar{d}) = \mathbf{T}$ for all $\bar{d} \in D^l$.

By $v_{\mathcal{P}}(P_1(\bar{y}_1)) = \mathbf{T}$ we also obtain

$$v_{\mathcal{P}}(P(\bar{x}) \leftrightarrow (P_1(\bar{y}_1) \vee P_2(\bar{y}_2))) = \mathbf{T}.$$

Therefore $v_{\mathcal{P}}(E) = \mathbf{T}$ for all conjuncts appearing in $\mathcal{E}(A)^0$ and therefore $v_{\mathcal{P}}(E(A)^0) = \mathbf{T}$. Again $\mathcal{P}$ is of the form $\mathcal{M}'$ in statement (I).

Finally, we must be sure to really obtain a single interpretation Ξ of predicate and function symbols which is independent of specific environments. Let U be the set of environments J such that $v_{\mathcal{M}_1}(A_1) = \mathbf{T}$ and V be the complement of U. By (I) we obtain two evaluation functions Φ_1, Φ_2 such that the (D, Φ_1, I) are models of $E(A_1)^0$ for $I \in U$ and the (D, Φ_2, J) are models of $E(A_2)^0$ for $J \in V$. By $U \cap V = \emptyset$ the construction above is consistent and we obtain interpretations $\mathcal{N}_I : (D, \Psi, I)$ such that

$$v_{\mathcal{N}_I}(\mathcal{E}_{A_1}^0 \wedge \mathcal{E}_{A_2}^0) = \mathbf{T} \text{ for all } I.$$

The assignments of Ξ on P are clearly independent of environments and all interpretations $\mathcal{P}_I : (D, \Xi, I)$ verify $E(A)^0$. This concludes the proof of case $k + 1$.   ◊

The definitional equivalences are simple formulas (they contain at most three atom formulas) which can easily be transformed into CNF and into clause form. We define

$\text{clf}(\delta_A(B)) = \emptyset$ if $\delta_A(B) = \text{TAUT}$ for some tautology TAUT.

$\text{clf}(\delta_A(B)) = \{\omega_A(B)^d \vee \omega_A(B_1), \ \omega_A(B)^d \vee \omega_A(B_2),$

$\qquad \omega_A(B_1)^d \vee \omega(B_2)^d \vee \omega_A(B)\}$ if $B = B_1 \wedge B_2$,

$\text{clf}(\delta_A(B)) = \{\omega_A(B)^d \vee \omega_A(B_1) \vee \omega_A(B_2), \ \omega_A(B_1)^d \vee \omega_A(B),$

$\omega_A(B_2)^d \vee \omega_A(B)\}$ if $B = B_1 \vee B_2$.

as the clause form of a definitional equivalence.

Definition 6.2.3. *Let A be a formula in negation normal form. Then the structural clause form* $\mathrm{cl}_{\mathrm{struc}}$ *is defined by*

$$\mathrm{cl}_{\mathrm{struc}}(A) = \{\omega_A(A)\} \cup \bigcup\{\mathrm{clf}(\delta_A(B)) \mid B \in \mathrm{subf}(A)\}.$$

Example 6.2.3. Let $A = P(x,y) \vee (Q(x) \wedge \neg R(y))$.
In Example 6.2.2 we have computed the definitional equivalences

$$(\forall x)(\forall y)(P_1(x,y) \leftrightarrow (P(x,y) \vee P_2(x,y))),$$

$$(\forall x)(\forall y)(P_2(x,y) \leftrightarrow (Q(x) \wedge \neg R(y))),$$

and TAUT for the literals in A. So we obtain

$\mathrm{cl}_{\mathrm{struc}}(A) = \{P_1(x,y),\ \neg P_1(x,y) \vee P(x,y) \vee P_2(x,y),\ \neg P(x,y) \vee P_1(x,y),$
$\neg P_2(x,y) \vee P_1(x,y),\ \neg P_2(x,y) \vee Q(x),\ \neg P_2(x,y) \vee \neg R(y),$
$\neg Q(x) \vee R(y) \vee P_2(x,y)\}$.

Theorem 6.2.1. *Let A be a formula in negation normal form. Then $\forall(A)$ is sat-equivalent to $\mathrm{cl}_{\mathrm{struc}}(A)$.*

Proof. By Lemma 6.2.1, $\forall(A)$ is sat-equivalent to its extension formula $E_A(A)$. But $\mathrm{cl}_{\mathrm{struc}}(A)$ is a clause form of a conjunctive normal form of $E_A(A)$ which is based on distributivity and thus preserves logical equivalence. Therefore $\forall(A) \sim_{sat} \mathrm{cl}_{\mathrm{struc}}(A)$. ◊

In Example 6.2.3 the structural clause form of A consists of seven clauses, while there are only two clauses in the standard clause form. Thus the example might suggest that structural clause forms are always more complicated. For the use in practice there are several techniques to optimize structural transformation with respect to the number of clauses obtained (we just mention [BT90]). In the sense of worst-case complexity structural normalization is clearly superior to the standard normalization; the former is polynomial (see Exercise 6.2.2), the latter exponential. For example let

$$A_n :\ (L_1^1 \wedge L_2^1) \vee \ldots \vee (L_1^n \wedge L_2^n)$$

be a sequence of disjunctive normal forms. The standard clause form of A_n contains $n2^n$ literals and thus is exponential in the length of A_n. If one focuses on proof complexity (instead of the length of representation) the difference can even be much stronger. The following theorem is based on a slightly different structural transformation from that we have given here (it starts from full PL-syntax, not from NNF).

Theorem 6.2.2. *There exists a sequence of unsatisfiable formulas $(F_n)_{n \in \mathbb{N}}$ such that for the structural clause forms $\mathrm{cl}_{\mathrm{struc}}(F_n)$ there are resolution refutations of length $\leq 2^{2^{2^{dn}}}$ for some constant d. For the nonstructural clause forms $\mathcal{D}_n$ every resolution refutation has a length $> s(n-1)$ for $s(0) = 1$, $s(n+1) = 2^{s(n)}$ for $n \in \mathbb{N}$.*

Proof. We omit the proof, but refer to [BFL94].   ◇

Note that the bound on the resolution refutations of the $\mathcal{D}_n$ in Theorem 6.2.2 is the same as for Statman's example (Example 6.2.1). Indeed the formulas F_n above are constructed via Statman's example by adding formulas containing the cut formulas of short proofs in the Gentzen calculus LK. Roughly speaking, the nonstructural transformation destroys the structure of the cut formulas and thus prevents short resolution proofs. But by extending the signature via structural normalization we can keep the information contained in the formulas even under transformation to conjunctive normal form.

While extension is necessary for the elimination of quantifiers in skolemization, it can be avoided in the construction of propositional normal forms. However, at least from the theoretical point of view, there is no reason to avoid extension in propositional normalization. The method of structural transformation into clause form is also applicable to nonclassical logics [Min93].

6.2.2 Functional Extension

The second way to use extension is to add extension rules to logical inference systems. A detailed discussion of such extended inference systems can be found in E. Eder's book [Ede92]; the basic idea, formulated by Tseitin for propositional logic, can be found in [Tsei83]. In Eder's extension method new predicate and function symbols have to be introduced, where the introduction is based on definitional equivalences as in Definition 6.2.2. In a slightly different and more restrictive way a functional extension rule is defined in [BL92]; it is based on shifting of quantifiers and reskolemization. Introducing new function symbols in a resolution calculus is somehow against the philosophy of resolution, where minimality of substitutions and terms is the key characteristic. On the other hand, clinging to minimality prevents formulation and use of substantial lemmas. So, in using extension methods, we must carefully preserve the computational power of resolution and its small search space (otherwise the extension calculus would not be a "computational" one). The function introduction rule defined in [BL92] preserves clause forms and the principle of most general unification and is restricted by the pattern of variable occurrences in the clauses.

Example 6.2.4 (Egly 1994).

$$\mathcal{C}_n = \{C_1,\ C_2,\ C_3^n\} \text{ for}$$
$$C_1 = P(a, g(a, b)),\ C_2 = \neg P(x, y) \vee P(f(x), y) \vee P(f(x), g(x, y)),$$
$$C_3^n = \neg P(f^{2^n}(a), z).$$

Every R-refutation of $\mathcal{C}_n$ is of length exponential in n [Egl94]. Much shorter proofs (of length linear in n) can be obtained if C_2 is subjected to quantifier shifting and subsequent (re)skolemization. By this operation we eliminate the variable y in the positive literals of C_2. First let us consider the formula F_2 representing C_2:

$$F_2 = (\forall x)(\forall y)(\neg P(x, y) \vee P(f(x), y) \vee P(f(x), g(x, y))).$$

Applying the valid schema

$$(S)\ \ (\forall y)(A(y) \vee B(y)) \to (\forall y)A(y) \vee (\exists y)B(y)$$

to F_2 we obtain the formula

$$F : (\forall x)((\forall y)\neg P(x, y) \vee (\exists y)(P(f(x), y) \vee P(f(x), g(x, y)))).$$

By skolemizing F we obtain the clause

$$C : \neg P(x, y) \vee P(f(x), m(x)) \vee P(f(x), g(x, m(x))).$$

The short refutation is then based on iterated self-resolution of C combined with iterated factoring. We illustrate the first step:
Resolving C with a renamed copy of itself (on the second literal) gives the clause

$$D : \neg P(x, y) \vee P(f(x), g(x, m(x))) \vee P(f^2(x), m(f(x))) \vee P(f^2(x), g(f(x), m(f(x)))).$$

By resolving D (on its second literal) with a renamed copy of C we get the clause:

$$E : \ \neg P(x, y) \vee P(f^2(x), m(f(x))) \vee P(f^2(x), g(f(x), m(f(x)))) \vee P(f^2(x), m(f(x)$$
$$P(f^2(x), g(f(x), m(f(x)))).$$

By factoring (even p-reduction suffices) in E we obtain

$$\neg P(x, y) \vee P(f^2(x), m(f(x))) \vee P(f^2(x), g(f(x), m(f(x)))).$$

Eventually (after $5n$ steps) the clause

$$C' : \neg P(x, y) \vee P(f^{2^n}(x), m(f^{2^n - 1}(x))) \vee P(f^{2^n}(x), g(f^{2^n - 1}(x), m(f^{2^n - 1}(x))))$$

is derived. The two positive literals in C' can be resolved with C_3 which gives us $C'' : \neg P(a, y)$. C'' and C_1 resolve to $\square$. The length of the whole refutation is linear in n. In trying to simulate the proof above, using C_2 instead of C, we obtain clauses of exponential length that cannot be factored. Thus the introduction of the new function symbol via the schema (S) made additional factoring possible and thus led to a much shorter refutation.

Of course, C is not R-derivable from C_n. We don't even have the validity of $F(C_n) \rightarrow F(\{C\})$, which could be called the strong correctness of inferring C. But still $C_n \cup \{C\} \sim_{sat} C_n$ holds, which guarantees refutational correctness. Note that the principle of logical equivalence has already been given up in skolemization.

There are many variants of function introduction depending on the type of inference rule (shifting of quantifiers, introduction of $\exists$-quantifiers) applied to clauses. For theoretical purposes it suffices to restrict the inference to the innermost quantifier.

Definition 6.2.4. *Let C be a set of clauses and $C \in \mathcal{C}$. Suppose that*

$$A: \quad (\forall \bar{x})(\forall y)(F_1 \vee F_2)$$

is a representation of the clause C in predicate logic where the range of the $\forall$-quantifiers in A is minimized (under the rules of quantifier shifting). Then the (skolemized) clause form of the formula

$$F(\mathcal{C}) \wedge (\forall \bar{x})((Qy)F_1 \vee (Q^d y)F_2)$$

(obtained via the transformation defined in Section 2.2) for $Q \in \{\forall, \exists\}$ is called a 1-F-extension (or simply F-extension) of $\mathcal{C}$.

Note that there are many different ways of representing a clause in the form A above and thus there are different F-extensions based on C. If $A = C_1 \vee C_2$ such that $V(C_1) \cap V(C_2) = \emptyset$ then there exists no F-extension based on this representation of C. Moreover F-extension is not just a rule applied to single clauses but has to be applied to the whole set of clauses $\mathcal{C}$. This is necessary because the newly introduced function symbol (generated by skolemization) may not occur in $\mathcal{C}$. Indeed the F-extension rule would be incorrect if $(\forall \bar{x})((Qy)F_1 \vee (Q^d y)F_2)$ were skolemized relative to itself (without respect to the other clauses in $\mathcal{C}$).

By shifting k quantifiers at once we obtain the principle of k-F-extension and by shifting until A is of the form $G_1 \vee G_2$ that of SF-extension (splitting F-extension). Although a single step of a k-F-extension cannot be represented by k 1-F-extensions, 1-F-extension polynomially simulates k-F-extension [Egl94]. Thus for the purpose of complexity analysis we may focus on F-extension only.

Suppose that Q in Definition 6.2.4 is $\forall$; then the F-extension is of the form

$$\mathcal{C} \cup \{C_1 \vee C_2\{y \leftarrow f(x_1, \ldots, x_k)\}\}.$$

For $Q = \exists$ we obtain

$$\mathcal{C} \cup \{C_1\{y \leftarrow f(x_1, \ldots, x_k)\} \vee C_2\}.$$

Note that (by definition of the form A in Definition 6.2.4)

$$y \in V(C_1) \cap V(C_2), \text{ but } y \notin V(C_1) \cap V(C_2\{y \leftarrow f(x_1, \ldots, x_k)\}).$$

The last property indicates that F-extension somehow works as "variable-decomposer" on clauses. In fact the connection of variables within a clause codes much of the logical structure of a problem. Note that without any variable connections we obtain sets of decomposed clauses (i.e., $V(L) \cap V(M) = \emptyset$ for any two different literals L, M in C); the corresponding class of clause sets is decidable (it is reducible to the Herbrand class via splitting)!

Because F-extension changes the term universe we cannot expect to preserve strong correctness (i.e. all clauses derivable from $\mathcal{C}$ logically follow from $\mathcal{C}$). But if $\mathcal{C} \cup \{C\}$ is an F-extension of $\mathcal{C}$ then $\mathcal{C} \cup \{C\} \sim_{sat} \mathcal{C}$ and (at least) we obtain refutational correctness.

Definition 6.2.5 (FR-deduction). *Let $\mathcal{C}$ be a set of clauses and C be a clause. A sequence $C_1, \ldots, C_n$ is called FR-deduction of C from $\mathcal{C}$ if the following conditions hold:*

(1) $C_n = C$.
(2) For all $i = 1, \ldots, n$:
   a) C_i is a resolvent of C_j, C_k for $j, k < i$ or

   b) $\mathcal{C} \cup_{j=1}^{i} \{C_j\}$ is an F-extension of $\mathcal{C} \cup_{j=1}^{i-1} \{C_j\}$.

The behavior of FR-deductions is different from that of R-deductions due to the fact that skolemization is a global rule. Thus if Γ and Δ are both FR-deductions from $\mathcal{C}$, Γ, Δ need not be an FR-deduction.

Example 6.2.5. $\mathcal{C} = \{P(x) \vee P(f(x)), \neg P(y)\}$.

$\Gamma : P(x) \vee P(f(c)), \neg P(y), \square$ is an FR-refutation of $\mathcal{C}$. Note that $\mathcal{C} \cup \{P(x) \vee P(f(c))\}$ is an F-extension of $\mathcal{C}$ and $P(x) \vee P(f(c))$ has $P(f(c))$ as factor. On the other hand there exists no nontrivial factor of the clause $P(x) \vee P(f(x))$. Thus one of the shortest R-refutations of $\mathcal{C}$ is

$$P(x) \vee P(f(x)), \ \neg P(y), \ P(x), \ \square.$$

The Herbrand complexity of $\mathcal{C}$ is 3, but

$$\mathrm{HC}(\mathcal{C} \cup \{P(x) \vee P(f(c))\}) = 2.$$

Even in this very simple case we see that F-extension may lead to proofs of a different type; we can make factoring possible, where in pure resolution it cannot be applied.

Theorem 6.2.3 (soundness and completeness of FR-deduction).
Let C be a set of clauses. C is unsatisfiable iff there exists an FR-refutation of C.

Proof. Let us assume that C is unsatisfiable.

If Γ is an R-refutation of C then Γ is also an FR-refutation. Thus (by the completeness of resolution shown in Theorem 2.7.2) there exists an FR-refutation of C.

To prove the other direction, let Γ be an FR-refutation of C. Then Γ is of the form $C_1, \ldots, C_n$ and $C_n = \square$. If C_i is a resolvent of C_j, C_k for $j, k < i$ and $C \cup \bigcup_{j=0}^{i-1} \{C_j\}$ is satisfiable then $C \cup \bigcup_{j=0}^{i} \{C_j\}$ is satisfiable too. On the other hand if $C \cup \bigcup_{j=0}^{i} \{C_j\}$ is satisfiable then $C \cup \bigcup_{j=0}^{i-1} \{C_j\}$ is also satisfiable. Therefore (in case of a resolvent) we obtain

$$C \cup \bigcup_{j=0}^{i-1} \{C_j\} \sim_{sat} C \cup \bigcup_{j=0}^{i} \{C_j\}$$

Now let us assume that $C \cup \bigcup_{j=0}^{i} \{C_j\}$ is an F-extension of $C \cup \bigcup_{j=0}^{i-1} \{C_j\}$. Then C_i is the clause form of a formula

$$B : (\forall \bar{x})((Qy)F_1 \vee (Q^d y)F_2),$$

where $A : (\forall \bar{x})(\forall y)(F_1 \vee F_2)$ is a minimal representation of a clause C_j for $j < i$ or of a $D \in C$ (again "minimal" means the range of quantifiers). But the formula

$$(\forall \bar{x})(\forall y)(F_1 \vee F_2) \rightarrow (\forall \bar{x})((Qy)F_1 \vee (Q^d y)F_2)$$

is logically valid and therefore

$$F(C \cup \bigcup_{j=0}^{i-1} \{C_j\}) \sim F(C \cup \bigcup_{j=0}^{i-1} \{C_j\}) \wedge B.$$

But then we also have

$$C \cup \bigcup_{j=0}^{i-1} \{C_j\} \sim_{sat} C \cup \bigcup_{j=0}^{i} \{C_j\}.$$

Thus a straightforward induction argument yields

$$(*) \quad C \sim_{sat} C \cup \bigcup_{j=0}^{i} \{C_j\} \text{ for } i = 1, \ldots, n$$

But $C_n = \square$ and therefore $C \cup \bigcup_{j=0}^{n} \{C_j\}$ is unsatisfiable. By $(*)$ C must be unsatisfiable too. $\diamond$

Although F-extension models a simple quantifier shifting rule, its effect can be very strong (in fact stronger than in Example 6.2.4).

Theorem 6.2.4 ([BL92]). *There exists a sequence of sets of clauses $(C_n)_{n \in \mathbb{N}}$ such that $C_n = C \cup \{\neg P_n\}$ (where the P_n's are atoms for $n \in \mathbb{N}$) and constants c, d such that*

(1) $l(\Gamma) \geq cs(n-1)$ for all R-refutations Γ of C_n,
(2) For every n there exists an FR-refutation Δ_n of C_n such that $l(\Delta_n) \leq 2^{dn}$

(s is defined by $s(0) = 1$, $s(n+1) = 2^{s(n)}$ as in Example 6.2.1).

An exact proof of Theorem 6.2.4 requires tools from combinatory logic and proof theory and is beyond the scope of this book. For a complete proof we refer to [Sta79] and [BL92]. However, we sketch the idea of the proof in order to give an insight into what methods have to be used.

To prove (1) in Theorem 6.2.4 we formulate a modified version of Statman's example having a nonelementary Herbrand complexity, i.e., $HC(C_n) \geq c's(n)$ for some constant c'. As R-deduction maximally gives an exponential speed-up over HC (Theorem 6.1.3) we obtain $l(\Gamma) \geq cs(n-1)$ for all R-refutations Γ of C_n (c is an appropriate constant).

For (2) formulate a short refutation in a full logic calculus (see Figure 6.4 and derive some clausal codifications of the skolemized cut formulas via F-extension. The structural coding of the complex cut formulas (the H_m in Example 6.2.1) is based on a representation of the schema of structural CNF-transformation for NNFs. The simulation of the short refutation by resolution and F-extension (only) requires exponential expense.

Theorem 6.2.4 states that functional extension can shorten proofs nonelementarily. The question remains whether FR-deduction can be of real computational value. It is hardly surprising that F-extension does not always reduce the search space. However, there have been several encouraging experiences in experiments [Egl90], [Pel94].

Frequently a beneficial effect results if F-extension leads to the decomposition of a clause. In such a case the set of clauses can be split into two parts such that each of them can be treated in parallel. Such a splitting can be interpreted as the application of a lemma in a proof [BL90].

Example 6.2.6 ([BL92]). Let C be the set of clauses $\{C_1, C_2, C_3, C_4\}$ for

$$C_1 = P(x, f(y)) \vee Q(x, y), \ C_2 = \neg P(x, f(y)) \vee Q(x, y),$$
$$C_3 = P(x, f(y)) \vee \neg Q(x, y), \ C_4 = \neg P(x, y) \vee \neg Q(x, f(y))$$

We first construct the FR-deduction

$$C_5, C_6 : \ \neg P(x, y) \vee \neg Q(x, f(g(x))), \neg P(a, y) \vee \neg Q(x, f(g(x))).$$

The clause C_6 is decomposed and consequently $C \cup \{C_5, C_6\}$ is unsatisfiable iff the sets of clauses

$$C_1 : \ \mathcal{C} \cup \{C_5\} \cup \{\neg P(a,y)\} \text{ and } C_2 : \ \mathcal{C} \cup \{C_5\} \cup \{\neg Q(x, f(g(x)))\}$$

are both unsatisfiable. Then Γ_1 is a refutation of C_1 for

$$\Gamma_1 : \neg P(a,y), \ P(u, f(v)) \lor Q(u,v), \ Q(a,v), \ P(x, f(y)) \lor \neg Q(x,y), \ \neg Q(a,y), \ \Box.$$

Γ_2 is a refutation of C_2:

$$\Gamma_2 : \neg Q(x, f(g(x))), \ P(x, f(y)) \lor Q(x,y), \ P(x, f(f(g(x)))),$$
$$\neg P(x, f(y)) \lor Q(x,y), \ \neg P(x, f(f(g(x)))), \ \Box.$$

In using $\neg P(a,y)$ and $\neg Q(x, f(g(x)))$ instead of the original clause C_4 we obtain two shorter refutations instead of a single longer one. The refutations Γ_1 and Γ_2 can be carried out independently of each other, which creates a beneficial effect with respect to proof search. Afterwards we can combine Γ_1 and Γ_2 and obtain a refutation Γ of $\mathcal{C}$. In such a combination we see the effect of lemmatization: $\mathcal{C}$ is "proved" out of C_1 and C_2. In using unrestricted resolution we cannot obtain a similar effect, as the clause $\neg P(a,y) \lor \neg Q(x, f(g(x)))$ is not derivable. Moreover no decomposable clause can be derived from $\mathcal{C}$ via resolution.

Note that F-extension is an inference method, i.e., we add a new clause to the original set of clauses. In general it is not allowed to *replace* a clause C by its F-extension, a rule which would destroy the completeness of the calculus. However, in particular cases it is possible to perform such a replacement [BL85] without losing completeness. In practice it may be useful to apply F-extensions which split clauses (as in Example 6.2.6) and to decompose the problem – even without further information [Egl90]; if we only search for a refutation (and do not try to decide the set of clauses) the loss of completeness cannot lead to incorrect results.

The introduction of new function symbols can be considered as a computational tool to simulate quantificational inference in clause logic, where quantifiers don't belong to the syntax. Although more research is required to obtain efficient applications, functional extension marks a starting point for stronger computational inference systems. Unlike the extension rules defined in [Ede92] function introduction is purely "predicate logical" and without significance to propositional logic (it simply does not exist there). While many methods in automated deduction owe their existence to prototypes in propositional logic, F-extension is of quantificational nature. It is weaker than Hilbert's ϵ-calculus [HB34] in which quantifiers are coded by terms under preservation of logical equivalence (instead of sat-equivalence only), but the calculi are intuitively related. In FR-deduction skolemization becomes a principle of inference instead of a principle of normalization in the preprocessing of formulas.

Exercises

Exercise 6.2.1. Let A be a formula in NNF such that $A = A_1 \circ A_2$ for $\circ \in \{\wedge, \vee\}$ and let $\mathcal{M}$ be an interpretation of A. Assume further that $\mathcal{M}_1$ is the restriction of $\mathcal{M}$ to A_1 and $\mathcal{M}_2$ the restriction of $\mathcal{M}$ to A_2. Let $\mathcal{N}_1$ ($\mathcal{N}_2$) be an extension of $\mathcal{M}_1$ ($\mathcal{M}_2$) to $E(A_1)^0$ ($E(A_2)^0$). Then $\mathcal{N}_1$ and $\mathcal{N}_2$ coincide on their common signature $\Sigma(E(A_1)) \cap \Sigma(E(A_2))$ (note that $\Sigma(E(A_1)) \cap \Sigma(E(A_2))$ may properly contain $\Sigma(A_1) \cap \Sigma(A_2)$).

Exercise 6.2.2. Prove that the structural transformation to clause form is polynomial, i.e., there exists a polynomial p such that for all formulas A in negation normal form

$$\| (\mathrm{cl}_{\mathrm{struc}}(A)) \| \leq p(\| A \|)$$

(where $\| A \|$ denotes the number of symbols occurring in A).

References

[AB70]   R. Anderson, W.W. Bledsoe: A linear format for resolution with merging and a new technique for establishing completeness. Journal of the ACM **17**(3), 525–534 (1970)

[Ack28]   W. Ackermann: Über die Erfüllbarkeit gewisser Zählausdrücke. Mathematische Annalen **100**, 638–649 (1928)

[AHU76]   A. Aho, J. Hopcroft, J. Ullman: The design and analysis of computer algorithms. Addison-Wesley (1975)

[And81]   P.B. Andrews: Theorem proving via general matings. Journal of the ACM **28**(2), 193–214 (1981)

[Ba87]   M. Baaz: Automatisches Beweisen für Logiksysteme, in denen Widersprüche behandelt werden können. Informatik-Fachberichte 151, Springer (1987)

[BF81]   A. Barr, E.A. Feigenbaum (eds.): The handbook of Artificial Intelligence, vol. I, chap. 2, "Search". Pitman Books (1981)

[BG94]   L. Bachmair, H. Ganzinger: Rewrite-based equational theorem proving with selection and simplification. Journal of Logic and Computation **4**(3), 1–31 (1994)

[Bib82]   W. Bibel: Automated theorem proving. Vieweg (1982)

[BJ74]   G.S. Boolos, R.C. Jeffrey: Computability and logic. Cambridge University Press (1974).

[BFL94]   M. Baaz, C. Fermüller, A. Leitsch: A non-elementary speed-up in proof length by structural clause form transformation. Proceedings LICS '94, IEEE Computer Science Press, 213–219 (1994)

[BL74]   W.S. Brainerd, L.H. Landweber: Theory of computation. John Wiley & Sons (1974)

[BL85]   M. Baaz, A. Leitsch: Die Anwendung starker Reduktionsregeln im automatischen Beweisen. Proc. of the Austrian Academy of Sciences II **194**(4-10), 287–307 (1985)

[BL90]   M. Baaz, A. Leitsch: A strong reduction method based on function introduction, ISSAC'90, ACM Press 30–37 (1990)

[BL92]   M. Baaz, A. Leitsch: Complexity of resolution proofs and function introduction. Annals of Pure and Applied Logic **57**, 181–215 (1992)

[BL94]   M. Baaz, A. Leitsch: On skolemization and proof complexity. Fundamenta Informaticae **20**(4), 353–379 (1994)

[Boe92]   E. Börger: Berechenbarkeit, Komplexität, Logik. Vieweg (1992)

[Boy71]   R.S. Boyer: Locking: a restriction of resolution. The University of Texas at Austin, Ph.D. dissertation (1971)

[BS28]   P. Bernays, M. Schönfinkel: Zum Entscheidungsproblem der Mathematischen Logik. Math. Annalen **99**, 342–372 (1928)

[BT90]   T. Boy de la Tour: Minimizing the number of clauses by renaming. Proc. CADE 10, 558–572, Springer (1990)

[Cha72]   C.L. Chang: Theorem proving with variable-constraint resolution. Infor-
            mation Sciences 4, 217–231
[Chu36]   A. Church: A note on the Entscheidungsproblem. Journal of Symbolic
            Logic 1, 40–44 (1936)
[CGT90]   S. Ceri, G. Gottlob, L. Tanca: Logic programming and databases.
            Springer (1990)
[CL73]   C.L. Chang, R.C.T. Lee: Symbolic logic and mechanical theorem proving.
            Academic Press (1973)
[CP96]   R. Caferra, N. Peltier: Decision procedures using model building tech-
            niques. Proc. CSL '95, Lecture Notes in Computer Science 1092, 130–
            144, Springer (1996)
[CZ92]   R. Caferra, N. Zabel: A method for simultaneous search for refutations
            and models by equational constraint solving. Journal of Symbolic Com-
            putation 13, 613–641 (1992)
[DLL62]   M. Davis, G. Logemann, D. Loveland: A machine program for theorem
            proving. Communications of the ACM 5(7), 394–397 (1962)
[DP60]   M. Davis, H. Putnam: A computing procedure for quantification theory.
            Journal of the ACM 7(3), 201–215 (1960)
[DL84]   L. Denenberg, H.R. Lewis: Logical syntax and computational complexity.
            Lecture Notes in Math. 1104, 101–115, Springer (1984)
[Ede85]   E. Eder: Properties of substitutions and unifications. J.Symbolic Com-
            putation 1, 31–46 (1985)
[Ede92]   E. Eder: Relative complexities of first-order calculi, Vieweg (1992)
[Egl90]   U. Egly: Problem reduction methods and clause splitting in automated
            theorem proving. Master thesis, Technische Universität Wien (1990)
[Egl91]   U. Egly: A generalized factorization rule based on the introduction of
            Skolem terms. Proc. Seventh Austrian Conference on Artificial Intelli-
            gence, Informatik-Fachberichte 287, Springer (1991)
[Egl92]   U. Egly: Shortening proofs by quantifier introduction. Proc. LPAR'92,
            Lecture Notes in AI 624, 148–159, Springer (1992)
[Egl94]   U. Egly: Methods of function introduction. Dissertation, TU Darmstadt
            (1994)
[Fer91]   C. Fermüller: Deciding classes of clause sets by resolution. Dissertation,
            Technische Universität Wien (1991)
[Fer91a]   C. Fermüller: A resolution variant deciding some classes of clause sets.
            CSL '90, Lecture Notes in Computer Science 533, 128–144, Springer
            (1991)
[Fit90]   M.C. Fitting: First-order logic and automated theorem proving. Springer
            (1990)
[FL93]   C. Fermüller, A. Leitsch: Model building by resolution. Proc. CSL '92,
            Lecture Notes in Computer Science 702, 134–148, Springer (1993)
[FL96]   C. Fermüller, A. Leitsch: Hyperresolution and automated model build-
            ing. Journal of Logic and Computation 6(2), 173–203 (1996)
[FLTZ93]   C. Fermüller, A. Leitsch, T. Tammet, N. Zamov: Resolution methods for
            the decision problem. Lecture Notes in AI 679, Springer (1993)
[Gen34]   G.Gentzen: Untersuchungen über das logische Schliessen I-II. Math.Z.
            39, 176–210, 405–431 (1934)
[GF92]   G. Gottlob, Ch. Fermüller: Removing redundancy from a clause. Artifi-
            cial Intelligence 61, 263–289 (1993)
[GH93]   D.M. Gabbay, C.J. Hogger (eds.): Handbook of logic in artificial intelli-
            gence and logic programming. Oxford University Press (1993)
[Gil60]   P.C. Gilmore: A proof method for quantification theory; its justification
            and realization. IBM J. Res. Develop. 4, 28–35 (1960)

[GJ79]   M.R. Garey, D.S. Johnson: Computers and intractability. Freeman (1979)
[GL85]   G. Gottlob, A. Leitsch: On the efficiency of subsumption algorithms. Journal of the ACM **32**(2), 280–295 (1985)
[GL85a]   G. Gottlob, A. Leitsch: Fast subsumption algorithms. Proc. EUROCAL '85, Lecture Notes in Computer Science **204**-II, 64–77, Springer (1985)
[Göd30]   K. Gödel: Die Vollständigkeit der Axiome des logischen Funktionenkalküls. Mh. Math. Phys. **37**, 349–360 (1930)
[Göd31]   K.Gödel: Über formal unentscheidbare Sätze der Principia Mathematica und verwandter Systeme. Mh. Math. Phys. **38**, 175–198 (1931)
[Göd32]   K. Gödel: Ein Spezialfall des Entscheidungsproblems der theoretischen Logik. Ergebn. math. Kolloq. **2**, 27–28 (1932)
[Got87]   G. Gottlob: Subsumption and implication. Information Processing Letters **24**, 109–111 (1987)
[Gur73]   Y. Gurevich: Formuly s odnim ∀ (formulas with one ∀). In Izbrannye voprosy algebry i logiki (Selected Questions in Algebra and Logics; in memory of A. Mal'cev). Nauka, Nowosibirsk, 97–110 (1973)
[HB34]   D. Hilbert, P. Bernays: Grundlagen der Mathematik II, Springer (1970)
[Her30]   J. Herbrand: Recherches sur la theorie de la démonstration. Travaux de la Societé des Sciences et des Lettres de Varsovie **33** (1930)
[Her31]   J. Herbrand: Sur le problème fondamentale de la logique mathématique, Sprawozdania z posiedzeń Towarzystwa Naukowego Warzawskiego, Wydział III **33** (1931)
[Hil01]   D. Hilbert: Probleme der Mathematik. Archiv der Mathematik und Physik **3**(1), 44–63, 213–237 (1901)
[Im85]   R. Imhoff: Subsumptionsalgorithmen und ihre Effizienz. Master's thesis, Dept. of Mathematics, Univ. of Linz Austria (1985)
[Joy73]   W.H. Joyner: Automated theorem proving and the decision problem. Ph.D. thesis, Harvard University (1973)
[Joy76]   W.H. Joyner: Resolution strategies as decision procedures. Journal of the ACM **23**(1), 398–417 (1976)
[KD68]   D.E. Knuth: The art of computer programming. Addison-Wesley (1968)
[KH94]   S. Klingenbeck, R. Hähnle: Semantic tableaus with ordering restrictions. Proc. of the CADE'94, Lecture Notes in AI **814**, 708–722, Springer (1994)
[KH69]   R. Kowalski, P. Hayes: Semantic trees in automatic theorem proving. In: B. Meltzer, D. Michie (eds.): Machine Intelligence **4**, 87–101, Elsevier (1969)
[Kow75]   R. Kowalski: A proof procedure using connection graphs. Journal of the ACM **22**, 572–595 (1975)
[KP88]   J. Krajicek, P. Pudlak: The number of proof lines and the size of proofs in first-order logic. Arch. Math. Logic **27**, 69–84 (1988)
[KMW61]   A.S. Kahr, E.F. Moore, Hao Wang: Entscheidungsproblem reduced to the ∀∃∀ case. Proc. Nat. Acad. Sci. USA **48**, 365–377 (1962)
[Lee67]   R.C.T. Lee: A completeness theorem and a computer program for finding theorems derivable from given axioms. Ph.D. thesis, University of California at Berkeley (1967)
[Leib]   G.W. Leibniz: Calculus ratiocinator. In: Sämtliche Schriften und Briefe edited by Preussische Akademie der Wissenschaften Darmstadt, Reichel (1923)
[Lei88]   A. Leitsch: Implication algorithms for classes of Horn clauses. In: W.H. Janko (ed.): Statistik Informatik + Ökonomie 172–189, Springer (1988).
[Lei89]   A. Leitsch: On different concepts of resolution. Zeitschr. für Math. Logik und Grundlagen der Math. **35**, 71–77 (1989)

[Lei90]   A. Leitsch: Deciding Horn classes by hyperresolution. Proc. of the CSL '89, Lecture Notes in Computer Science **440**, 225–241, Springer (1990)

[Lei93]   A. Leitsch: Deciding clause classes by semantic clash resolution. Fundamenta Informaticae **18**, 163–182 (1993)

[Lei93a]   A. Leitsch: Resolution theorem proving. AILA preprint **15**, Associazione Italiana di Logica e sue Applicazioni (1993)

[LG90]   A. Leitsch, G. Gottlob: Deciding clause implication problems by ordered semantic resolution. In: F. Gardin and G. Mauri (eds.): Computational Intelligence II, 19–26, North-Holland (1990)

[Lew79]   H.R. Lewis: Unsolvable classes of quantificational formulas. Addison-Wesley (1979)

[Lov70]   D.W. Loveland: A linear format for resolution. Proc. IRIA Symposium on Automatic Demonstration, Lecture Notes in Mathematics **125**, 147–162, Springer (1970)

[Lov78]   D.W. Loveland: Automated theorem proving – a logical basis. North-Holland (1978)

[Llo87]   J.W. Lloyd: Foundations of logic programming. Springer (2nd ed. 1987)

[Loew15]   L. Löwenheim: Über Möglichkeiten im Relativkalkül. Math. Ann. **68**, 169–207 (1915)

[LP92]   S.J. Lee, D. Plaisted: Eliminating duplication with the hyperlinking strategy. Journal of Automated Reasoning **9**(1), 25–42 (1992)

[Luc70]   D. Luckham: Refinements in resolution theory. Proc. IRIA Symposium on Automatic Demonstration, Lecture Notes in Mathematics **125**, 163–190, Springer (1970).

[MB88]   R. Manthey, F. Bry: Satchmo: a theorem prover implemented in Prolog. CADE 9, Lecture Notes in Computer Science **310**, 415–434, Springer (1988)

[Mas64]   S.Y. Maslov: An inverse method of establishing deducibilities in the classical predicate calculus. Dokl. Akad. Nauk SSSR **159**, 1420–1424 (1964)

[Mas68]   S.Y. Maslov: The inverse method of establishing deducibility for logical calculi. Proc. Steklov Inst. Math. **98**, 25–96 (1968)

[Mat93]   R. Matzinger: On lock resolution and decision methods for clause classes. Master's thesis, Dept. of Computer Science, Technische Universität Wien (1993)

[McC90]   W. McCune: Otter 2.0 users guide. Argonne National Laboratory, Argonne (1990)

[Min93]   G.E. Mints: Gentzen-type systems and resolution rules. Part II: Predicate Logic. In: Logic Colloquium 1990, Lecture Notes in Logic **2**, 163–190, Springer (1993)

[MM82]   A. Martelli, U. Montanari: An efficient unification algorithm. ACM Transactions on Programming Languages and Systems **4**(2), 258–282 (1982)

[MP92]   J. Marcinkowski, L. Pacholski: Undecidability of the Horn clause implication problem. Rapport de Recherche 1992-5, Groupe de recherche algorithmique et logique, University of Caen, France (1992)

[MR85]   N.V. Murray, E. Rosenthal: Path resolution and semantic graphs. Proc. of the EUROCAL '85, Lecture Notes in Computer Science **204**, 50–63, Springer (1985)

[Niv96]   H. de Nivelle: Resolution games and non-liftable orderings. Collegium Logicum, Annals of the Kurt Gödel Society **2**, 1–20 (1996)

[Pea84]   J. Pearl: Heuristics: Intelligent search strategies for computer problem solving. Addison-Wesley (1984)

[Pel94]   J. Pellizzari: Methods of function introduction. Master's thesis, Dept. of Computer Science, Technische Universität Wien (1994)

[Pra69]   D. Prawitz: Advances and problems in mechanical proof procedures. Machine Intelligence **4**, 59–71, Elsevier (1969)

[Pra71]   D. Prawitz: Ideas and results in proof theory. In: J.E. Fenstad (ed.): Proc. of the 2nd Scandinavian Logic Symposium, 235 –308, North-Holland (1971)

[PW78]   M.S. Paterson, M.N. Wegman: Linear unification. Journal of Computer and System Sciences **16**(2), 158–167(1978)

[Rob65]   J.A. Robinson: A machine oriented logic based on the resolution principle. Journal of the ACM **12**(1), 23–41 (1965)

[Rob65a]   J.A. Robinson: Automatic deduction with hyperresolution. Intern. Journal of Comp. Math. **1**, 227–234 (1965)

[Rob79]   J.A. Robinson: Logic: form and function. North-Holland (1979)

[Rud92]   V. Rudenko: Decision of a Horn clause implication class with ≤ 2 variables in the head. Yearbook of the Kurt Gödel Society 1992, 81–91

[Sko20]   T. Skolem: Logisch-kombinatorische Untersuchungen über die Erfüllbarkeit und Beweisbarkeit mathematischer Sätze nebst einem Theoreme über dichte Mengen. In: J.E. Fenstad (ed.): Selected works in logic by Th. Skolem, Universitetsforlaget Oslo-Bergen-Tromsö 103–136 (1970)

[Sla67]   J.R. Slagle: Automatic theorem proving with renamable and semantic resolution. Journal of the ACM **14**(4), 687–697 (1967)

[Sla92]   J. Slaney: Finder (finite domain enumerator): notes and guide. Technical Report TR-ARP-1/92, Australian National University Automated Reasoning Project, Canberra (1992)

[Sla93]   J. Slaney: Scott: A model-guided theorem prover. Proc. of the IJCAI '93 vol. 1, 109–114, Morgan Kaufmann (1993)

[SS76]   E. Specker, V. Strassen: Komplexität von Entscheidungsproblemen. Lecture Notes in Computer Science **43**, Springer (1976)

[SS88]   M. Schmidt-Schauss: Implication of clauses is undecidable. Theoretical Computer Science **59**, 287–296 (1988)

[Sta79]   R. Statman: Lower bounds on Herbrand's theorem. Proc. AMS **75**, 104–107 (1979)

[Ste71]   S. Stenlund: Combinators, λ-terms and proof theory. Reidel (1971)

[Sti73]   R.B. Stillman: The concept of weak substitution in theorem proving. Journal of the ACM **20**(4), 648–667 (1973)

[Tak75]   G. Takeuti: Proof theory. North-Holland (1975)

[Tam91]   T. Tammet: Using resolution for deciding solvable classes and building finite models. Lecture Notes in Computer Science **502**, 33–64, Springer (1991)

[Tsei83]   G.S. Tseitin: On the complexity of derivation in propositional calculus. In: J. Siekmann, G. Wrightson (eds.): Automation of reasoning, 466–483, Springer (1983)

[Tur36]   A. Turing: On computable numbers with an application to the Entscheidungsproblem. Proc. of the London Math. Soc. Ser. 2 **42**, 230–265 (1936/37)

[WRC65]   L. Wos, G.A. Robinson, D.F. Carson: Efficiency and completeness of the set of support strategy in theorem proving. Journal of the ACM **12**(4), 536–541 (1965)

[Zam72]   N.K. Zamov: On a bound for the complexity of terms in the resolution method. Proc. of the Steklov Math. Inst. **128**, 5–13 (1972)

Notation Index

$<_A$ 99
α normalization 12
AS atom set 24
AT atom formulas 6

β $\exists$-elimination 13
$\square$ empty clause 18
BS 230
BSALG 236
BSH 230
BSH* 231
BS* 231

$\mathcal{C}_+$ 132
$\sim_C$ 190
$C \setminus L$ 44
CL 89
$\text{Clos}(\mathcal{T})$ 36
C_N 130
CNF conjunctive normal form 17
$comp$ 150
$CORR(,)$ 64
C_P 130
C_r reduced clause 43
CS constant symbols 5
CS_H 155
CS_l 155
$CS_{<_A^s}$ 167
CS_{SULI} 172
cl_{struc} 279
CS_{ULI} 172
CS_x 154
CS_{xsT} 198
CS_{xt} 157

$<_d$ 100
DEC 187
Δ_{SULI} 172
Δ_{ULI} 172
δ 169

δ_{SULI} 172
δ_{ULI} 172
DIFF(,) difference set 65
dom domain 10
d_x 154
d_{xr} 178

E( ) edges 31
ϵ empty substitution 60
$=_s$ 61

$F()$ formula operator 23
FS function symbols 5

GI 253

HC 253
$H(\mathcal{C})$ Herbrand universe 24

K_∞ 217

l length 76
$\|\|\|$ symbolic length 256
$\leq_s$ generality relation 61
LIT literals 6
$<_1$ 224
$<_2$ 224

MB 250
MON 223
MON* 225

N_c 96
N_l 109
N_o 95, 96
NOD( ) nodes 31
N_r 95, 96
N_s 96
N_v 95, 96
N_{vl} 109

φ 92

PL predicate logic   5
$\prec_q$ ordering of quantifiers   12
PS predicate symbols   5
PVD   233
PVD_r 233

$R_{<_A}$ 103
$R_{c\mathcal{M}}$ 142
$R_\emptyset$ 92
$\mathcal{R}es()$ 92
$\mathcal{R}es(,)$ resolvents   91
rg range   10
R_H 132
ρ_x 92
$\rho_{<_A}$ 101
$\rho_{c\mathcal{M}}$ 142
ρ_H 132
ρ_l 111
$\rho_{\mathcal{M}}$ 142
ρ_N 98
$\rho_{<_A}^0$ 226
$\hat{\rho}_x$ 152
R_H^+ 132
R_l 111
R_{lT} 199
$R_{\mathcal{M}}$ 142
R_N 98
$R_{<_A}^0$ 226
ROOT   31
$R_{\emptyset_t}$ 156
R_x refinement operator   92
R_{xs} 163
R_{xsT} 197
R_{xT} 196

s 270
$\sim_{sat}$ sat-equivalence   10

$\leq_{sc}$ 105
$\leq_{scs}$ 137
$sf(,)$ 163
S_x^i 92
S_H^i 132
S_{Hr}^i 175
$S_{\emptyset_t}^i$ 156
S_{xr}^i 174
$\leq_{sl}$ 112
$\leq_{ss}$ subsumption   158
SSIMP   190
ST   182
ST0ST_0 182
stn 186
STP   187
$stpn$ 188
sub 174
SUBST($\mathcal{C}$)   60
SUBST_Σ 60
SUBST substitutions   60

T   245
T terms   5
$\mathcal{T}$ semantic tree   31
τ term depth   6
$\tau_{\max}$ 100
TAUTEL   156, 196

$UN(W)$ 62

V variables   5
$\sim_v$ 73
V1C   187
VAR1   215
VAR1C   216
$\sim vc$ 190
VPROP   255

Subject Index

A-ordering   99
A-resolvent   101
algorithm
– BSALG   236
– KCIA   206
– MB   250
– of Stillman   182
atom   6
– resolved   101
– resolved upon   73
– set   24
atom representation   139

B-refinement   196
backtracking   152
branch   32

clash
– semantic   142
– sequence   131
clause   18
– center   117
– condensed   96
– connected component   190
– copying   266
– empty   18
– Horn   19
– – implication problem   204
– implication   200
– – class 1VCI   207
– – class H0CI   208
– indexed   109
– Krom   19
– – implication problem   205
– normalization   95
– PN-form   130
– powers   209
– r-equivalent   255
– reduced   43
– relevant   119

– side   117
– simple   187
– top   117
– variant   73
clause form   18
– structural   279
completeness
– relative   169
component
– constant   227
– functional   227
– v-   227
computational logic   77
condensation   96
connection method   268
connectives   5
constant symbol   5
corresponding pairs   64

Davis-Putnam   43
– completeness   48
– correctness   48
– method of   43
– rules   44
decision class
– Ackermann   212, 215, 223
– Bernays-Schönfinkel   51
– Bernays-Schönfinkel   212, 230
– BS   230
– BSH   230
– dyadic   212
– Gödel   212
– Herbrand   213
– K   217
– monadic   212, 223
– PS2   215
– PVD   233
– VAR1   214, 215
– VAR1C   216, 223
decision problem   212

deduction
- A-ordering 102
- Chang-Lee- 86
- FR- 283
-- completeness 284
-- soundness 284
- LI- 125
-- unrestricted 169
- linear input 125
- lock 111
-- completeness 113
- LR- 117
-- completeness 123
- NR- 98
- RA- 102
-- completeness 105
- Robinson- 83
- STULI- 199
- SULI- 169
- ULI- 169
difference set 65
DP-
- decision tree 48
- tree 46

environment 8
equivalence
- sat- 15–17
equivalent
- interpretations 8
- logically 10
- sat- 10
expression 60
- length 69
extension 272
- F- 282
- formula 274
- functional 280

F-extension 282
fact 126
factor 75
- l- 117
- lock 110
- nontrivial 75
- S- 82
- selected 226
- standard 84
failure node 35
finitely controllable 212
formula 6
- atom 6
- closed 7

- open 7
- prenex 40
- satisfiable 10
- standard form 7
- sub 7
- valid 10
function symbol 5

G-instance 75
generality 61
Gilmore 41
- method of 41
goal 126
GR-deduction 54
- completeness 55
ground
- projection 260
ground instance 24

H-interpretation 25
- atom representation 139
- corresponding 27
H-model
- least 145
Herbrand
- complexity 253
  interpretation 25
- theorem of 38, 41
- universe 24
heuristics 150
Horn logic 126
hyperresolution 130
- completeness 136
- with replacement 178
hyperresolvent 131

instance 60
interpretation 8
- equivalence 8
- function 8

lifting lemma 79
lifting theorem 79
literal 6
- indexed 109
locking 108
logically isomorphic 256

m.g.u. 62
matrix 40, 215
maximal path 31
merging low 109
model 9
- Herbrand

– – least   145
model building   139, 238
– algorithm MB   250

normal form
– clausal   18
– conjunctive   17
– negation   17

one-variable class   215
operator
– A-ordering   103
– clause ordering   96
– clause reduction   96
– hyperresolution   132
– lock   111
– N-resolution   98
– replacement-   174
– standard normalization   96
– T   245
– unrestricted resolution   92
– variable
– – standardization   96
ordering
– a posteriori   102
– a priori   102

p-reduct   53
p-resolvent   53
pair
– corresponding   64
– irreducible   64
permutation   61
PR-deduction   54
– soundness   54
Prawitz's method   266
predicate symbol   5
prefix   40
prenex form   40
projection
– ground   260
– propositional   260
proof search   149
propositional
– projection   260
– renaming   255
– skeleton   255

quantifiers   5

R-deduction   76
– more general   77
refinement   90
– A-ordering   102

– – completeness   105
– – with subsumption   165
– B-   196
– complete   90
– forward subsumption-   163
– hyperresolution   132
– – completeness   136
– linear   117
– – completeness   123
– lock   111
– – completeness   113
– model-based   141
– operator   92
– – complete   93
– semantic   142
– semantic clash   142
– – completeness   144
refutation
– -sequence   174
relation
– strictly stronger   161
– stronger   161
renaming   61
replacement   172
– completeness of   178
– minimal refutation depth   178
– sequence   174
– – convergent   174
– – divergent   174
resolution
– completeness   80
– deduction   76
– – length   76, 259
– general   75
– propositional   52
– refinement   90
– set of support   116
resolvent   75
– A-ordering   101
– binary   73
– Chang-Lee-   84
– hyper-   131
– lock   110
– LRM-   117
– N-   97
– PRF   130
– propositional   53
– Robinson-   82
– semantic   142
rule   126

S-factor   82
satisfiable   10

saturation   223
search   149
– breadth-first   151
– complexity   154, 167, 172
– – ULI-deduction   172
– depth-first   152
semantic clash   142
semantic tree   31
– closed   35
– complete   32
sentence   7
setting   133
sign-renaming   132
signature   6
skeleton   255
Skolem   14
skolemization   14
splitting   285
ST-tree   182
stable   239
Statman's example   270
subformula   7
– immediate   7
substitution   10
– domain   10
– ground   11, 24, 60
– range   10
– total   262
subsumption   158
– -reduced   162, 173
– algorithm
– – DC   192
– – SSIMP   190
– – ST   182
– backward   162
– forward   162
symbolic length   256

taut-equivalent   40
taut-reduced   48
TAUTEL   196
tautology
– elimination   156, 196
term   5
– depth   6
– dominating   224
· functional   6
– ground   6
– similar   224
theorem
– of Lee   203
transformation
– nonstructural   272
– structural   272
tree
– semantic   31
– ST-   182

UAL   66
unification   59
– algorithm   66
– complexity   70
– theorem   67
unifier   61
– most general   62
universal closure   8
unlocking   113

V-resolution   269
valid   10
variable   5
– assignment   8
– depth   100
– standardization   96

Texts in Theoretical Computer Science – An EATCS Series

J. L. Balcázar, J. Díaz, J. Gabarró
Structural Complexity I
2nd ed. (see also overleaf, Vol. 22)

M. Garzon
Models of Massive Parallelism
Analysis of Cellular Automata and
Neural Networks

J. Hromkovič
**Communication Complexity
and Parallel Computing**

A. Leitsch
The Resolution Calculus

A. Salomaa
Public-Key Cryptography
2nd ed.

K. Sikkel
Parsing Schemata
A Framework for Specification and
Analysis of Parsing Algorithms

Monographs in Theoretical Computer Science – An EATCS Series

C. Calude
Information and Randomness
An Algorithmic Perspective

K. Jensen
Coloured Petri Nets
Basic Concepts, Analysis Methods
and Practical Use, Vol. 1
2nd ed.

K. Jensen
Coloured Petri Nets
Basic Concepts, Analysis Methods
and Practical Use, Vol. 2

A. Nait Abdallah
The Logic of Partial Information

Former volumes appeared as
EATCS Monographs on Theoretical Computer Science

Vol. 5: W. Kuich, A. Salomaa
Semirings, Automata, Languages

Vol. 6: H. Ehrig, B. Mahr
Fundamentals of Algebraic Specification 1
Equations and Initial Semantics

Vol. 7: F. Gécseg
Products of Automata

Vol. 8: F. Kröger
Temporal Logic of Programs

Vol. 9: K. Weihrauch
Computability

Vol. 10: H. Edelsbrunner
Algorithms in Combinatorial Geometry

Vol. 12: J. Berstel, C. Reutenauer
Rational Series and Their Languages

Vol. 13: E. Best, C. Fernàndez C.
Nonsequential Processes
A Petri Net View

Vol. 14: M. Jantzen
Confluent String Rewriting

Vol. 15: S. Sippu, E. Soisalon-Soininen
Parsing Theory
Volume I: Languages and Parsing

Vol. 16: P. Padawitz
Computing in Horn Clause Theories

Vol. 17: J. Paredaens, P. DeBra, M. Gyssens,
D. Van Gucht
**The Structure of the Relational
Database Model**

Vol. 18: J. Dassow, G. Páun
**Regulated Rewriting in Formal Language
Theory**

Vol. 19: M. Tofte
Compiler Generators
What they can do, what they might do,
and what they will probably never do

Vol. 20: S. Sippu, E. Soisalon-Soininen
Parsing Theory
Volume II: LR(k) and LL(k) Parsing

Vol. 21: H. Ehrig, B. Mahr
Fundamentals of Algebraic Specification 2
Module Specifications and Constraints

Vol. 22: J. L. Balcázar, J. Díaz, J. Gabarró
Structural Complexity II

Vol. 24: T. Gergely, L. Úry
First-Order Programming Theories

R. Janicki, P. E. Lauer
**Specification and Analysis of Concurrent
Systems**
The COSY Approach

O. Watanabe (Ed.)
**Kolmogorov Complexity
and Computational Complexity**

G. Schmidt, Th. Ströhlein
Relations and Graphs
Discrete Mathematics for Computer Scientists

S. L. Bloom, Z. Ésik
Iteration Theories
The Equational Logic of Iterative Processes

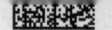